Methods in Molecular Biology

Series Editor
John M. Walker
School of Life and Medical Sciences
University of Hertfordshire
Hatfield, Hertfordshire, UK

For further volumes:
http://www.springer.com/series/7651

For over 35 years, biological scientists have come to rely on the research protocols and methodologies in the critically acclaimed *Methods in Molecular Biology* series. The series was the first to introduce the step-by-step protocols approach that has become the standard in all biomedical protocol publishing. Each protocol is provided in readily-reproducible step-by-step fashion, opening with an introductory overview, a list of the materials and reagents needed to complete the experiment, and followed by a detailed procedure that is supported with a helpful notes section offering tips and tricks of the trade as well as troubleshooting advice. These hallmark features were introduced by series editor Dr. John Walker and constitute the key ingredient in each and every volume of the *Methods in Molecular Biology* series. Tested and trusted, comprehensive and reliable, all protocols from the series are indexed in PubMed.

Computational Methods for Estimating the Kinetic Parameters of Biological Systems

Edited by

Quentin Vanhaelen

Insilico Medicine Hong Kong Ltd, Pak Shek Kok, Hong Kong

Humana Press

Editor
Quentin Vanhaelen
Insilico Medicine Hong Kong Ltd
Pak Shek Kok, Hong Kong

ISSN 1064-3745 ISSN 1940-6029 (electronic)
Methods in Molecular Biology
ISBN 978-1-0716-1766-3 ISBN 978-1-0716-1767-0 (eBook)
https://doi.org/10.1007/978-1-0716-1767-0

This Humana imprint is published by the registered company Springer Science+Business Media, LLC, part of Springer Nature.
The registered company address is: 1 New York Plaza, New York, NY 10004, U.S.A.

Preface

One of the biggest challenges encountered when building quantitative mechanistic models to study the dynamics of a biological system is to obtain quantitative information about the kinetics of the biochemical processes involved. Depending on the system, one could look for information about the kinetic rates of enzymatic reactions, the binding and unbinding kinetics of a ligand to its receptor, the association and dissociation rates characterizing the interactions between proteins or between a drug and its target, or even the kinetic parameters needed to simulate the dynamics of a large gene regulatory network. Mechanistic models are often developed following a two-step methodology. Firstly, hypothesis-driven approaches are used for the assembly of the dynamical model. This step makes an extensive use of state-of-the-art knowledge from various fields including chemistry, biophysics, and biology. The purpose of this step is to build a formal description of the different mechanisms one wants to incorporate within the dynamical model. The level of detail of the mathematical description depends on the scope of the study as well as on the information that can be gathered about the system. Secondly, one relies on experimental data available and/or data-driven methods for obtaining quantitative information about the parameters that appear within the mathematical formulation of the model. This includes kinetic parameters, concentrations, and time delays for instance. The assembly of relatively large and accurate quantitative models of biological systems is often made difficult because of the lack of experimental data for many of these parameters. This is especially true for kinetic parameters. These parameters are complicated to measure and their values strongly vary in function of the experimental conditions considered. This situation has triggered a large interest within the scientific community for the development of computational methods that can be applied for the estimation of kinetic parameters in various contexts.

The aim of this book is to provide an overview of various classes of computational techniques, including machine learning techniques, commonly used for evaluating kinetic parameters of biological systems. The book focuses on three types of situations: the prediction of the kinetics of enzymatic reactions, the prediction of the kinetics of protein-protein or protein-ligand interactions (binding rates, dissociation rates, binding affinities), and the prediction of relatively large set of kinetic rates of reactions usually found in quantitative models of large biological networks. Each chapter has been written by scientists whose research focuses on developing and using such techniques. In each chapter, the authors have used their experience to describe in a comprehensive and accessible manner a specific method. Furthermore, they explain the necessary steps required for its implementation and successful use. In addition, review chapters have been integrated in order to introduce from a larger perspective the main characteristics of some of these methods.

Chapter 1 is a review of the current approaches for assembling mechanistic drug-target binding models. Chapter 2 describes a model of drug-receptor interaction and its associated kinetic and binding properties. Chapter 3 covers a method based on Bayesian inference for the analysis of enzyme kinetics. Chapter 4 presents a method for parameter selection and estimation based on multiobjective optimization. Chapter 5 discusses how to apply data-driven normalization of the simulation to ODE-based dynamic models of systems biology using the framework PEPSSBI (Parameter Estimation Pipeline for Systems and Synthetic Biology). Chapter 6 describes an optimization approach for parameter estimation based on the Gauss-Newton method and Levenberg-Marquardt method. The algorithm is applied to a Microalgae Growth Model. Chapter 7 presents the software ENZO designed for the automatic assembly and calibration of ODE-based models of enzymatic reactions. Chapter 8 describes an approach to estimate kinetic parameters of the reduction and oxidation of redox switches using time-series data. Chapter 9 provides a review of current computational methods for structure-based drug design using systems biology. Chapter 10 gives a comprehensive tutorial on how to use QM-based methods to evaluate the kinetics of enzyme reactions. Chapter 11 focuses on the characterization of dissociation rates by Surface Plasmon Resonance (SPR) and the associated challenges. Chapter 12 provides an overview of the current computational methods for binding free energy prediction. Chapter 13 presents a dynamic model based on the Molecular Mechanics Generalized Born Surface Area (MM/GBSA) approach to analyze how T cell receptor, peptide, and the major histocompatibility complex (MHC) act together to trigger celiac disease. Chapter 14 describes an umbrella sampling-based method that can be used to compute ligand-binding affinity. Chapter 15 presents a method combining molecular dynamics and network analysis to investigate the ligand-protein binding landscape. Chapter 16 reviews several computational methods for estimating the strength of protein-protein interactions. The authors present the method BADOCK that can be used to predict binding energy without using the structure of the protein complex. Finally, Chapter 17 presents a method designed to evaluate the kinetic parameters subject to a variety of explicit and implicit constraints of models describing the dynamics of ion channels.

There is currently a large and diverse set of methods to estimate kinetic parameters. These methods have their specific characteristics and requirements, for example in terms of data needed to perform the computational analysis. Thus, the most appropriate type of method depends on multiple factors such as the nature of the system of interest, the scope of the study, and the amount of a priori knowledge needed. For scientists interested in using such techniques, identifying the best type of algorithm can be challenging. Hopefully, the methods presented in this book will be of great interest for scientists who would like to use or develop a computational method for the estimation of kinetic parameters.

I would like to thank the Series Editor Prof. John Walker for giving me the opportunity to edit this volume. This book would not exist without the work of all authors who collaborated on it. I deeply acknowledge the authors for their contributions, in-depth knowledge, and professional presentation. I would like to express my gratitude to the collaborators of Springer Nature for their support and for their efficient and professional handling of this project.

Pak Shek Kok, Hong Kong ***Quentin Vanhaelen***

Contents

Contributors

PIA ABEL ZUR WIESCH • *Department of Pharmacy, Faculty of Health Sciences, UiT The Arctic University of Norway, Tromsø, Norway; Department of Biology, Eberly College of Science, The Pennsylvania State University, University Park, PA, USA; Center for Infectious Disease Dynamics, Huck Institutes of the Life Sciences, The Pennsylvania State University, University Park, PA, USA; Centre for Molecular Medicine Norway, Nordic EMBL Partnership, Blindern, Oslo, Norway*

MARZIE AMIRSHENAVA • *Division of Biological Sciences, University of Missouri, Columbia, MO, USA*

FERNANDO ANTUNES • *Departamento de Química e Bioquímica and Centro de Química Estrutural, Faculdade de Ciências, Universidade de Lisboa, Lisboa, Portugal*

YADIRA BOADA • *Synthetic Biology and Biosystems Control Lab, I.U. de Automatica e Informatica Industrial (ai2), Universitat Politecnica de Valencia, Valencia, Spain; Centro Universitario EDEM, Escuela de Empresarios, Valencia, Spain*

TOMASZ BOROWSKI • *Jerzy Haber Institute of Catalysis and Surface Chemistry, Polish Academy of Sciences, Kraków, Poland*

PATRICIA BOTA • *Structural Bioinformatics Lab (GRIB-IMIM), Department of Experimental and Health Science, University Pompeu Fabra, Barcelona, Catalonia, Spain; Department of Biosciences, U Science Tech, Universitat de Vic-Universitat Central de Catalunya, Vic, Catalonia, Spain*

PAULA BRITO • *Departamento de Microbiologia e Imunologia and iMED.ULisboa, Faculdade de Farmácia, Universidade de Lisboa, Lisboa, Portugal*

BOSEUNG CHOI • *Department of National Statistics, Korea University Sejong Campus, Sejong, South Korea*

ANDREA DEGASPERI • *MRC Cancer Unit, Hutchison/MRC Research Centre, University of Cambridge, Cambridge, UK; Department of Medical Genetics, University of Cambridge, Addenbrooke's Hospital, Cambridge, UK; Wellcome Sanger Institute, Hinxton, Cambridge, UK; Systems Biology Ireland, School of Medicine, University College Dublin, Dublin, Ireland*

ALEX DICKSON • *Department of Biochemistry and Molecular Biology, Michigan State University, East Lansing, MI, USA; Department of Computational Mathematics, Science and Engineering, Michigan State University, East Lansing, MI, USA*

TOM DIXON • *Department of Biochemistry and Molecular Biology, Michigan State University, East Lansing, MI, USA; Department of Computational Mathematics, Science and Engineering, Michigan State University, East Lansing, MI, USA*

AYKUT ERBAŞ • *UNAM National Nanotechnology Research Center and Institute of Materials Science Nanotechnology, Bilkent University, Ankara, Turkey*

NARCIS FERNÁNDEZ-FUENTES • *Institute of Biological, Environmental and Rural Sciences, Aberystwyth University, Aberystwyth, UK; Department of Biosciences, U Science Tech, Universitat de Vic-Universitat Central de Catalunya, Vic, Catalonia, Spain*

DIRK FEY • *Systems Biology Ireland, School of Medicine, University College Dublin, Dublin, Ireland; Conway Institute of Biomolecular and Biomedical Research, University College Dublin, Dublin, Ireland*

MICHAŁ GLANOWSKI • *Jerzy Haber Institute of Catalysis and Surface Chemistry, Polish Academy of Sciences, Kraków, Poland*
COLIN HEMEZ • *Graduate Program in Biophysics, Harvard University, Boston, MA, USA*
MILAN HODOŠČEK • *National Institute of Chemistry, Ljubljana, Slovenia*
HYUKPYO HONG • *Department of Mathematical Sciences, Korea Advanced Institute of Science and Technology, Daejeon, Republic of Korea*
FATIH INCI • *UNAM National Nanotechnology Research Center and Institute of Materials Science Nanotechnology, Bilkent University, Ankara, Turkey*
SITI S. JAMAIAN • *Department of Mathematics and Statistics, Faculty of Applied Sciences and Technology, Universiti Tun Hussein Onn Malaysia, Muar, Johor, Malaysia*
DUŠANKA JANEŽIČ • *Faculty of Mathematics, Natural Sciences and Information Technologies, University of Primorska, Koper, Slovenia*
SANGITA KACHHAP • *Jerzy Haber Institute of Catalysis and Surface Chemistry, Polish Academy of Sciences, Kraków, Poland*
AMAN CHANDRA KAUSHIK • *Wuxi School of Medicine, Jiangnan University, Wuxi, China*
BORIS N. KHOLODENKO • *Systems Biology Ireland, School of Medicine, University College Dublin, Dublin, Ireland; Conway Institute of Biomolecular and Biomedical Research, University College Dublin, Dublin, Ireland; Department of Pharmacology, Yale University School of Medicine, New Haven, CT, USA*
JAE KYOUNG KIM • *Department of Mathematical Sciences, Korea Advanced Institute of Science and Technology, Daejeon, Republic of Korea*
JINGYI LIANG • *Department of Pharmacy, Faculty of Health Sciences, UiT The Arctic University of Norway, Tromsø, Norway; Department of Biology, Eberly College of Science, The Pennsylvania State University, University Park, PA, USA; Center for Infectious Disease Dynamics, Huck Institutes of the Life Sciences, The Pennsylvania State University, University Park, PA, USA*
KIM S. LING • *Department of Mathematics and Statistics, Faculty of Applied Sciences and Technology, Universiti Tun Hussein Onn Malaysia, Muar, Johor, Malaysia*
SAMUEL D. LOTZ • *Department of Biochemistry and Molecular Biology, Michigan State University, East Lansing, MI, USA*
ALBERTO MESEGUER • *Structural Bioinformatics Lab (GRIB-IMIM), Department of Experimental and Health Science, University Pompeu Fabra, Barcelona, Catalonia, Spain*
LORIN S. MILESCU • *Department of Biology, University of Maryland, College Park, MD, USA*
MIRELA MILESCU • *National Institute of Neurological Disorders and Stroke, National Institutes of Health, Bethesda, MD, USA*
MARCO A. NAVARRO • *Division of Biological Sciences, University of Missouri, Columbia, MO, USA*
SON TUNG NGO • *Laboratory of Theoretical and Computational Biophysics, Ton Duc Thang University, Ho Chi Minh City, Vietnam*
LAN K. NGUYEN • *Systems Biology Ireland, School of Medicine, University College Dublin, Dublin, Ireland; Department of Biochemistry and Molecular Biology, School of Biomedical Sciences and Biomedicine Discovery Institute, Monash University, Clayton, VIC, Australia*
BALDO OLIVA • *Structural Bioinformatics Lab (GRIB-IMIM), Department of Experimental and Health Science, University Pompeu Fabra, Barcelona, Catalonia, Spain*
LAMBERTUS A. PELETIER • *Mathematical Institute, Leiden University, Leiden, The Netherlands*

MINH QUAN PHAM • *Graduate University of Science and Technology, Vietnam Academy of Science and Technology, Hanoi, Vietnam; Institute of Natural Products Chemistry, Vietnam Academy of Science and Technology, Hanoi, Vietnam*

JESÚS PICÓ • *Synthetic Biology and Biosystems Control Lab, I.U. de Automatica e Informatica Industrial (ai2), Universitat Politecnica de Valencia, Valencia, Spain*

LINQIONG QIU • *Shanghai Key Laboratory of Green Chemistry and Chemical Processes, School of Chemistry and Molecular Engineering, State Key Laboratory of Precision Spectroscopy, East China Normal University, Shanghai, China*

MARIA M. REIF • *Physics Department (T38), Technische Universität München, Garching, Germany*

SHAKTI SAHI • *School of Biotechnology, Gautam Buddha University, Greater Noida, India*

AUTOOSA SALARI • *Department of Molecular and Cellular Biology, University of California, Berkeley, CA, USA*

JIANING SONG • *NYU-ECNU Center for Computational Chemistry, NYU Shanghai, Shanghai, China*

JURE STOJAN • *Institute for Biochemistry, Faculty of Medicine, University of Ljubljana, Ljubljana, Slovenia*

MACIEJ SZALENIEC • *Jerzy Haber Institute of Catalysis and Surface Chemistry, Polish Academy of Sciences, Kraków, Poland*

VI NGOC-NHA TRAN • *Department of Pharmacy, Faculty of Health Sciences, UiT The Arctic University of Norway, Tromsø, Norway*

ALEJANDRO VIGNONI • *Synthetic Biology and Biosystems Control Lab, I.U. de Autoamatica e Infortiatica Industrial (ai2), Universitat Politecnica de Valencia, Valencia, Spain*

DONG-QING WEI • *Department of Bioinformatics and Biostatistics, College of Life Sciences and Biotechnology, The State Key Laboratory of Microbial Metabolism, Shanghai Jiao Tong University, Shanghai, China*

MARTIN ZACHARIAS • *Physics Department (T38), Technische Universität München, Garching, Germany*

JOHN Z. H. ZHANG • *Shanghai Key Laboratory of Green Chemistry and Chemical Processes, School of Chemistry and Molecular Engineering, State Key Laboratory of Precision Spectroscopy, East China Normal University, Shanghai, China; NYU-ECNU Center for Computational Chemistry, NYU Shanghai, Shanghai, China; Department of Chemistry, New York University, New York, NY, USA*

FATHUL H. ZULKIFLI • *Department of Mechanical Engineering Technology, Faculty of Engineering Technology, Universiti Tun Hussein Onn Malaysia, Muar, Johor, Malaysia*

Chapter 1

Current Approaches of Building Mechanistic Pharmacodynamic Drug–Target Binding Models

Jingyi Liang, Vi Ngoc-Nha Tran, Colin Hemez, and Pia Abel zur Wiesch

Abstract

Mechanistic pharmacodynamic models that incorporate the binding kinetics of drug–target interactions have several advantages in understanding target engagement and the efficacy of a drug dose. However, guidelines on how to build and interpret mechanistic pharmacodynamic drug–target binding models considering both biological and computational factors are still missing in the literature. In this chapter, current approaches of building mechanistic PD models and their advantages are discussed. We also present a methodology on how to select a suitable model considering both biological and computational perspectives, as well as summarize the challenges of current mechanistic PD models.

Key words Pharmacodynamics, Drug–target binding kinetics, Occupancy–efficacy relationship, Mechanistic models, Mathematical biology, Differential equations, Model selection, Parameter estimation, PK-PD simulation tools

1 Introduction

Pharmacokinetic (PK) and pharmacodynamic (PD) modeling studies the dose–exposure–response relationships of drugs. PK/PD models have been widely used in preclinical and clinical studies since the 1960s to interpret observed data and predict optimal dosing strategies [1, 2].

PD studies the relationship between the local concentration of drug at the target site and its biochemical and physiological effects. Traditional PD models predict optimum dosing regimens from preclinical data based on a Hill-based function that links the drug dose to efficacy (Fig. 1a). Recently, mechanistic PD models that incorporate the binding kinetics of drug–target interactions have gained increasing attention (Fig. 1b). Compared with traditional Hill-based PD models, these mechanistic models have shown

Jingyi Liang and Vi Ngoc-Nha Tran contributed equally with all other contributors.

Quentin Vanhaelen (ed.), *Computational Methods for Estimating the Kinetic Parameters of Biological Systems*, Methods in Molecular Biology, vol. 2385, https://doi.org/10.1007/978-1-0716-1767-0_1,

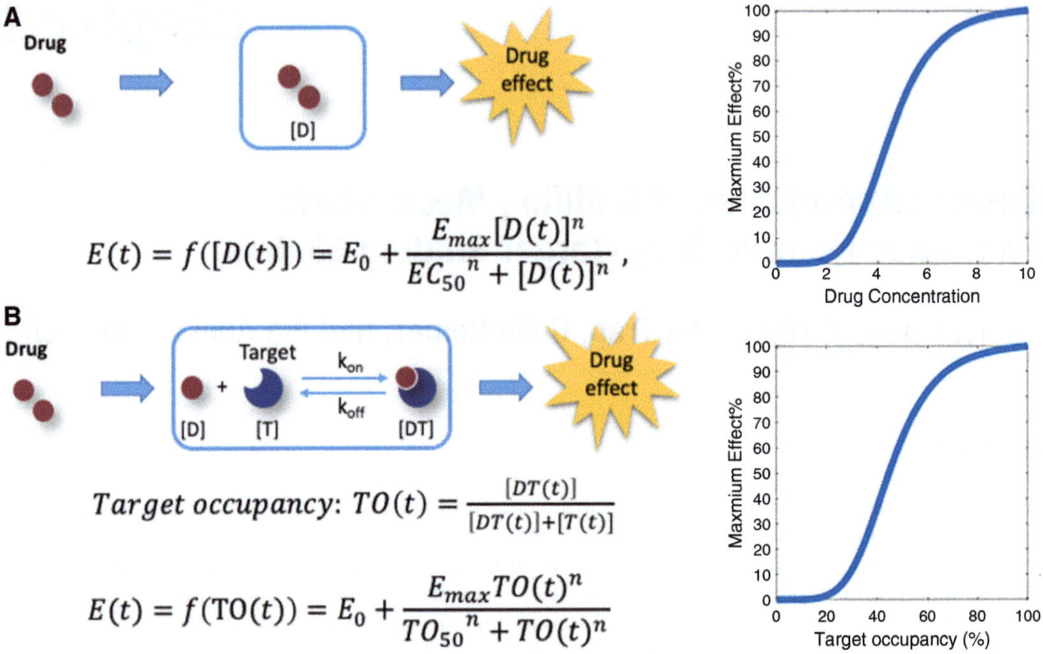

Fig. 1 Schematic overview of a traditional PD model (**a**) and a PD model that incorporates drug–target binding kinetics (**b**). Note that under certain assumptions (see below), the two models can be equivalent

advantages in understanding drug action and target engagement and are useful for the selection of drug candidates and dosing regimens for clinical trials.

In this chapter, we first introduce traditional PD models and summarize the current approaches of building mechanistic PD models that integrate the drug–target binding kinetics.

We then list several advantages of mechanistic PD models compared with traditional methods. We also discuss the computational and mathematical methods that are involved in mechanistic PD modeling. We conclude the chapter with the current challenges we are facing in developing mechanistic PD models.

2 Modeling Approaches for Drug–Target Binding Kinetics

The interaction between drug and target molecules is often viewed as a reversible process and can be expressed using the law of mass action. The number of bound target depends on the concentration of free drug, the concentration of free target, and the association and dissociation rates of the drug–target complex. In the simplest setting, one step-reversible reaction between drug and target molecules can be described as

$$[D] + [T] \underset{k_{\text{off}}}{\overset{k_{\text{on}}}{\rightleftharpoons}} [DT] \tag{1}$$

where $[D]$, $[T]$, and $[DT]$ are the concentration of free drug, free target, and drug–target complex, respectively. k_{on} and k_{off} represent the association and dissociation rate constants of the drug–target complex, respectively. All molecules are assumed to be able to freely diffuse within the medium, and the chemical and physical properties of the drug and target molecules are not altered after dissociation from the complex.

This approach can yield the same results as traditional PD models under certain assumptions, as described in the following. At equilibrium, i.e., at infinity and with constant drug and target molecule concentrations,

$$k_{\text{on}}[D][T] = k_{\text{off}}[DT],$$

and the dissociation constant K_d can be calculated as

$$K_d = \frac{k_{\text{off}}}{k_{\text{on}}} = \frac{[D][T]}{[DT]}. \tag{2}$$

K_d describes the affinity between a drug and a target molecule. A smaller K_d indicates a tighter binding to the drug target (higher binding affinity).

The fractional occupancy ($f_{\text{occupancy}}$) of drug targets describes the fraction of binding sites occupied at a particular drug concentration. At equilibrium, the fractional occupancy is as follows:

$$f_{\text{occupancy}} = \frac{[DT]}{[DT] + [T]} = \frac{[\text{D}]}{[\text{D}] + K_d}. \tag{3}$$

In pharmacology, drug effects are believed to depend on the fractional occupancy of targets, therefore,

$$\text{Effect} = E_{\max} * \text{f}_{\text{occupancy}}, \tag{4}$$

where $E_{\max}$ is the maximum effect where 100% of targets are occupied.

The binding between free drug and free target molecules is often assumed to be a reversible and rapid equilibrium process.

3 Traditional Pharmacodynamic Models

The drug–target binding model introduced above can be easily transferred into a traditional PD model or Hill function—in fact, traditional PD models were derived from drug–target binding. In the traditional PD model, the effect of a drug is described as a function of the local drug concentration (Fig. 1a). Because it is difficult to measure drug concentration at the target site, the plasma drug concentration that can be easily accessed is used as an approximation. Although simple dose–response functions (e.g., fixed, linear, and log-linear) have been used, most PD models employ a Hill-based function developed from receptor binding theory. The Hill function was initially developed by Hill to describe the binding of oxygen to hemoglobin [3] and modified by Wagner [4] to be applied to PD modeling in the 1960s.

Most PD models are developed based on a number of assumptions (Table 1) [5, 6]. They assume the binding between drug and target molecules occurs at equilibrium, and therefore, the concentration of drug–target complex at equilibrium is:

$$[DT] = [T_0]\frac{[D]}{[D] + K_d'}$$

where $[T_0]$ is the total concentration of target that represents the maximum binding capacity ($[T_0] = [T] + [DT]$).

The drug effect (E) is directly and immediately reflected by the concentration of the drug–target complex $[DT]$, and two common forms of E_{max} model are basic E_{max} model:

$$E = E_0 + \frac{E_{max}[D]}{EC_{50} + [D]}, \tag{5}$$

and sigmoidal E_{max} model:

$$E = E_0 + \frac{E_{max}[D]^n}{{EC_{50}}^n + [D]^n}, \tag{6}$$

where E is the drug effect, E_0 is the baseline effect in the absence of a drug, E_{max} is the maximum effect that a drug can induce, EC_{50} is the drug concentration that produces 50% of the maximal effect, and n is the Hill coefficient. In the sigmoidal E_{max} model, n as a slope factor indicates the steepness of the drug concentration–effect relationship.

4 Mechanistic Pharmacodynamic Models that Include Drug–Target Binding

Since not all drug–target interactions are in rapid equilibrium, for example, those with a slow dissociation rate of the drug–target complex, the duration of drug target binding also affects duration of the drug effect [15]. Therefore, the traditional PD model may underpredict the drug effects, and it is more realistic to include binding kinetics for drugs for which binding does not occur much faster than the half-life of the drug. This may be either drugs with a slow drug–target dissociation rate or drugs with a very short half-life. Compared with traditional PD models, the assumptions listed in Table 1 are relaxed in mechanistic PD models. The benefits of studying binding kinetics in drug discovery have been well-reviewed in [16, 17]; in this section, we focus on the development

Table 1
Common assumptions in traditional PD model (left column) and examples of mechanistic PD models with corresponding relaxed assumptions discussed in Subheading 4 (right column)

Assumptions in traditional PD model	Mechanistic PD model with relaxed assumptions
Receptor independence	
Homogeneous distribution of receptors	
Reversible binding consequence	[7]
Constant number of target molecules	[8]
No bacteria growth or death	[7]
Equivalent target or receptors	[7]
Equal drug concentration inside and outside the cell	[7, 9]
Constant drug concentration during equilibrium period	[7, 8, 10–14]
Fast association and disassociation rate	[7, 8, 10–14]

of mechanistic pharmacodynamic models that incorporate drug–target binding kinetics. There also exist PK models that integrate drug–target binding, which are beyond the focus of this chapter, but have been reviewed in [5].

To include binding kinetics into the PD model, for example, in an ordinary deterministic model, the concentrations of free drug, free target, and drug–target complex are calculated separately by including explicit terms of association and dissociation into the system:

$$\begin{aligned} \frac{d[D]}{dt} &= -k_{\text{on}}[D][T] + k_{\text{off}}[DT], \\ \frac{d[T]}{dt} &= -k_{\text{on}}[D][T] + k_{\text{off}}[DT], \\ \frac{d[DT]}{dt} &= k_{\text{on}}[D][T] - k_{\text{off}}[DT]. \end{aligned} \tag{7}$$

4.1 Factors that Influence Binding Kinetics

Many factors at both PK and PD levels will affect the binding kinetics and therefore influence drug effects. To make accurate predictions of target occupancy, it is important to take into consideration these factors, which lead to PD models with different complexity. At the PK level, kinetic processes that determine the concentration profile of drugs at the site of action should be taken into consideration. At the PD level, processes such as nonspecific binding, endogenous competition, and target turnover have been shown to influence the target occupancy differently.

4.1.1 Concentration Gradient Between Intracellular and Extracellular Space

For drug targets that localize inside the cellular compartment, the drug availability to the target is significantly affected by the drug permeability across the cell membrane. Especially for bacteria such as *Mycobacterium tuberculosis* (TB), the causative agent of tuberculosis disease, the thick cell wall limits the access of drugs to the site of action and causes large concentration gradients between extracellular and intracellular drugs. The process is managed through various mechanisms (e.g., transporter proteins, efflux pumps, passive diffusion), which strongly depend on the nature of the cell envelope and characteristics of the drug molecules. Therefore, the concentration of extracellular and intracellular drug concentration can be calculated separately and terms that cover the diffusion mechanisms will be added to each one of them. Since the diffusion process is rather complicated and key parameter values may be difficult to obtain from experiments, the diffusion can be simply simulated as an overall effect, which is calculated based on a permeability factor and the concentration gradient of drugs [7]. For instance, to calculate the contribution of diffusion to the intracellular drug concentration, the diffusion term can be written as $pV_{\text{ratio}}([D_e] - [D_i])$, where p is the permeability factor, V_{ratio} is the volume ratio between extracellular and intracellular, and $([D_e] - [D_i])$ represents the concentration gradient between extracellular ($[D_e]$) and intracellular drugs ($[D_i]$). If the concentration at equilibrium is different intracellularly than in the surrounding tissue, this equation can be further modified to $pV_{\text{ratio}}([D_e]\alpha - [D_i])$, where α is the ratio of the equilibrium concentration outside and inside of the cells. The parameter α will be larger than 1 if the drug accumulates intracellularly and smaller than 1 if the intracellular concentration can never reach the external concentration (e.g., constant action of efflux pumps).

4.1.2 Nonspecific Binding

Nonspecific binding can affect target occupancy by influencing free drug concentration. Nonspecific binding can serve as a reservoir of drug molecules. When free drug decreases, nonspecifically bound drug with low binding affinity can be released and, therefore, increase the availability of free drugs in the system and lead to increased formation of specific binding complexes [7]. Similar to Eq. 7, nonspecific binding can be simulated by adding the following equations to the system:

$$\begin{aligned} \frac{d[U]}{dt} &= -k_{u,\text{on}}[D][U] + k_{u,\text{off}}[DU], \\ \frac{d[DU]}{dt} &= k_{u,\text{on}}[D][U] - k_{u,\text{off}}[DU], \end{aligned} \tag{8}$$

where $[U]$ and $[DU]$ are the concentration of nonspecific binding sites and nonspecific binding complex, respectively. $k_{u,\text{ on}}$ and $k_{u,\text{ off}}$ are the nonspecific binding and dissociation rates, respectively. Therefore, the free drug concentration $[D]$ is

$$\frac{d[D]}{dt} = -k_{\text{on}}[D][T] + k_{\text{off}}[DT] - k_{u,\text{on}}[D][U] + k_{u,\text{off}}[DU]. \tag{9}$$

4.1.3 Endogenous Competition

The competition between drugs and endogenous compounds which also bind to the drug target is an important factor that affects the drug binding kinetics. de Witte et al. developed a mechanistic model of the dopamine D_2 receptor by incorporating competitive binding of both endogenous dopamine and the antagonist [18]. Their model has shown the dissociation rate of the endogenous ligand from the target is one of the factors that limits fast drug–target dissociation. They suggest that the fluctuation in the concentration of endogenous ligand is required to understand and simulate the influence of drug–target binding kinetics and drug efficacy.

In the simplest setting, the interaction between endogenous ligand and target can be simulated as

$$\begin{aligned} \frac{d[L]}{dt} &= -k_{l,\mathrm{on}}[L][T] + k_{l,\mathrm{off}}[LT], \\ \frac{d[LT]}{dt} &= k_{l,\mathrm{on}}[L][T] - k_{l,\mathrm{off}}[LT], \end{aligned} \tag{10}$$

where $[L]$ and $[T]$ are the concentration of endogenous ligand and its ligand–target complex, respectively. $k_{l,\ \mathrm{on}}$ and $k_{l,\ \mathrm{off}}$ are the binding and dissociation rates of the endogenous ligand, respectively. So the concentration of free target is

$$\frac{d[T]}{dt} = -k_{\mathrm{on}}[D][T] + k_{\mathrm{off}}[DT] - k_{l,\mathrm{on}}[L][T] + k_{l,\mathrm{off}}[LT]. \tag{11}$$

4.1.4 Target Turnover

The abundance and turnover of target molecules are important factors that affect target occupancy. For targets with a fast turnover rate, target occupancy is determined by not only the drug–target binding kinetics but also the rates of target synthesis and degradation. It is especially important to take into consideration the turnover of the drug target for irreversible drug–target interactions. For instance, Hong et al. proposed a PD model of antiplatelet effects of aspirin and ibuprofen [8]. The model simulates the irreversible inhibition by aspirin and reversible binding of ibuprofen on the cyclo-oxygenase-1 (COX-1) enzyme. It incorporates the turnover of COX-1, and the model is able to describe the long duration of inhibition of proaggregatory thromboxane A_2 production, as a result of irreversible acetylation of platelet COX-1.

4.2 Drug Efficacy as a Function of Target Occupancy

Drug efficacy can be surprisingly difficult to define and measure. Clearly, drugs should minimize morbidity and mortality, but quantitative molecular readouts can remain elusive. Designing effective methods to measure drug efficacy will speed up the drug discovery process. PD biomarkers have been used as promising predictors for disease diagnosis and tracking of disease progression in vitro and in vivo, in combination with PK-PD modeling approaches [13]. A safe, reliable, sensitive, and specific biomarker and effective

Table 2
Types of occupancy–efficacy functions. E_0 denotes the baseline effect in the absence of a drug

Function type	Equations	Model examples
Simple linear	$E = E_0 + k * [DT]$ or $E = E_0 + k * TO$	[8, 19–21]
Linear with condition	$E = \begin{cases} \{E_{01} + k_1 * [DT], & \text{condition 1} \\ \{E_{02} + k_2 * [DT], & \text{condition 2} \\ & \ldots\ldots \end{cases}$	[7, 22]
Sigmoid function	$E = E_0 + \frac{E_{max} TO^n}{TO_{50}{}^n + TO^n}$	[18, 23]

techniques to measure the level of biomarkers are required for revealing the alteration by the drug and making accurate predictions [14].

Mechanistic PD models that incorporate binding kinetics offer the opportunity to analyze the relationship between drug concentration, drug target occupancy, and effect. This is especially important for drugs with a slow drug–target dissociation rate, since the rapid equilibrium between free drug and drug–target complex will not be achieved, and thus, the traditional PD model is not applicable. Therefore, identifying the relationship between target occupancy and drug efficacy is critical in predicting the drug effect. In this section, we discuss current PD approaches that employ detailed binding kinetic scheme and occupancy–efficacy functions in predicting drug effect (Table 2).

4.2.1 Linear Function

In the simplest approach, the relationship between target occupancy and effect is assumed to be linear [8, 19–21]. In these approaches, the pharmacological responses of drugs are predicted through a linear transduction function based on the level of drug–target complex or the apparent fractional receptor occupancy when the total amount of targets at the site of action is uncertain [20] (line a in Fig. 2a). In the model of paLpxC inhibitor in animal studies of *Pseudomonas* infection by Walkup et al. [22], they assume that target binding only increases the killing rate of bacteria and that this killing is linearly associated with the target occupancy within a certain range (line b in Fig. 2a). The lower threshold (minimum target occupancy) of this range is defined as the target occupancy when the first measurable response is shown in experiments and the upper threshold (maximum target occupancy) is the target occupancy when the maximum killing rate is reached due to the saturation limit of complex-induced killing of bacteria. A similar approach can be found in the model by Abel zur Wiesch et al. [7]. Translation inhibitors such as tetracycline only inhibit bacterial growth without increasing bacterial death. They assume that the

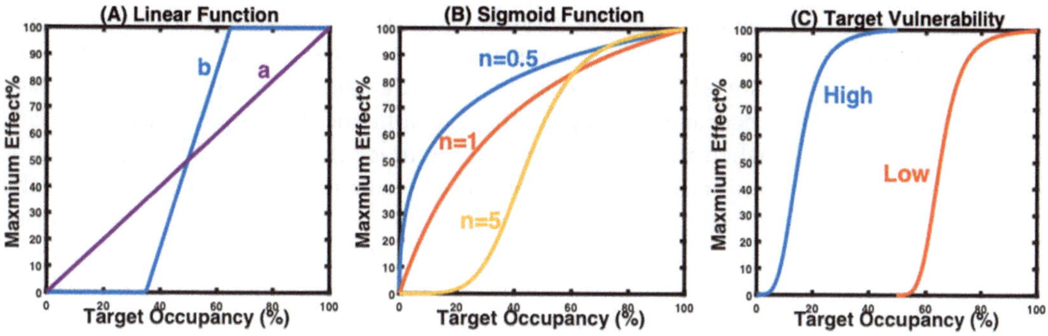

Fig. 2 Target occupancy–efficacy relationship. The occupancy–efficacy function can be linear (**a**) and sigmoidal (**b**). The occupancy–efficacy plot is a great way to show the target vulnerability (**c**)

replication rate of bacteria is linearly associated with the fraction of free ribosomes above a critical threshold, below which there is no growth of bacteria. The critical threshold can be defined as the equilibrium fraction of free targets at the minimal antibiotic concentration of drugs [9].

4.2.2 Sigmoidal Function

Mechanistic PD models must use an adequate occupancy–efficacy function or so-called target vulnerability function that can capture target engagement in inducing the desired drug response. Targets with lower vulnerability (Fig. 2c) require a higher level of target occupancy to generate the desired responses [16]. However, the relationship between target occupancy and drug efficacy is often more complicated than linear based on experimental observations [24]. Sigmoidal type occupancy–efficacy functions (Table 2), which are very similar to the form of sigmoidal $E_{\max}$ function of the traditional model (Eqs. 5 and 6), are used in several models to define the level of drug–target complex and the pharmacological effects ([18, 23, 25]). To study the cellular response for competition binding between antagonists and dopamine to D_2 receptor, de Witte et al. developed a mechanistic model that contains detailed binding kinetics of competitive binding between D_2 receptor antagonists and endogenous dopamine, and the downstream response of cellular cyclic adenosine monophosphate (cAMP) [18]. A combination of Hill functions is used to define how the production rate of cAMP is affected by both the concentration of the D_2-receptor-antagonist complex and the receptor-dopamine complex. Clarelli et al. [25] published a quantitative model to predict bacterial antibiotic efficacy. Since the exact mechanism underlying the bacteriostatic and bactericidal effects induced by certain antibiotics may not be certain, they use sigmoidal functions to describe the occupancy–efficacy relationship between bound targets and growth/death rates of bacteria, respectively.

4.2.3 Signal Transduction Network

Mechanistic PD models can further incorporate detailed downstream signaling transduction pathways. This can include mechanisms that are induced by the drug–target binding and/or feedback mechanisms which in turn will influence binding kinetics. Especially when the direct occupancy–efficacy relationship is complicated or uncertain, this type of model framework is useful to estimate the drug responses from the PD/PK data and to explore their relationship between target occupancy [26]. However, it may increase the complexity of the model structure and bring difficulties in solving and analyzing of model systems.

4.3 Advantages of Mechanistic PD Models

Mechanistic PD models that incorporate binding kinetics have demonstrated their power in making more accurate predictions of drug efficacy compared with the traditional approach and offer opportunities in understanding the physiological responses induced by drug–target binding kinetics.

4.3.1 Population Heterogeneity

Population heterogeneity can be observed among cell populations such as viruses and bacteria. Subpopulation groups show diverse properties in terms of cell shape/volume, drug permeability, replication/death rate, total content of targets/endogenous agonists/antagonists, etc. and consequently lead to heteroresistant responses to drugs. By employing the target occupancy–efficacy function, mechanistic PD models link chemical reaction kinetics and bacterial population biology [7]. The diversity among subpopulations can be modeled by assigning different values to corresponding parameters in detailed mechanistic models [27]. Consequently, these models are beneficial in investigating the heterogeneity among the target cell population and gaining insights into improving the overall drug efficacy for the whole population.

4.3.2 Prolonged Drug Action

Prolonged drug action after dosing (such as the post-antibiotic effect) is a phenomenon in which cells continue being affected by the drug even after the free drug around targets escapes or dissipates. These phenomena thus extend the time course of pharmacological effects by drugs. Prolonged drug action can result from long residence time or drug rebinding [28]. Nonspecific binding can also serve as a drug reservoir that may contribute to prolonged drug action. Mechanistic PD models that incorporate explicit association and dissociation terms of binding between specific/nonspecific targets can describe the prolonged drug action more precisely [7, 9, 22].

4.3.3 Synergistic and Antagonistic Action of Drugs

Combination drug therapy is a widely used practice that shows promising results for treating various diseases. Depending on the interaction of drugs with their targets, combinations can lead to synergistic or antagonistic effects. Due to the fact that drugs interact with their target in a network manner, it remains challenging to fully understand the mechanisms of drug synergy or antagonism, and therefore, rationally design suitable drug combinations for treatment. There are several approaches proposed for studying the synergistic/antagonistic effect of multiple drug treatments. For instance, a mechanistic model of antimicrobial combination by Baeder et al. [29] uses a multi-hit framework to link the level of individual components of bacteria–inhibitor interactions to the resulting pharmacodynamic function. Through simulation, the model provides guidance for selecting a suitable additivity term for synergy or antagonism of antimicrobials. In addition, the mechanistic PD model incorporating downstream signal transduction pathways after drug–target interaction offers a framework to investigate the interaction response of drug combination in a more systematic way [30].

5 Methodology

In the previous sections, we discussed the advantages of mechanistic models in explaining pharmacodynamic effects. In this section, we offer guidelines for selecting a suitable model according to two main considerations: biological and computational factors.

5.1 How to Select a Suitable Model

5.1.1 Selecting Traditional or Mechanistic Models to Describe the Process

From a biological perspective, whether choosing a traditional or mechanistic model depends on how much the mechanisms of drug action are known. There are two main factors to decide whether the mechanism is known: the availability of model parameters or coefficients and the model assumptions. The model parameters and coefficients can be found with preclinical data or early clinical data such as drug–target binding kinetics. The model assumptions are listed in Table 1, where an example of a traditional model is the Hill-based PK/PD model [2, 3, 31] and examples of mechanistic models are the PD models used in [7, 9]. Depending on the complexity of each mechanistic model, a subset of these assumptions is released. The traditional model is simpler and mostly used when the mechanisms of drug action is unknown. The mechanistic models have an increased complexity and can explain more details of drug binding and pharmacodynamic effects.

5.1.2 Selecting the Mathematical Modeling Approaches to Represent the Chosen Model

From a computational perspective, different types of models require different solvers or mathematical approaches to describe the drug–target binding process. The traditional model can be represented by a Hill-function such as [2, 3, 31]. The mechanistic models can be modeled and simulated by deterministic approaches or stochastic approaches.

The deterministic group of modeling approaches includes ordinary differential equations (ODE) systems such as mechanistic PD models [7, 9], master equation—a complex form of ODE, or partial differential equations (PDE) such COMBAT PD models [25]. Another group of modeling approaches employs stochastic simulations such as the Gillespie algorithm [32, 33]. Gillespie simulations are also used in modeling bacterial populations when multi-step processes of bactericidal and bacteriostatic effects are required [27].

Mechanistic PD models with different compartments [7, 9, 34] have been implemented with ODE systems that can predict the bacterial population for each compartment and population in total over time as well as predict the number of targets and target occupancy. The ODE implementation requires that the bacterial population of each compartment is represented by one differential equation. Therefore, the number of equations depends on the possible number of bound targets.

Modeling reaction kinetics of drug–target binding for multi-step processes of bactericidal and bacteriostatic effects are more challenging and require a more complex ODE system which is called a "master equations" approach [27]. The multi-step processes are the processes whereby the antibiotic–target binding can produce toxic byproducts that affect the kill rate of the cells [35]. This master equations approach, however, requires a large number of differential equations (specifically, the number of equations required is the product of the possible number of bound targets and the possible number of toxic byproducts in the system). This leads to a significant computational complexity, and it is challenging for the solver to provide meaningful results in a reasonable time to test and evaluate the models [27].

The challenge of inflated numbers of differential equations can be solved by using PDEs [25] or stochastic models such as Gillespie simulation algorithm [32, 33].

Models represented by PDEs can predict full pharmacodynamic profiles including drug efficacy and target molecule occupancy. They can also be used to predict the drug concentration ranges of the resistance selection window. PDE-based models are different from ODE-based models since they require a single equation to describe all bacterial populations with different target occupancy states. The full PD model with PDE can describe the sum of bacteria within all target occupancy states over time by a time–kill curve and therefore facilitates the fitting of this time–kill curve to experimental data while still being able to predict the bacterial populations based on occupancy states.

The binding kinetic process can also be modeled and solved by stochastic models such as Gillespie simulations. Martinecz et al. [27] use Gillespie simulations to predict bacterial population dynamics while modeling multi-step intracellular processes with bactericidal and bacteriostatic rate diversity. The Gillespie stochastic algorithm runs one simulation for each cell in the population and requires the same number of running simulations as the number of cells in the population. Their study [27] proves that using Gillespie stochastic algorithm in modeling multi-step intracellular processes is effective to model organisms with slow replication rates. However, since Gillespie simulations neglect bacterial replication, it is not suited for organisms with fast replication rates.

5.2 Estimation of Uncertain Parameters

The accuracy of mathematical models in systems biology in general and drug–target binding kinetics in particular depends on the accuracy of model parameters. Unknown model parameters can be measured experimentally or can be estimated from experimental data using model fitting techniques [36]. The most common fitting techniques are regression techniques [37], Bayesian-based techniques or their advance forms such as Monte Carlo techniques [38], or ensemble methods [39, 40].

The group of regression analyses includes maximum likelihood estimation and least-squares fitting where parameters have normal distribution. If the parameter distribution is not normal, Bayesian inference can be applied. This study review provides a guide to data regression in data-driven mechanistic modeling of systems biology [37]. Bayesian-based methods can retrieve meaningful information from uncertain data of stochastic systems and are popularly applicable in biological systems. However, Bayesian-based techniques usually rely on heavy computational workload.

Recently, computational biology models have been using control theory for parameter fitting and estimation [41–46]. In the case where experimental data is noisy and limited, combined and control-based methods have been shown to estimate the model parameters with high accuracy. The Kalman filtering procedure is one of the control-based approaches that combines control theory with optimization techniques for model parameter selection [36].

5.3 Simulation Tools

Software and tools that create PK-PD models are available in the literature. The traditional models can be implemented with the population analysis software summarized by [47] such as NONMEM, S-PLUS, SAS Proc NLIN. The PD mechanistic models of bacterial antibiotic–target binding can be created and visualized by vCOMBAT [34], an interactive web-based platform for nonquantitative scientists and healthcare providers. There also exist web-based tools that support the visualization of PK models such as a UCSF TB-Lesion tool, which simulates drug concentration over time of seven drugs on different lesions during the treatment of TB [48].

6 Challenges of Current Mechanistic PD Models

Although mechanistic PD models with drug–target binding kinetics have shown increasing value in drug discovery, there are certain challenges that have emerged when developing the models. In the mechanistic PD model, association and dissociation rate constants are critical parameters, and it is important to apply suitable values to them based on their nature to make accurate predictions. The binding kinetics of drug–target interaction are mostly measured in vitro. In vivo measurement methods are available; however, their validity remains unclear [49]. The binding kinetics in vitro and in vivo are influenced by different factors and are measured based on different methodologies. It remains a challenge to reliably translate the binding kinetics estimates in vitro into in vivo ones.

When incorporating binding kinetics and relevant pathways, model complexity increases greatly and leads to high-dimensional variables and parameters. This will bring more uncertainty when the parameter values are not well known as well as make the modeling process become time-consuming. Therefore, it is important to identify and include the most relevant pathways and most important parameters into the system. Both in vitro and in vivo experiments may be required to obtain reliable drug- and pathway-specific parameters for the model. Data generated from experiments will be particularly beneficial during model development, and therefore, a rapid cycle of communication between experimentalists and modelers is indispensable.

On the other hand, when there are several unknown parameters in the model, it is very often that several sets of parameter values can fit the model equally, making the model nonidentifiable. There are approaches available that aim to solve the issues of large parameter spaces while still being able to handle sparse and noisy data [50–52]. For example, the Kalman filtering procedure is one of the approaches for model parameter selection that combines control and optimization techniques [36].

Model simulation time depends on simulated time length, the number of model variables, and how the model is implemented. The longer the simulated time length is, the longer the computational time is. The higher number of model variables (e.g., number of ODEs) also means more required computation. The solvers not only need to find variable values but also to balance and assure their precision. Therefore, choosing the suitable solver and model implementation is one solution to reduce simulation time. For example, the vCOMBAT study and tool has improved their simulation time significantly when implementing their model in C programming languages instead of R [34].

References

1. Levy G (1964) Relationship between elimination rate of drugs and rate of decline of their pharmacologic effects. J Pharm Sci 53:342
2. Gesztelyi R, Zsuga J, Kemeny-Beke A, Varga B, Juhasz B, Tosaki A (2012) The Hill equation and the origin of quantitative pharmacology. Arch Hist Exact Sci 66(4):427–438. https://doi.org/10.1007/s00407-012-0098-5
3. Hill A (1910) The possible effects of the aggregation of the molecules of haemoglobin on its dissociation curves. J Physiol 4:4–7
4. Wagner J (1968) Kinetics of pharmacologic response I. Proposed relationships between response and drug concentration in the intact animal and man. J Theor Biol 20(2):173–201
5. Clarelli F, Liang J, Martinecz A, Heiland I, Zur Wiesch PA (2020) Multi-scale modeling of drug binding kinetics to predict drug efficacy. Cell Mol Life Sci 77(3):381–394
6. Salahudeen M, Nishtala P (2017) An overview of pharmacodynamic modelling, ligand-binding approach and its application in clinical practice. Saudi Pharm J 25:165–175
7. Abel zur Wiesch P, Abel S, Gkotzis S, Ocampo P, Engelstadter J, Hinkley T, Magnus C, Waldor MK, Udekwu K, Cohen T (2015) Classic reaction kinetics can explain complex patterns of antibiotic action. Sci Transl Med 7(287):287ra273. https://doi.org/10.1126/scitranslmed.aaa8760
8. Hong Y, Gengo FM, Rainka MM, Bates VE, Mager DE (2008) Population pharmacodynamic modelling of aspirin- and Ibuprofen-induced inhibition of platelet aggregation in healthy subjects. Clin Pharmacokinet 47 (2):129–137. https://doi.org/10.2165/00003088-200847020-00006
9. Abel Zur Wiesch P, Clarelli F, Cohen T (2017) Using chemical reaction kinetics to predict optimal antibiotic treatment strategies. PLoS Comput Biol 13(1):e1005321. https://doi.org/10.1371/journal.pcbi.1005321
10. Hammarlund-Udenaes M, Paalzow LK, de Lange EC (1997) Drug equilibration across the blood-brain barrier--pharmacokinetic considerations based on the microdialysis method. Pharm Res 14(2):128–134
11. Calvetti D, Cheng Y, Somersalo E (2015) A spatially distributed computational model of brain cellular metabolism. J Theor Biol 376:48–65. https://doi.org/10.1016/j.jtbi.2015.03.037
12. Magnus C (2013) Virus neutralisation: new insights from kinetic neutralisation curves. PLoS Comput Biol 9(2):e1002900. https://doi.org/10.1371/journal.pcbi.1002900
13. Jackson RC (2012) Pharmacodynamic modelling of biomarker data in oncology. ISRN Pharmacol 2012:590626–590626. https://doi.org/10.5402/2012/590626
14. de Lange EC, van den Brink W, Yamamoto Y, de Witte WE, Wong YC (2017) Novel CNS drug discovery and development approach: model-based integration to predict neuro-pharmacokinetics and pharmacodynamics. Expert Opin Drug Discovery 12 (12):1207–1218
15. Copeland RA, Pompliano DL, Meek TD (2006) Drug–target residence time and its implications for lead optimization. Nat Rev Drug Discov 5(9):730–739
16. Tonge PJ (2018) Drug–target kinetics in drug discovery. ACS Chem Neurosci 9(1):29–39
17. Schuetz DA, de Witte WEA, Wong YC, Knasmueller B, Richter L, Kokh DB, Sadiq SK, Bosma R, Nederpelt I, Heitman LH (2017) Kinetics for drug discovery: an industry-driven effort to target drug residence time. Drug Discov Today 22(6):896–911
18. de Witte WEA, Versfelt JW, Kuzikov M, Rolland S, Georgi V, Gribbon P, Gul S, Huntjens D, van der Graaf PH, Danhof M, Fernandez-Montalvan A, Witt G, de Lange ECM (2018) In vitro and in silico analysis of the effects of D2 receptor antagonist target binding kinetics on the cellular response to fluctuating dopamine concentrations. Br J Pharmacol 175(21):4121–4136. https://doi.org/10.1111/bph.14456
19. Shimada S, Nakajima Y, Yamamoto K, Sawada Y, Iga T (1996) Comparative pharmacodynamics of eight calcium channel blocking agents in Japanese essential hypertensive patients. Biol Pharm Bull 19(3):430–437
20. Yassen A, Olofsen E, Kan J, Dahan A, Danhof M (2007) Animal-to-human extrapolation of the pharmacokinetic and pharmacodynamic properties of buprenorphine. Clin Pharmacokinet 46(5):433–447. https://doi.org/10.2165/00003088-200746050-00005
21. Abelo A, Holstein B, Eriksson UG, Gabrielsson J, Karlsson MO (2002) Gastric acid secretion in the dog: a mechanism-based pharmacodynamic model for histamine stimulation and irreversible inhibition by omeprazole. J Pharmacokinet Pharmacodyn 29 (4):365–382
22. Walkup GK, You Z, Ross PL, Allen EK, Daryaee F, Hale MR, O'Donnell J, Ehmann

DE, Schuck VJ, Buurman ET, Choy AL, Hajec L, Murphy-Benenato K, Marone V, Patey SA, Grosser LA, Johnstone M, Walker SG, Tonge PJ, Fisher SL (2015) Translating slow-binding inhibition kinetics into cellular and in vivo effects. Nat Chem Biol 11 (6):416–423. https://doi.org/10.1038/nchembio.1796

23. Singh AP, Guo L, Verma A, Wong GG, Shah DK (2019) A cell-level systems PK-PD model to characterize in vivo efficacy of ADCs. Pharmaceutics 11(2):98. https://doi.org/10.3390/pharmaceutics11020098

24. Haraguchi K, Ito K, Kotaki H, Sawada Y, Iga T (1997) Prediction of drug-induced catalepsy based on dopamine D1, D2, and muscarinic acetylcholine receptor occupancies. Drug Metab Dispos 25(6):675–684

25. Clarelli F, Palmer A, Singh B, Storflor M, Lauksund S, Cohen T, Abel S, Abel Zur Wiesch P (2020) Drug-target binding quantitatively predicts optimal antibiotic dose levels in quinolones. PLoS Comput Biol 16(8):e1008106. https://doi.org/10.1371/journal.pcbi.1008106

26. Yin N, Pei J, Lai L (2013) A comprehensive analysis of the influence of drug binding kinetics on drug action at molecular and systems levels. Mol BioSyst 9(6):1381–1389. https://doi.org/10.1039/c3mb25471b

27. Martinecz A, Clarelli F, Abel S, Abel Zur Wiesch P (2019) Reaction kinetic models of antibiotic heteroresistance. Int J Mol Sci 20 (16):3965. https://doi.org/10.3390/ijms20163965

28. Vauquelin G, Charlton SJ (2010) Long-lasting target binding and rebinding as mechanisms to prolong in vivo drug action. Br J Pharmacol 161(3):488–508

29. Baeder DY, Yu G, Hoze N, Rolff J, Regoes RR (2016) Antimicrobial combinations: bliss independence and Loewe additivity derived from mechanistic multi-hit models. Philos Trans R Soc Lond, B, Biol Sci 371(1695):20150294

30. Yin N, Ma W, Pei J, Ouyang Q, Tang C, Lai L (2014) Synergistic and antagonistic drug combinations depend on network topology. PLoS One 9(4):e93960

31. Holford N (2017) Pharmacodynamic principles and the time course of immediate drug effects. Transl Clin Pharmacol 25(4):157–161

32. Gillespie DT, Hellander A, Petzold LR (2013) Perspective: Stochastic algorithms for chemical kinetics. J Chem Phys 138 (17):170901–170901. https://doi.org/10.1063/1.4801941

33. Gillespie DT (1977) Exact stochastic simulation of coupled chemical reactions. J Phys Chem 81(25):2340–2361. https://doi.org/10.1021/j100540a008

34. Tran VN-N, Shams A, Ascioglu S, Martinecz A, Liang J, Clarelli F, Mostowy R, Cohen T, Abel zur Wiesch P (2020) vCOMBAT: a novel tool to create and visualize a COmputational M odel of Bacterial Antibiotic Target-binding. bioRxiv:2020.2008.2005.236711. https://doi.org/10.1101/2020.08.05.236711

35. Belenky P, Ye JD, Porter CB, Cohen NR, Lobritz MA, Ferrante T, Jain S, Korry BJ, Schwarz EG, Walker GC, Collins JJ (2015) Bactericidal antibiotics induce toxic metabolic perturbations that lead to cellular damage. Cell Rep 13(5):968–980. https://doi.org/10.1016/j.celrep.2015.09.059

36. Lillacci G, Khammash M (2010) Parameter estimation and model selection in computational biology. PLoS Comput Biol 6(3): e1000696. https://doi.org/10.1371/journal.pcbi.1000696

37. Jaqaman K, Danuser G (2006) Linking data to models: data regression. Nat Rev Mol Cell Biol 7(11):813–819. https://doi.org/10.1038/nrm2030

38. Brooks S (1998) Markov chain Monte Carlo method and its application. J R Stat Soc 47 (1):69–100. https://doi.org/10.1111/1467-9884.00117

39. Brown KS, Sethna JP (2003) Statistical mechanical approaches to models with many poorly known parameters. Phys Rev E 68

40. Battogtokh D, Asch DK, Case ME, Arnold J, Schüttler HB (2002) An ensemble method for identifying regulatory circuits with special reference to the qa gene cluster of Neurospora crassa. Proc Natl Acad Sci 99(26):16904. https://doi.org/10.1073/pnas.262658899

41. Hulhoven X, Wouwer AV, Bogaerts P (2003) Hybrid extended Luenberger-asymptotic observer for bioprocess state estimation. In: 2003 European Control Conference (ECC), 1–4 Sept. 2003. pp 2535–2540. https://doi.org/10.23919/ECC.2003.7085347

42. Wells CH (1971) Application of modern estimation and identification techniques to chemical processes. AICHE J 17(4):966–973. https://doi.org/10.1002/aic.690170433

43. Lillacci G, Valigi P (2007) State observers for the estimation of mRNA and protein dynamics. In: 2007 IEEE/NIH life science systems and applications workshop, 8–9 Nov. 2007. pp 108–111. https://doi.org/10.1109/LSSA.2007.4400896

44. Lillacci G, Valigi P (2008) State estimation for a model of gene expression. In: 2008 IEEE International Symposium on Circuits and Systems, 18–21 May 2008. pp 2046–2049. https://doi.org/10.1109/ISCAS.2008.4541850
45. Boizot N, Busvelle E, Gauthier J-P (2010) An adaptive high-gain observer for nonlinear systems. Automatica 46(9):1483–1488. https://doi.org/10.1016/j.automatica.2010.06.004
46. Dochain D (2003) State and parameter estimation in chemical and biochemical processes: a tutorial. J Process Control 13(8):801–818. https://doi.org/10.1016/S0959-1524(03)00026-X
47. Duffull SB, Wright DFB, Winter HR (2011) Interpreting population pharmacokinetic-pharmacodynamic analyses—a clinical viewpoint. Br J Clin Pharmacol 71(6):807–814. https://doi.org/10.1111/j.1365-2125.2010.03891.x
48. Strydom N, Gupta SV, Fox WS, Via LE, Bang H, Lee M, Eum S, Shim T, Barry CE III, Zimmerman M, Dartois V, Savic RM (2019) Tuberculosis drugs' distribution and emergence of resistance in patient's lung lesions: a mechanistic model and tool for regimen and dose optimization. PLoS Med 16(4): e1002773. https://doi.org/10.1371/journal.pmed.1002773
49. de Witte WE, Wong YC, Nederpelt I, Heitman LH, Danhof M, van der Graaf PH, Gilissen RA, de Lange EC (2016) Mechanistic models enable the rational use of in vitro drug-target binding kinetics for better drug effects in patients. Expert Opin Drug Discovery 11 (1):45–63
50. Wilkinson DJ (2007) Bayesian methods in bioinformatics and computational systems biology. Brief Bioinform 8(2):109–116. https://doi.org/10.1093/bib/bbm007
51. McAdams HH, Arkin A (1999) It's a noisy business! Genetic regulation at the nanomolar scale. Trends Genet 15(2):65–69. https://doi.org/10.1016/s0168-9525(98)01659-x
52. Toni T, Welch D, Strelkowa N, Ipsen A, Stumpf MPH (2009) Approximate Bayesian computation scheme for parameter inference and model selection in dynamical systems. J R Soc Interface 6(31):187–202. https://doi.org/10.1098/rsif.2008.0172

Chapter 2

An Extended Model Including Target Turnover, Ligand–Target Complex Kinetics, and Binding Properties to Describe Drug–Receptor Interactions

Lambertus A. Peletier

Abstract

Since the beginning of this century, target-mediated drug disposition has become a central concept in modeling drug action in drug development. It combines a range of processes, such as turnover, protein binding, internalization, and non-specific elimination, and often serves as a nucleus of more complex pharmacokinetic models. It is simple enough to comprehend but complex enough to be able to describe a wide range of phenomena and data sets. However, the complexity comes at a price: many parameters. In this chapter, we present an overview of the temporal development of the compounds involved after different types of drug doses and offer convenient handles for dissecting data sets in a sophisticated manner in order to estimate the values of these parameters, such as rate constants and pertinent concentrations.

Key words Target, Receptor, Antibodies, Drug disposition, Michaelis–Menten, Quasi-equilibrium, Quasi-steady state

1 Introduction

1.1 Target-Mediated Drug Disposition (TMDD)

The disposition of a drug (ligand) is the sum of all distributional processes, such as binding, partitioning, and elimination (metabolism and excretion). *Target-mediated drug disposition* (TMDD) occurs when drug binds to a target with high affinity and specificity to such an extent that this affects the pharmacokinetic characteristics (cf. Levy [1] and [2]). It often includes elimination of ligand as ligand–target complex. This is particularly apparent when ligand and target occur in equimolar concentrations and exhibit nonlinear binding and clearance of ligand. Target-mediated drug disposition has become central to many drug discovery projects (cf. [3–12]).

1.2 The Basic TMDD Model and Its Equations

The basic conceptual TMDD model is shown in Fig. 1: ligand is supplied to the system and binds the target, often a receptor, to form a ligand–target complex. Target is synthesized and degraded

Quentin Vanhaelen (ed.), *Computational Methods for Estimating the Kinetic Parameters of Biological Systems*, Methods in Molecular Biology, vol. 2385, https://doi.org/10.1007/978-1-0716-1767-0_2,

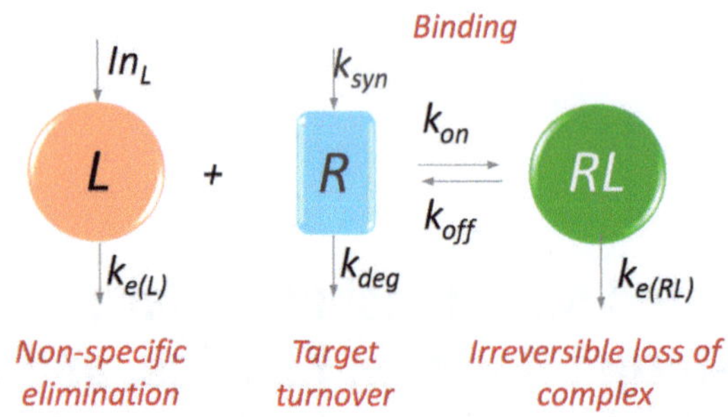

Fig. 1 Schematic rendering of the basic model for target-mediated drug disposition (TMDD): Ligand (L) binds reversibly the target (R) with a second-order rate constant k_{on} and a first-order rate constant k_{off} and forms the ligand–target complex RL, while target is synthesized with a zeroth-order rate constant k_{syn} and degraded with a first-order rate constant k_{deg}. In addition, ligand and complex are eliminated with first-order rate constants, respectively, $k_{e(L)}$ and $k_{e(RL)}$

by a turnover process, and both ligand and complex are eliminated by first-order processes.

Mathematically, the TMDD model shown in Fig. 1 can be formulated as a system of three nonlinear ordinary differential equations for the concentrations of ligand (L), receptor (R), and ligand–receptor complex (RL):

$$\begin{cases} \dfrac{dL}{dt} &= k_{\text{infus}} - k_{\text{on}} L \cdot R + k_{\text{off}} RL - k_{e(L)} L, \\ \dfrac{dR}{dt} &= k_{\text{syn}} - k_{\text{deg}} R - k_{\text{on}} L \cdot R + k_{\text{off}} RL, \\ \dfrac{dRL}{dt} &= k_{\text{on}} L \cdot R - (k_{\text{off}} + k_{e(RL)}) RL. \end{cases} \tag{1}$$

Here, $k_{\text{infus}} = In_L / V_c$, where In_L denotes the amount of ligand (e.g., antibody, peptide, protein or low molecular compound) flowing into the central compartment per unit time and V_c the volume of this compartment. In this system, k_{syn} denotes the zeroth-order receptor synthesis rate, k_{deg} the first-order receptor degradation, and $k_{e(L)}$ the first-order non-specific ligand elimination rate constant. The formation of drug–receptor complex RL involves a second-order on-rate constant k_{on} and a first-order off--rate constant k_{off}. This complex is irreversibly lost through a first-order internalization process with rate constant $k_{e(RL)}$.

The TMDD model embodies a complex of *four* different processes: (*i*) ligand–target binding, (*ii*) target turnover, (*iii*) ligand non-specific elimination, and (*iv*) ligand–target internalization, each with its own time scale. In practice, this results in ligand *vs.* time graphs with typical characteristics that can be used for diagnostic purposes, yielding information about the different processes and concentrations involved.

Balance Equations: In light of the rapid binding of ligand and receptor, the total amounts of free and bound ligand and receptor,

$$L_{\text{tot}} = L + RL \qquad \text{and} \qquad R_{\text{tot}} = R + RL, \tag{2}$$

are very useful quantities in the analysis of the TMDD model. They satisfy the following balance equations:

$$\begin{cases} \dfrac{dL_{\text{tot}}}{dt} &= k_{\text{infus}} - k_{e(L)}L - k_{e(RL)}RL, \\ \dfrac{dR_{\text{tot}}}{dt} &= k_{\text{syn}} - k_{\text{deg}}R - k_{e(RL)}RL. \end{cases} \tag{3}$$

These equations can readily be derived from the full system (1): the first one by adding the equations involving L and the second one by adding the equations involving R.

Note that in the absence of ligand the baseline concentrations are given by

$$L = 0, \qquad R = R_0 \stackrel{\text{def}}{=} \frac{k_{\text{syn}}}{k_{\text{deg}}}, \qquad RL = 0. \tag{4}$$

A simple corollary of the target balance equation in (3) is the following invariance property:

Corollary 1 *If* $k_{\text{deg}} = k_{e(RL)}$, *then* $R_{\text{tot}}(t) = R_0$ *for all* $t \geq 0$.

Indeed, if $k_{\text{deg}} = k_{e(RL)}$, then the target balance equation simplifies to

$$\frac{dR_{\text{tot}}}{dt} = k_{\text{deg}}(R_0 - R_{\text{tot}}). \tag{5}$$

The initial values in (4) imply that $R_{\text{tot}}(0) = R_0$. Therefore, $R_{\text{tot}}(t) = R_0$ for all $t \geq 0$.

In this chapter, we offer ways of interpreting the different characteristics of the kinetic time courses of ligand, target, and ligand–target complex, revealing different properties of the constituent processes. We do this by analyzing data sets obtained after different *iv* bolus doses (Subheading 2), data sets during and after *iv* constant-rate infusions over a finite period of time (Subheading 3), the corresponding steady states (Subheading 4), and the way clearance CL depends on the ligand concentration at steady state (Subheading 5). In practice, the data may not always be sufficiently detailed to determine the values of all the six parameters in the system of Eqs. 1. For such situations, alternative, simpler, models have been proposed (Subheading 6). This chapter closes with a brief discussion of some of the contexts and generalizations in which the basic TMDD model has featured (Subheading 7).

Numerical computations and simulations presented in this chapter are made with MATLAB R2018a.

2 Characterization of TMDD Disposition After an *iv* Bolus Administration of Ligand

Initially, a bolus dose D of ligand is administered *intravenously* (*iv*) into the central compartment (volume V_c), while the system is at baseline (cf. (4)). This results in the following initial concentrations of ligand, receptor, and ligand–receptor complex will be given by

$$L(0) = L_0 \stackrel{\text{def}}{=} \frac{D}{V_c}, \qquad R(0) = R_0, \qquad RL(0) = 0. \qquad (6)$$

It is well established that after a large *iv* bolus dose, ligand *vs.* time curves display a characteristic shape (cf. e.g. [10–13]). Figure 2 gives a schematic illustration of such ligand *vs.* time data in plasma obtained after a large *iv* bolus dose. Following [11], the kinetics can be divided into four phases $A, \ldots, D$, each associated with one or more specific processes.

Counting from the time of administration, during the first phase (A), which is often brief, ligand and target bind complex is formed, and at the end of this phase, assuming the dose is large enough, the target is saturated. During the second phase (B), the target remains saturated, and since the target route has limited capacity, for higher doses ligand is primarily eliminated via the non-specific route. In the third phase (C), the target is no longer saturated and ligand is also eliminated via the target route, i.e., as part of the ligand–target complex and, as in phase B, the non-specific route. Finally, the fourth phase (D) is the terminal

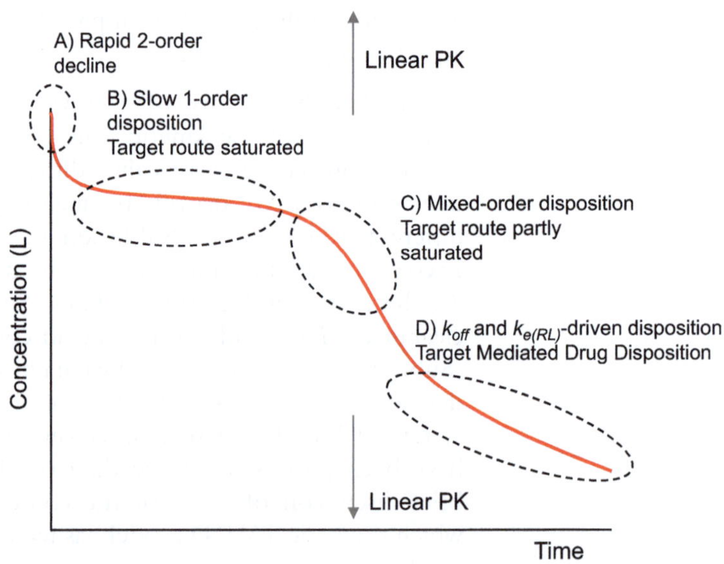

Fig. 2 Structure of a typical ligand *vs.* time data set after a large *iv* bolus dose for TMDD kinetics together with the associated processes. Ligand concentration is measured on a logarithmic scale, while time is measured on a linear scale

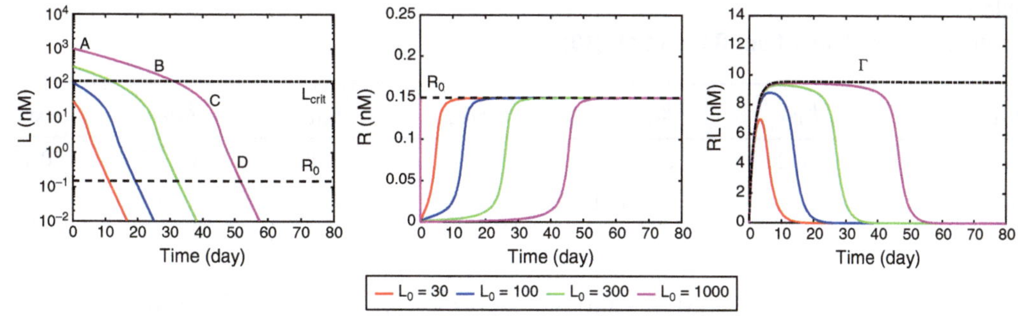

Fig. 3 Graphs of the concentrations of ligand L, receptor R, and ligand–receptor RL vs. time for initial concentrations given by (8) and parameters listed in Table 1. In the left panel, the letters A, B, C, and D along the top curve indicate the four phases, and the dash-dotted line marks the ligand concentration L_{crit} defined by (18), which indicates the *onset of nonlinearity*. In the right panel, the curve Γ is the graph of the approximate ligand–receptor complex concentration $RL_{\text{approx}}(t)$ defined in (9)

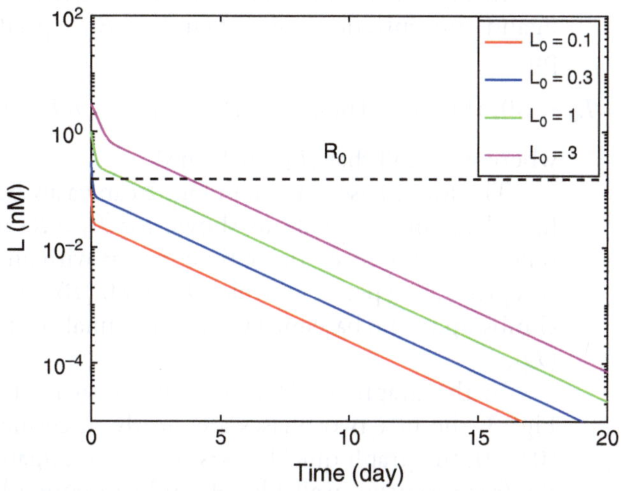

Fig. 4 Graphs of the free ligand concentration, L, vs. time for initial concentrations $L_0 = 0.3$, 1, 3, 10 nM and $R(0) = R_0$ and $RL(0) = 0$ nM and parameters shown in Table 1

phase in which $R \approx R_0$, the kinetics is approximately linear, and ligand, target, and complex converge to their baseline values (4).

For convenience, we denote the approximate times of the transitions between these four phases by T_1 (A/B), T_2 (B/C), and T_3 (C/D).

In Figs. 3 and 4, we show simulations of the TMDD model (1) for parameters taken form Betts et al. [10], quoted in Table 1. The concentration *vs.* time curves exhibit the characteristic features seen in TMDD data.

For the parameter values in Table 1, the dissociation constant K_d and the Michaelis–Menten constant K_m are

Table 1
Parameter values taken from Betts et al. [10]

k_{on}	k_{off}	k_{syn}	k_{deg}	$k_{e(L)}$	$k_{e(RL)}$	R_0
316	16.21	5.9	39	0.051	0.613	0.15
$(nM)^{-1}$/day	1/day	(nM)/day	1/day	1/day	1/day	nM

$$K_d = \frac{k_{off}}{k_{on}} = 0.051 \text{ nM} \quad \text{and} \quad K_m = \frac{k_{off} + k_{e(RL)}}{k_{on}} = 0.053 \text{ nM}, \tag{7}$$

i.e., the affinity of the ligand to the receptor is high.

For the ligand *vs.* time graphs, it makes a difference whether $L_0 > R_0$ or $L_0 < R_0$. For the larger doses, the graphs have the four-phase structure discussed above. For the smaller values, the structure is simpler and close to bi-exponential.

In Fig. 3, we show simulations for larger values of the initial ligand concentration, i.e., when $L_0 > R_0$. Specifically, at $t = 0$ we put

$$L = L_0 = 30, 100, 300, 1000, \qquad R = R_0, \qquad RL = 0 \text{ nM}, \tag{8}$$

where we recall that $R_0 = 0.15$ nM.

The four phases shown in Fig. 2 can easily be recognized in the ligand *vs.* time simulations shown for $L_0 > R_0$ in Fig. 3. They are reflected in the receptor *vs.* time graphs with an initial drop of free receptor (A), a period in which $R(t) \approx 0$ (B), a period in which $R(t)$ climbs up to the baseline (C), and the final stretch when $R(t) \approx R_0$ (D).

In the graph of $RL(t)$ *vs.* time, shown in the right panel of Fig. 3, the first two phases are clearly discernible. Starting at $RL(0) = 0$, the graph quickly rises to R_0, i.e., ligand quickly binds all the free receptors available (A) and subsequently increases as long as $R(t) \approx 0$, i.e., the kinetics is in phase (B) while following the graph Γ of the function

$$R_{approx}(t) = R_* + (R_0 - R_*) \cdot e^{-k_{e(RL)}t}, \qquad R_* = \frac{k_{syn}}{k_{e(RL)}}. \tag{9}$$

Subsequently, $RL(t)$ enters a quick descent toward $RL = 0$. This period coincides, first with (C) and then, when $RL(t)$ is very small, with (D).

Figure 4 shows simulations for smaller initial doses. The initial ligand concentrations $L_0 = 0.1$, 0.3, 1, 3 nM straddle the target baseline concentration $R_0 = 0.15$ nM. Thus, R_0 is seen to be a threshold: For $L_0 > R_0$, the structure of the graphs is as shown in Fig. 3, with four phases $A \ldots D$, while for $L_0 < R_0$ the ligand *vs.* time curves become increasingly bi-exponential as $L_0 \to 0$.

Below, we successively describe the kinetics of ligand, target, and ligand–receptor in the four phases seen when $L_0 > R_0$. Specifically, we focus on two questions:

(i) Which parameters determine the properties, such as slope and duration, of the phases $A \ldots D$?

(ii) How can initial estimates of the model parameters be obtained from experimental data?

2.1 Phase A

The shape and duration of *Phase A* after an *iv* bolus dose are mainly determined by the loss of ligand L and receptor R due to a second-order binding reaction $k_{\text{on}} \times L \times R$. The second-order rate constant k_{on}, in combination with the initial concentrations of target R_0 and ligand L_0, will be seen to be critical parameters during *Phase A*.

Specifically, if (i) K_d is small (e.g., $K_d \ll R_0$), (ii) the first-order rate constants $k_{e(L)}$, $k_{e(RL)}$, and k_{deg} are all comparable in magnitude to k_{off}, i.e., none of them is exceptionally small or large in comparison with k_{off}, and (iii) if $L_0 > R_0$, then it is possible to show (cf. [11]) that if $L \gg K_m$, then [1]

$$L(t) \searrow L_0 - R_0, \quad R(t) \searrow 0 \quad \text{and} \quad RL(t) \nearrow R_0 \quad \text{as} \quad t \to T_1,^2 \tag{10}$$

where T_1 denotes the time when *Phase A* runs over into *Phase B*. The half-life of this limit is approximately given by

$$t_{1/2} = \frac{\log(2)}{k_{\text{on}}(L_0 - R_0)}. \tag{11}$$

Thus, the half-life is inversely proportional to k_{on} and, for larger values of the initial dose, with L_0. Note that when we take $T_1 \approx 4 \times t_{1/2}$, use the parameter values of Table 1, and put $L_0 = 100$ nM, then *Phase A* spans approximately 10^{-4} days.

Remark: From the balance equation for L in (3), it can be seen that because *Phase A* is usually short, very little ligand gets lost during this phase (cf. [11]), i.e., in light of the initial concentration of L and RL,

$$L(t) + RL(t) \approx L_0 \qquad \text{for} \qquad 0 < t < T_1. \tag{12}$$

Parameter Estimates: The half-life given by Eq. 11 yields a handle on the important parameter k_{on} when L_0 and R_0 are known. Since *Phase A* is often very short, the data need to have a high degree of granularity.

[1] Here $F \nearrow L$ means F approaches L from below and $G \searrow L$ means G approaches L from above.

2.2 Phase B

During *Phase B*, the receptor is close to saturation, so that little free receptor is left, i.e., $R \approx 0$ and $R_{\text{tot}} \approx RL$. Therefore, during this phase, the balance Eqs. 3 are well approximated by the following equations:

$$\begin{cases} \dfrac{d}{dt}(L + RL) = -k_{e(L)}L - k_{e(RL)}RL, \\ \dfrac{dRL}{dt} = k_{\text{syn}} - k_{e(RL)}RL. \end{cases} \tag{13}$$

Subtracting the second equation from the first then yields

$$\frac{dL}{dt} = -k_{e(L)}L - k_{\text{syn}}. \tag{14}$$

Thus, during *Phase B*, ligand is eliminated via two routes: (i) a first-order non-specific route with rate constant $k_{e(L)}$ and (ii) a zeroth-order target route with rate constant k_{syn}.

Equation 14 can be solved analytically and so yields an explicit approximation of the ligand *vs.* time graph during *Phase B*:

$$L_{\text{approx}}(t) = \left(L_0 - R_0 + \frac{k_{\text{syn}}}{k_{e(L)}}\right)e^{-k_{e(L)}t} - \frac{k_{\text{syn}}}{k_{e(L)}} \qquad \text{for} \quad 0 < t < T_2. \tag{15}$$

Since the exponential function in the expression for $L_{\text{approx}}(t)$ tends to zero as $t \to \infty$, it follows that the function $L_{\text{approx}}(t) < 0$ for t large enough and therefore vanishes at a finite time $t = T_2^0 > 0$. This time can be computed explicitly:

$$T_2^0 = \frac{1}{k_{e(L)}} \log\left\{\frac{k_{e(L)}}{k_{\text{syn}}}(L_0 - R_0) + 1\right\}. \tag{16}$$

The time T_2^0 can serve as an estimate for the time T_2 at which *Phase B* ends.

Note that the function $L_{\text{approx}}(t)$ is determined by the initial target and ligand concentrations, the synthesis rate k_{syn}, and the non-specific ligand elimination rate $k_{e(L)}$. Hence, in *Phase B*, non-specific elimination and target synthesis are the dominant processes.

The second equation in the system (13) can be solved for $RL(t)$. Thanks to rapid initial binding, we may take $RL(0) = R_0$ and so obtain

$$RL(t) \approx RL_{\text{approx}}(t) = R_* + (R_0 - R_*)\, e^{-k_{e(RL)}t} \tag{17}$$

as long as $R(t) \approx 0$, i.e., in *Phase B*. The graph of $RL_{\text{approx}}(t)$ is shown in the third panel of Fig. 3 and is denoted by Γ.

Clearly, $RL(t) \to R_*$ as $t \to \infty$. Since $R(t) \approx 0$ in *Phase B*, it follows that $R_{\text{tot}}(t) \to R_*$. Thus, for sufficiently large doses, the amount of receptor *accumulates* during *Phase B* by a factor $R_*/R_0 = k_{\text{deg}}/k_{e(RL)}$ nM. For the values in Table 1, $R_* = 9.62$ nM and hence target accumulation amounts to about 63.6.

Parameter Estimates:
From the ligand graphs:

(i) The slope of the "linear" part of the graphs yields an estimate for $k_{e(L)}$.

(ii) Fitting the graph of $L_{\text{approx}}(t)$ to the ligand data yields an estimate for k_{syn}.

From the ligand–receptor graphs:

(iii) Fitting the graph of $RL_{\text{approx}}(t)$ to the data of the receptor–ligand complex yields estimates for $k_{e(RL)}$ and $R_* = k_{\text{syn}}/k_{e(RL)}$ and hence of k_{syn}.

2.3 Phase C

Phase C covers the transition of first-order kinetics of L in *Phase B*, down to first-order kinetics in *Phase D*. Thus, *Phase C* begins when the receptor ceases to be saturated, and $L(t)$ drops down to values that are of the same order as K_m. Thus, elimination via the target route then starts to contribute.

An alternative marker for the *Onset of nonlinearity* has been proposed by Stein et al. in [14]; it states that *Phase C* begins when L drops below a critical value L_{crit} defined by

$$L_{\text{crit}} \stackrel{\text{def}}{=} \frac{k_{\text{syn}}}{k_{e(L)}}. \tag{18}$$

It is a marker of practical value because in *Phase C* the *receptor occupancy* (RO) drops below 100% and the therapeutic value may become less than optimal.

In Fig. 3, it is seen that the graph of $L(t)$ shifts to the right with increasing doses. This shift is easy to measure and can yield information about the system (1). Specifically, let D_1 and D_2 be two doses so numbered that $D_1 > D_2$, and let the corresponding ligand *vs.* time graphs be denoted by $L(t;D_1)$ and $L(t;D_2)$. Then, when the doses are large enough, the graphs of these functions cross the level L_{crit}, say at times t_1 and t_2, i.e.,

$$L(t_1; D_1) = L_{\text{crit}} \quad \text{and} \quad L(t_2; D_2) = L_{\text{crit}}.$$

It can be shown that $t_1 > t_2$ and (cf. [11])[2]

$$t_1 - t_2 \sim \frac{1}{k_{e(L)}} \{\log(D_1) - \log(D_2)\} \quad \text{as} \quad D_1,\ D_2 \to \infty. \tag{19}$$

[2] We write $f(x) \sim A \cdot g(x)$ as $x \to 0$ (∞) if: $f(x)/g(x) \to A$ as $x \to 0$ (∞).

Parameter Estimates: The top of *Phase C* yields an estimate for L_{crit}, i.e., for the quotient of two rate constants involved (cf. (18)); and the rate with which the profile translates with the drug dose D gives an estimate for $k_{e(L)}$ (cf. (19)).

2.4 Phase D

Phase D is the terminal phase in which $R(t) \approx R_0$. This means that the second-order term $k_{\mathrm{on}} L \times R$ can be approximated by $k_{\mathrm{on}} L \times R_0$. The resulting linear system has three negative eigenvalues of which the smallest one is the terminal slope λ_z. It is shown in Appendix 7 of [11] that if the condition

$$k_{e(RL)} \ll k_{\mathrm{on}} R_0 \tag{20}$$

is satisfied, then

$$\lambda_z \approx k_{e(RL)}. \tag{21}$$

For the parameter values of Table 1, $k_{e(RL)}/(k_{\mathrm{on}} R_0) = 0.0129$, i.e., the condition (20) is satisfied.

Parameter Estimates: Because λ_z can be expressed in an explicit formula, the terminal slope yields a relation between the parameters involved in this formula. In fact, if the condition (20) is satisfied, Eq. 21 can be used to obtain a value for the parameter $k_{e(RL)}$.

2.5 Conclusions

The table below summarizes the processes, the associated parameters, and the data characteristics, such as the slope or shift or the duration of the phase, of the ligand concentration *vs.* time, which are important in the different phases *A*...*D*. The slopes are of graphs with concentrations measured on a logarithmic scale (Table 2).

The slope in *Phase D* is estimated subject to the condition stated in (20).

Note that provided the experimental ligand data are sufficiently granular, with the exception of k_{off}, all the parameters in the system (1) can be estimated from the ligand *vs.* time graphs after large *iv* bolus data. However, sufficiently granular data are often not available. For such situations, periodic *iv* dosing or drug administration

Table 2
Summary of processes, parameters, and data characteristics

Phase	Time span	Kinetics	Parameters	Slopes	Shifts
A	$(0, T_1)$	*Ligand–target binding*	k_{on}, R_0	–	–
B	(T_1, T_2)	*Linear first-order kinetics*	$k_{e(L)}, k_{e(RL)}, k_{\mathrm{syn}}, R_0$	$-k_{e(L)}$	$1/k_{e(L)}$
C	(T_2, T_3)	*Michaelis–Menten kinetics*	K_m	–	–
D	(T_3, ∞)	*Linear first-order kinetics*	$k_{e(L)}, k_{e(RL)}$	$-k_{e(RL)}$	$1/k_{e(L)}$

by an *iv* infusion may yield good estimates from properties of transient behavior and steady states. Kinetics during such an administration will be discussed in the next section that will be followed by an analysis of steady states.

3 Characterization of TMDD Disposition After an *iv* Constant-Rate Infusion of Ligand

Here, drug is administered by means of an *iv* constant-rate infusion followed, after a time t_{washout}, by washout. This is a strong experimental approach for discriminating between different kinetic phenomena. For this type of drug administration, the infusion rate $k_{\text{infus}}(t)$ is defined by

$$k_{\text{infus}}(t) = \begin{cases} k_f & 0 < t < t_{\text{washout}} \quad \textit{Uptake phase}, \\ 0 & t_{\text{washout}} < t < \infty \quad \textit{Washout phase}, \end{cases} \tag{22}$$

where k_f denotes the rate of infusion. Initially, there is no ligand in the system, and the target concentration is at baseline level, i.e.,

$$L(0) = 0, \qquad R(0) = R_0 \quad \text{and} \quad RL(0) = 0. \tag{23}$$

As in the previous section, the parameters are taken from Table 1.

The kinetics for this regime is shown in Fig. 5 for the infusion rates and the time of washout

$$k_f = 0.5,\ 3,\ 6,\ 10 \quad \text{nM/h}; \qquad t_{\text{washout}} = 100 \text{ days}. \tag{24}$$

The ligand concentrations are shown on a semi-logarithmic scale, and the concentrations of free receptor and receptor–ligand complex are shown on a linear scale.

It is seen that in the uptake phase the ligand concentrations increase and converge monotonically toward constant states through phases that are comparable to *Phase A* in which ligand binds the receptor and *Phase B* in which the receptor is saturated,

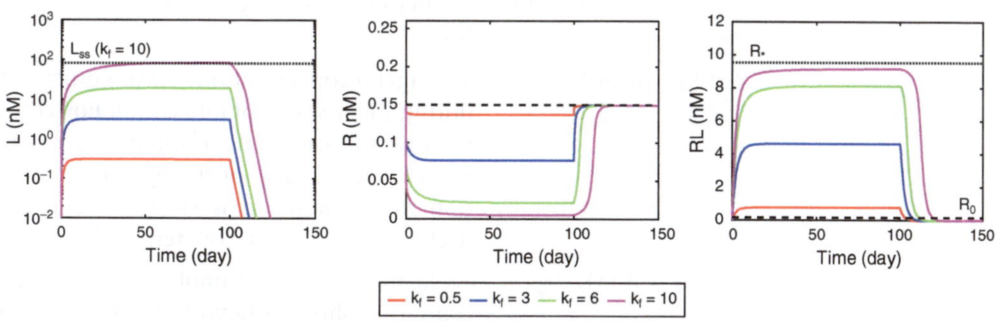

Fig. 5 Concentrations of ligand, free target, and ligand–target complex vs. time for infusion rates k_f given by (24), which straddle the synthesis rate $k_{\text{syn}} = 5.9$ (nM)/h. The parameter values are taken from Table 1 with $R_0 = 0.15$ nM, $R_* = k_{\text{syn}}/k_{e(RL)} = 9.6$ nM, and $t_{\text{washout}} = 100$ days

while in the washout phase the ligand graphs descend to zero through phases that are comparable to *Phases B, C,* and *D*, as seen after an *iv* bolus dose.

During the uptake phase, the ligand concentrations increase—first rapidly (*Phase A*) and then more slowly—and in the end converge toward a steady state $L_{ss}(k_f)$. The washout time has been chosen long enough so that L, R, and RL have stabilized at $t = T$.

For infusion rates that exceed the synthesis rate ($k_f > k_{syn}$), as for $k_f = 6$ *&* 10 (nM)/h, the ligand concentrations are so large that the receptor is quickly saturated and eventually ceases to play a significant role in the clearance. What remains is the linear non-specific clearance of ligand, so that after *Phase A*, the ligand kinetics is well described by the equation

$$\frac{dL}{dt} = k_f - k_{syn} - k_{e(L)}L, \tag{25}$$

so that

$$L(t) \to L_\infty \stackrel{\text{def}}{=} \frac{k_f - k_{syn}}{k_{e(L)}} \quad \text{as } t \to \infty \qquad \text{and} \qquad t_{1/2} = \frac{\log(2)}{k_{e(L)}}. \tag{26}$$

Hence, for large doses, the half-life no longer depends on the infusion rate. For the parameter values given in Table 1, this yields a half-life of $t_{1/2} = 13.6$ day and a time to steady state of $4 \times t_{1/2} = 54.4$ days, which agrees with the first graph on the left in Fig. 5.

For larger infusion rates ($k_f > k_{syn}$), the **washout kinetics** is very similar to the kinetics after an *iv* bolus dose, as described in Subheading 2. However, since ligand and target have formed a complex during the uptake phase, *Phase A* is now absent. Plainly, for larger infusion rates ($k_f > k_{syn}$), the characteristic phases, seen after a bolus dose, *B, C,* and *D* can be recognized in the post-washout kinetics.

We describe the ligand graphs for the four infusion rates one by one; the first two rates are below the synthesis rate and the second two infusion rates are above:

(a) $\mathbf{k_f = 0.5}$nM/day: The ligand concentration shows a quick jump during the first day of infusion, followed by a rise toward steady state of about 4 days. There is then a rapid drop of the ligand concentration as the infusion is stopped. The rapid decline approaches a slow terminal phase.

(b) $\mathbf{k_f = 3}$nM/day: As in the previous example, the ligand concentration shows a rapid initial rise but then approaches the steady state more slowly than in Case (a). The washout behavior is similar to Case (a).

(c) $k_f = 6$nM/day: After an initial jump in ligand concentration as in Cases (a) and (b), ligand approaches the steady state more slowly and levels off. The washout profile in Case (c) exhibits a slight nonlinearity until the ligand concentration approaches the terminal phase.

(d) $k_f = 10$nM/day: The initial quick jump in ligand exposure is also present, but the steady state is reached still more slowly than in Case (c). The washout profile displays a more pronounced nonlinear pattern until it reaches the terminal phase.

It is seen that as the infusion rate increases, the time to steady state increases toward about 4 times the half-life given in (26). The reason is that for small infusion rates, the ligand concentration is low and ligand is cleared along both routes, the target route and the non-specific route. For large infusion rates, target is saturated and plays a limited role in ligand clearance so that the total clearance is smaller (cf. Fig. 8).

Summarizing: As the infusion rate k_f increases, the ligand *vs.* time graphs show a complex collection of profiles, distinguished by: (i) an increasing sequence of limit values, i.e., steady-state values $L_{ss}(k_f)$, (ii) varying times to steady state, and (iii) complex post-washout profiles. In the next section, we derive explicit expressions for the way L_{ss} depends on k_f and show among other things that

$$L_{ss}(k_f) \sim \frac{k_f - k_{syn}}{k_{e(L)}} \quad \text{as} \quad k_f \to \infty, \tag{27}$$

which is consistent with the observation made in (26).

Parameter Estimates:

Uptake Period

(a) For larger infusion rates, the half-life of the rate with which ligand converges toward L_∞ yields an estimate for $k_{e(L)}$.

(b) Once $k_{e(L)}$ has been estimated, the expression for L_∞ yields an estimate for k_{syn}.

Washout Period

(c) As in Subheading 2, if $k_{e(RL)} \ll k_{on}R_0$, then the terminal slope λ_z is found to be $\lambda_z \approx k_{e(RL)}$ and thus can be used to estimate $k_{e(RL)}$.

4 Equilibrium Relationships

Whereas in the previous section, the dynamical properties of concentration *vs.* time graphs were exploited for diagnostic purposes, in this section we focus on two types of characterizations of the steady-state concentrations of the three compounds:

(i) Graphs of steady-state concentrations of ligand L_{ss}, receptor R_{ss}, and ligand–receptor complex RL_{ss} versus the infusion rate k_{infus}

(ii) Graphs of steady-state concentrations of receptor R_{ss}, and ligand–receptor complex RL_{ss} versus the steady-state ligand concentration L_{ss}

In Subheading 4.1, we analyze graphs of the first type. It will be demonstrated that graphs of steady-state concentrations of ligand, receptor, and complex, L_{ss}, R_{ss}, and RL_{ss}, versus the infusion rate k_{infus} yield interesting and complex figures. We show how geometric properties of these graphs such as plateaus and slopes are related to the values of the parameters in the TMDD model (1) and yield estimates of those parameters.

In Subheading 4.2, we derive simple analytical expressions for R_{ss} and RL_{ss} as functions of the steady-state ligand concentration L_{ss} and introduce the ***in vivo potency*** EC_{50}, which displays succinctly how the target degeneration rate k_{deg}, the internalization rate $k_{e(RL)}$, and the Michaelis–Menten binding constant K_m each affect the potency of the drug.

For simplicity, we omit the subscript "ss" for the remainder of this section.

4.1 L, R, and RL vs. the Ligand Dosing Rate

The three steady-state concentrations can be computed as functions of the infusion rate k_{infus} by solving the three equations in the TMDD system (1) while putting the time derivatives on the left-hand side equal to zero. The resulting expressions for L, R, and RL as functions of $k_{infus} = k_f$ are graphed in Fig. 6 for the parameter values quoted in Table 1. The right panel is a blow-up for small concentrations.

Both graphs in Fig. 6 are complex and harbor much information. Since the graphs can also be computed explicitly, any information they contain can be extracted analytically.

Qualitative Properties: As the ligand infusion rate increases, more ligand binds the receptor, depleting the free receptor pool. Thus, R decreases causing new receptor to be synthesized. However, once k_{infus} exceeds the synthesis rate, the receptor becomes saturated, $R \approx 0$, and RL has reached its maximum value R_* (cf. Eq. 29).

It is interesting to note that once the receptor is saturated, the system has lost one clearance route for ligand, leaving only the non-specific route, causing the ligand concentration to climb faster

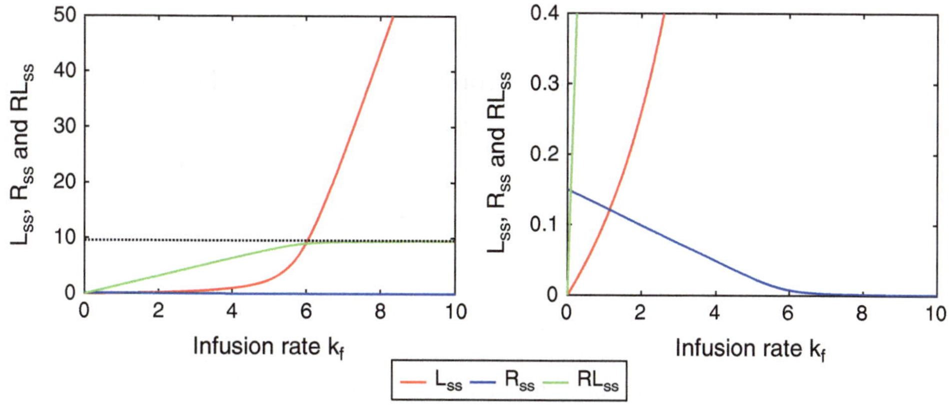

Fig. 6 Steady-state concentrations L_{ss}, R_{ss}, and RL_{ss} graphed versus the constant infusion rate $k_{infus} = k_f$ (*Left*). A blow-up of the small-concentration range, shown on the right, adds detail to the target concentration that starts at $R = R_0 = 0.15$ nM. Parameter values are taken from Table 1

with increasing infusion rate. This is seen in Fig. 6, where the ligand curve has a sharp kink when $k_{\text{infus}} = k_f \approx k_{\text{syn}}$.

Quantitative Properties: At steady state, the two balance Eq. 3 yield the following relations between the concentrations of the three compounds:

$$\begin{cases} L &= \dfrac{1}{k_{e(L)}}(k_{\text{infus}} - k_{e(RL)}RL), \\ R &= \dfrac{1}{k_{\text{deg}}}(k_{\text{syn}} - k_{e(RL)}RL). \end{cases} \tag{28}$$

Evidently, because $R \geq 0$, the second equation implies that

$$RL \leq R_* = \frac{k_{\text{syn}}}{k_{e(RL)}}, \tag{29}$$

as is seen in Fig. 6. According to the first equation in (28), once RL has reached its ceiling R_*, then L increases linearly, with a slope given by $(k_{e(L)})^{-1}$, as the infusion rate increases further.

Mathematical Analysis: The pair of equations in (28) states that L and R can be expressed in terms of RL, the concentration of the ligand–target complex. Using these equations in the third equation of the system (1) to eliminate L and R, we obtain a quadratic equation in terms of a single variable RL. This equation can be solved explicitly and yields the following solution:

$$RL = \frac{1}{2k_{e(RL)}}\left(a - \sqrt{a^2 - 4b}\right), \tag{30}$$

in which the constants a and b are combinations of the parameters in the model (1). They are defined by

$$a \stackrel{\text{def}}{=} k_{\text{infus}} + k_{\text{syn}}(1 + \mu \cdot \kappa_m) \qquad \text{and} \qquad b \stackrel{\text{def}}{=} k_{\text{infus}} k_{\text{syn}}. \tag{31}$$

The dimensionless constants, μ and κ_m, in these definitions are

$$\mu = \frac{k_{e(L)}}{k_{e(RL)}} \qquad \text{and} \qquad \kappa_m = \frac{K_m}{R_0}. \tag{32}$$

They have a pharmacokinetic significance: μ relates the two elimination time scales and κ_m relates the Michaelis–Menten binding constant to the baseline target concentration.

Whereas the steady-state concentration of ligand–target complex RL is given by (30) and (31), the steady-state concentrations of ligand and target can now be obtained by substituting RL into the two balance equations of the system (28).

For small and large infusion rates, the three concentrations are found to have the following limiting behavior:

Ligand–Target Complex:

$$\begin{cases} RL \to \dfrac{k_{\text{syn}}}{k_{e(RL)}} = R_* & \text{as} \quad k_{\text{infus}} \to \infty, \\[2ex] RL \sim \dfrac{1}{k_{e(RL)}} \dfrac{R_0}{R_0 + \dfrac{k_{e(L)}}{k_{e(RL)}} K_m} \cdot k_{\text{infus}} & \text{as} \quad k_{\text{infus}} \to 0. \end{cases} \tag{33}$$

At large infusion rates, the target will become saturated and the ligand–target complex approaches its a priori upper limit R_* (cf. (29)). At small infusion rates, RL increases linearly with k_{infus}.

Ligand:

$$\begin{cases} L \sim \dfrac{1}{k_{e(L)}} (k_{\text{infus}} - k_{\text{syn}}) & \text{as} \quad k_{\text{infus}} \to \infty, \\[2ex] L \sim \dfrac{1}{k_{e(L)}} \dfrac{\dfrac{k_{e(L)}}{k_{e(RL)}} K_m}{R_0 + \dfrac{k_{e(L)}}{k_{e(RL)}} K_m} \cdot k_{\text{infus}} & \text{as} \quad k_{\text{infus}} \to 0. \end{cases} \tag{34}$$

Target:

$$\begin{cases} R \to 0 & \text{as} \quad k_{\text{infus}} \to \infty, \\[2ex] R \sim R_0 \left(1 - \dfrac{R_0}{R_0 + \dfrac{k_{e(L)}}{k_{e(RL)}} K_m} \cdot \dfrac{k_{\text{infus}}}{k_{\text{syn}}} \right) & \text{as} \quad k_{\text{infus}} \to 0. \end{cases} \tag{35}$$

Plainly, the two dimensionless quantities μ and κ_m introduced in (32) are critical in the limiting behavior of RL, L, and R as $k_{\text{infus}} \to 0$. Notice that they appear together, as a product. This can be understood as follows: The parameter μ compares the rates of the two ligand elimination routes; when μ is large, non-specific elimination is faster, and when it is small, the target route is faster. On the other hand, κ_m says something about the target: when it is small, ligand will have a stronger tendency to bind the target and be eliminated through internalization. Thus, together μ and κ_m determine which of the elimination routes is dominant: when $\varepsilon \stackrel{\text{def}}{=} \mu \cdot \kappa_m$ is large, more ligands will go through the non-specific route, while when it is small, most of the ligands go through the target route. For the parameters from Table 1, μ, κ_m, and ε become

$$\mu = 0.083, \quad \kappa_m = 0.35, \quad \varepsilon = 0.029.$$

More generally, the value of these numbers determines (i) the shapes of the curves in Fig. 6, (ii) the small slope of the ligand curve for small infusion rates $k_{\text{infus}} < k_{\text{syn}}$, and (iii) the sharp transitions near when the infusion rate equals the synthesis rate: $k_{\text{infus}} = k_{\text{syn}}$. For instance, if $\varepsilon = \mu \cdot \kappa_m \ll 1$, as is the case for parameter values from Table 1, then for ligand the small infusion limit becomes

$$L(k_{\text{infus}}) \sim \frac{\varepsilon}{k_{e(RL)}} k_{\text{infus}} \quad \text{as} \quad k_{\text{infus}} \to 0 \tag{36}$$

as is demonstrated in Fig. 6.

4.2 R_{ss} and RL_{ss} Versus the Ligand Concentration L_{ss}

In order to express RL in terms of L, we use the ligand balance equation from (28) to eliminate k_{infus} from the first equation of the system (1) and subsequently use the receptor balance equation in (28) to eliminate R. This results in the equation

$$k_{e(RL)} RL \cdot \left(\frac{k_{\text{on}}}{k_{\text{deg}}} L + 1 + \frac{k_{\text{off}}}{k_{e(RL)}} \right) = k_{\text{on}} R_0 \cdot L, \tag{37}$$

which involves L and RL only. When we now divide by k_{on} and multiply by k_{deg}, we obtain the desired expression:

$$RL = \frac{k_{\text{syn}}}{k_{e(RL)}} \frac{L}{L + L_{50}} \quad \text{where} \quad L_{50} = \frac{k_{\text{deg}}}{k_{e(RL)}} \cdot K_m = \frac{k_{\text{syn}}}{k_{e(RL)}} \frac{K_m}{R_0}. \tag{38}$$

From the target balance equation in (28), we deduce that

$$R = \frac{k_{\text{syn}}}{k_{\text{deg}}} \cdot \frac{L_{50}}{L + L_{50}} = R_0 \cdot \frac{L_{50}}{L + L_{50}}. \tag{39}$$

The parameter L_{50} in Eqs. 38 and 39 is the ligand concentration resulting in the half-maximal concentration of complex RL and target R at steady state. It is called the *in vivo potency*, typically denoted by EC_{50} in the pharmacological literature (cf. [15]).

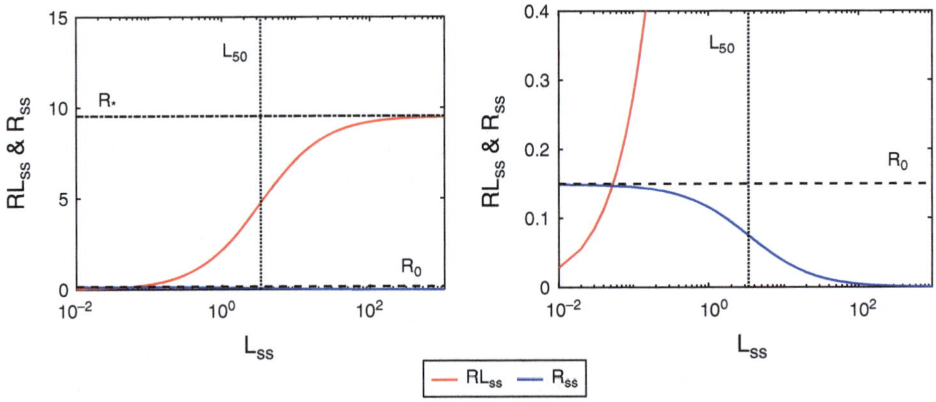

Fig. 7 Graphs of RL_{ss} and R_{ss} versus L_{ss} according to Eqs. 38 and 39 for the parameter values of Table 1. In the right figure, the concentrations are enlarged in order to see the graph of R_{ss} *vs.* L_{ss} more clearly. For $L_{ss} = L_{50}$, the graphs of R_{ss} and RL_{ss} both attain half their respective maximum values, R_0 and R_*. The value of L_{50} is here 3.385 nM

Figure 7 displays the graphs of the functions $RL(L)$ and $R(L)$ with L on a logarithmic scale, while RL and R are given on a linear scale.

4.3 *The* In Vivo *Potency* EC_{50}

The in vivo potency is interesting because it shows how in an in vivo environment, target turnover is relevant for the formation of ligand–target complex, and hence the potency of the drug. Specifically,

$$EC_{50} = \frac{k_{\text{deg}}}{k_{e(RL)}} \cdot \frac{k_{\text{off}} + k_{e(RL)}}{k_{\text{on}}} = \frac{k_{\text{deg}}}{k_{\text{on}}} \cdot \left(1 + \frac{k_{\text{off}}}{k_{e(RL)}}\right). \tag{40}$$

Thus, the in vivo potency is composed of two factors: (i) a factor that reflects the turnover rate *vs.* the internalization rate, $k_{\text{deg}}/k_{e(RL)}$ and (ii) a factor reflecting the binding properties, K_m.

The fact that the in vivo potency is dependent on target turnover, i.e., often enzyme turnover, is of great practical importance. It means that the optimal dose of a medicine may well depend on the personal characteristics of the patient: a slow turnover means a lower dose is sufficient and a rapid dose requires a higher dose.

Qualitative Properties: From the analytical expression of EC_{50} in (40), we deduce the following properties:

(i) When the turnover rate k_{deg} increases, then the in vivo potency also increases.

(ii) When the internalization rate $k_{e(RL)}$ increases, then the in vivo potency decreases and actually tends to zero as $k_{e(RL)} \to \infty$.

And for large and small values of the internalization rate, relative to the dissociation rate, the in vivo potency can be approximated as follows:

(iii) If $k_{e(RL)} \gg k_{\text{off}}$, i.e., when the system is almost *irreversible*, then

$$EC_{50} \approx \frac{k_{\text{deg}}}{k_{\text{on}}}. \tag{41}$$

(iv) If $k_{e(RL)} \ll k_{\text{off}}$, i.e., when the system is practically *reversible*, then

$$EC_{50} \approx \frac{k_{\text{deg}}}{k_{e(RL)}} \cdot \frac{k_{\text{off}}}{k_{\text{on}}} = \frac{k_{\text{deg}}}{k_{e(RL)}} \cdot K_d. \tag{42}$$

Remark: We mention here an alternative potency metric, the AFIR,[3] introduced by Stein et al. (cf. [16]) and based on the free target concentration relative to the baseline target concentration (R/R_0). For large drug concentrations ($L \gg K_d$), it is expressed in terms of three quantities: (i) the average drug concentration, (ii) the binding affinity, and (iii) the "target accumulation" (R_*/R_0). In Ahmed et al. [17], it is demonstrated how AFIR and a related metric AFTIR can be used for dose selection.

5 Clearance of Ligand *L*

The total clearance of the system CL is defined as the constant of proportionality between the rate of elimination and the plasma concentration (cf. [18] and [19]). Thus, at steady state,

$$CL = \frac{In_L}{L_{ss}}. \tag{43}$$

As in Subheading 4, we omit the subscript "ss" in this section.

According to the ligand balance equation in (3), the total clearance can be written as

$$CL = CL_{(L)} + CL_{(RL)}, \tag{44}$$

where $CL_{(L)}$ and $CL_{(RL)}$ refer to the two elimination routes, the non-specific route and the target route:

$$CL_{(L)} = k_{e(L)} V_c \quad \text{and} \quad CL_{(RL)} = k_{e(RL)} \frac{RL}{L} V_c. \tag{45}$$

By Eq. 38, we have

$$RL = R_* \cdot \frac{L}{L + L_{50}}, \tag{46}$$

where R_* is given by (29) and L_{50} is given by (38).

[3] *time-Averaged Free Target concentration to Initial target concentration Ratio.*

Substituting this expression for RL into the formula for $CL_{(RL)}$ in (45), we obtain

$$CL_{(RL)} = k_{e(RL)} \cdot \frac{R_*}{L + L_{50}} V_c. \tag{47}$$

Note that, using the expressions for R_* and L_{50} given by, respectively, (29) and (38), we obtain the following limits for small and large ligand concentrations:

$$CL_{(RL)} \to \begin{cases} k_{e(RL)} \dfrac{R_0}{K_m} \cdot V_c & \text{as} \quad L \to 0, \\ 0 & \text{as} \quad L \to \infty. \end{cases} \tag{48}$$

In light of the definition of $CL_{(L)}$ in Eq. 45, it follows that for the total clearance the limits are

$$\begin{aligned} CL(L) &= CL_{(L)} + CL_{(RL)} \\ &\to \begin{cases} \left(k_{e(L)} + k_{e(RL)} \dfrac{R_0}{K_m}\right) \cdot V_c & \text{as} \quad L \to 0, \\ k_{e(L)} \cdot V_c & \text{as} \quad L \to \infty. \end{cases} \end{aligned} \tag{49}$$

For the numerical values from Table 1, the limits of $CL(L)/V_c$ are 1.79 (1/day) for $L \to 0$ and 0.051 (1/day) for $L \to \infty$, i.e., substantially lower.

Figure 8 shows a graph of the total clearance CL as a function of the ligand concentration L.

6 Reduced Models

From the early studies of TMDD on, the complexity of the TMDD model, with its many parameters, has led to the design of simpler models, partly empirical and partly based on assumptions about the underlying processes. Below we describe a few.

6.1 The Parallel Linear and Michaelis–Menten Elimination Model

This is a single-compartment model that is frequently used to fit data that are similar to those sketched in Fig. 2, as in Bauer et al. [20]. In this model, ligand is eliminated via a linear non-specific route as well as a saturable route modeled by Michaelis–Menten kinetics. Thus, mathematically, it takes the form

$$\frac{dL}{dt} = -k_{e(L)} L - V_{\max} \frac{L}{L + K_M} \tag{50}$$

in which $k_{e(L)}$ denotes the non-specific elimination rate, $V_{\max}$ the maximal elimination rate along the saturable route, and K_M the "Michaelis–Menten constant."

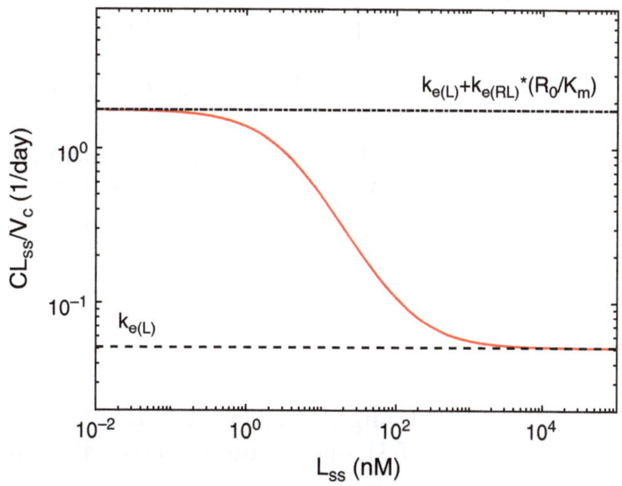

Fig. 8 Total clearance $CL(L)$ graphed versus the ligand concentration L_{ss} on a log–log scale for parameter values from Table 1. For small and large ligand concentrations, CL is seen to converge toward the limits given in (49), while in the transition region the relation is nonlinear

Note that the terminal slope of the ligand *vs.* time graph is here

$$\lambda_z = k_{e(L)} + \frac{V_{\max}}{K_M}. \tag{51}$$

To what extent, and under which conditions, this model captures the features of the TMDD model has been a challenging question that has been taken up by several authors, such as Yan et al. [21], Ma [22], Gibiansky et al. [7], Marathe et al. [23], and Peletier and Gabrielsson [9], using singular perturbation theory [24].

6.2 The Constant Target Pool Hypothesis

When no information about the target is available, it is sometimes assumed that the total concentration of free and bound target together is constant, i.e.,

$$R_{\text{tot}}(t) = R(t) + RL(t) = R_0 \qquad \text{for} \qquad t \geq 0. \tag{52}$$

By Corollary 1, this identity holds if $k_{\text{deg}} = k_{e(RL)}$ (cf. [11]).

Mager and Jusko [3] first studied this case (see also [9] and [22]). It is particularly interesting from an instructional point of view because the identity (52) makes it possible to eliminate one of the three variables from the full system (1) and so reduce it to a two-dimensional system. For instance, if R is eliminated from the first and third equations of (1), one obtains the system

$$\begin{cases} \dfrac{dL}{dt} &= -(k_{\text{on}}R_0 + k_{e(L)})L + (k_{\text{on}}L + k_{\text{off}})RL, \\ \dfrac{dRL}{dt} &= k_{\text{on}}R_0L - (k_{\text{on}}L + k_{\text{off}} + k_{e(RL)})RL. \end{cases} \tag{53}$$

This system is formally identical to the classical, closed Michaelis–Menten system (cf. [25–28]), where R and L stand for, respectively, the enzyme and the substrate concentration. It is well known (cf. [9, 29] Appendix A) that if $L_0 \gg R_0$, then the second equation in the system (53) is in quasi-steady state, so that

$$RL = R_0 \cdot \frac{L}{L + K_m}. \tag{54}$$

Using this expression to eliminate RL from the first equation of (53), one obtains the classical Michaelis –Menten equation:

$$\frac{dL}{dt} = -k_{e(L)}L - V_{\max}\frac{L}{L + K_m} \qquad \text{where} \qquad V_{\max} = k_{e(RL)} \cdot R_0, \tag{55}$$

as given in (50).

6.3 Quasi-Equilibrium (QE) and Quasi-Steady State (QSS) hypotheses

In practice, *Phase A* is often very short and difficult to obtain data for. Therefore, and also to simplify the model, hypotheses are made about the kinetics in the subsequent phases:

(i) The *quasi-equilibrium (QE) hypothesis*, first proposed by Mager and Krzyzanski [5]), assumes that within a brief initial period, ligand, target, and drug–target complex reach quasi-equilibrium, i.e.,

$$L \cdot R = K_d\, RL \qquad \text{where} \qquad K_d = \frac{k_{\text{off}}}{k_{\text{on}}} \tag{56}$$

and from then on remain in quasi-equilibrium.

(ii) The *quasi-steady state (QSS) hypothesis*, used by Gibiansky et al [7], is similar but assumes that ligand, target, and internalization quickly reach a quasi-steady state in which

$$L \cdot R = K_m\, RL \qquad \text{where} \qquad K_m = \frac{k_{\text{off}} + k_{e(RL)}}{k_{\text{on}}}, \tag{57}$$

and that they remain in this quasi-steady state throughout later times.

Once in quasi-equilibrium or quasi-steady state, it follows that

$$RL = R_{\text{tot}}\frac{L}{L + K} \qquad \text{where} \quad K = K_d \ \text{ or } \ K = K_m. \tag{58}$$

Because rapid binding of drug to target is assumed, the QE and QSS approximations focus on the total drug concentration

$L_{tot} = L + RL$ and the total receptor concentration $R_{tot} = R + RL$ as principal variables. They satisfy the system (3). When we eliminate $R = R_{tot} - RL$ and subsequently write $RL = L_{tot} - L$, we obtain the system

$$\begin{cases} \dfrac{dL_{tot}}{dt} &= k_{infus} + (k_{e(RL)} - k_{e(L)})L - k_{e(RL)}L_{tot} \\ \dfrac{dR_{tot}}{dt} &= k_{syn} + (k_{e(RL)} - k_{deg})(L - L_{tot}) - k_{deg}R_{tot}, \end{cases} \tag{59}$$

where $K = K_d$ is the QE approximation and $K = K_m$ is the QSS approximation.

Finally, L is related to L_{tot} and R_{tot} through the implicit relation

$$L_{tot} = L + RL = L + R_{tot}\frac{L}{L+K}. \tag{60}$$

Because for each value of R_{tot} the right-hand side of (60) is a strictly increasing function of L, we may solve for L in terms of L_{tot} and R_{tot} to obtain the expression

$$L = F(L_{tot}, R_{tot}) \stackrel{\text{def}}{=} \frac{1}{2}\left\{L_{tot} - R_{tot} - K + \sqrt{(L_{tot} - R_{tot} - K)^2 + 4KL_{tot}}\right\}. \tag{61}$$

When we substitute $L = F(L_{tot}, R_{tot})$ into the system (59), we obtain two equations that only involve the total concentrations of ligand L_{tot} and receptor R_{tot}:

$$\begin{cases} \dfrac{dL_{tot}}{dt} &= k_{infus} + (k_{e(RL)} - k_{e(L)})F - k_{e(RL)}L_{tot}, \\ \dfrac{dR_{tot}}{dt} &= k_{syn} + (k_{e(RL)} - k_{deg})(F - L_{tot}) - k_{deg}R_{tot}. \end{cases} \tag{62}$$

What has been gained through this reduction is that the on- and off-rates of ligand and target no longer feature individually in the model, but only in combination, through K_d or K_m, in the expression for the function $F(L_{tot}, R_{tot})$.

6.4 QE or QSS and Constant Target Pool Hypothesis Combined

Since in this reduction, R_{tot} is constant and, in light of the initial conditions (4), equal to R_0, it follows that (58) yields an expression for RL in terms of L. Using this in the ligand equation in the system (59), we obtain an equation in L only

$$\frac{d}{dt}\left(L + R_0\frac{L}{L+K}\right) = k_{infus} - k_{e(L)}L - k_{e(RL)}R_0\frac{L}{L+K}, \tag{63}$$

or

$$\frac{dL}{dt} = \frac{k_{infus} - k_{e(L)}L - k_{e(RL)}R_0\dfrac{L}{L+K}}{1 + \dfrac{R_0K}{(L+K)^2}}. \tag{64}$$

This equation was first derived by Wagner in 1971 (cf. [30] and [31]). Note that if $R_0 \ll K$, then

$$\frac{R_0 K}{(L+K)^2} \ll 1 \tag{65}$$

and Wagner's equation 64 reduces to the Michaelis–Menten equation 50 when one puts $V_{\max} = k_{e(RL)} R_0$.

The terminal slope of Wagner's equation is seen to be given by

$$\lambda_z = k_{e(L)} + k_{e(RL)} \cdot \frac{R_0}{K + R_0}. \tag{66}$$

Evidently, the terminal slope λ_z of the classical Michaelis–Menten model, given by (51), is larger than that of the Wagner equation because of the additional R_0 in the numerator of the latter.

Validity of the QE and the QSS Hypotheses: According to the local analysis of the initial concentration graphs (*Phase* A), when $L_0 > R_0$, then initially the identity $L \cdot R = K_m \cdot RL$ holds approximately, i.e., the QSS hypothesis holds, and remains valid for much of *Phase* B (cf. [9]). However, numerical evidence shown in [11] suggests that in *Phases* C and D the QSS hypothesis is no longer valid.

7 Discussion

Since the basic TMDD model has been formulated, it has found a place in many different contexts. Below we discuss a few.

The in vivo Michaelis–Menten Model: Whereas the classical Michaelis–Menten (MM) model is a *closed* model, in many pharmacological applications, this model is used in an in vivo environment in which enzyme turnover is present. When enzyme turnover is incorporated in the classical MM model, the resulting model has much in common with a TMDD model, with enzyme featuring as the target and ligand as the substrate. However, there is one important difference: after internalization, all of the enzymes are returned to the free enzyme pool, while the substrate is eliminated [29]. This can lead to large swings of the free enzyme concentration, which is rarely seen in TMDD (cf. Aston et al. [32]).

Multiple Targets and Ligands: The TMDD model has been extended in many different directions. For instance to systems in which drugs bind to two or more targets (cf. [33] and [34]). For example, when a single ligand chooses between two targets R_1 and R_2, then, at steady state, the corresponding receptor occupancies R_1L and R_2L can each be expressed in terms of the ligand concentration in a formula that is comparable to the one for a single target given in (38). They are (cf. [33])

$$R_1L = R_{1;*} \cdot \frac{L_{50;1}}{L + L_{50;1}} \qquad \text{and} \qquad R_2L = R_{2;*} \cdot \frac{L}{L + L_{50;2}}, \tag{67}$$

where $R_{1;*}$ and $R_{2;*}$ are the maximum concentrations of the ligand–receptor complexes R_1L, and R_2L and $L_{50;i}$ and $L_{50;2}$ the in vivo potencies. They are given by

$$R_{i;*} = \frac{k_{\text{syn};i}}{k_{e(R_iL)}} \qquad \text{and} \qquad L_{50;i} = \frac{k_{\text{deg};i}}{k_{e(R_iL)}} \cdot K_{m;i} \quad (i = 1, 2) \tag{68}$$

in which $K_{m;i} = (k_{\text{off};i} + k_{e(R_iL)})/k_{\text{on};i}$ $(i = 1, 2)$.

Alternatively, two ligands, with concentrations L_1 and L_2, may compete for a single target site. For instance, one of them is a drug that is supplied in order to inhibit the effect caused by an endogenous ligand (cf. Benson et al. [35, 36]). Identities can then be derived for the two complexes RL_1 and RL_2:

$$\begin{cases} RL_1 &= R_1^* \dfrac{L_1}{L_1 + \theta \cdot L_2 + L_{50;1}}, \\ RL_2 &= R_2^* \dfrac{L_2}{L_2 + \theta^{-1} \cdot L_1 + L_{50;2}}, \end{cases} \tag{69}$$

where $L_{50;1}$ and $L_{50;2}$ are given by

$$L_{50;i} = \frac{k_{\text{deg}}}{k_{e(RL_i)}} \cdot K_{m;i}, \qquad R_i^* = \frac{k_{\text{deg}}}{k_{e(RL_i)}} \quad (i = 1, 2) \qquad \text{and} \qquad \theta = \frac{L_{50;1}}{L_{50;2}}. \tag{70}$$

Thus, since here L_1 is the concentration of the drug and L_2 the concentration of the endogenous ligand, the EC_{50} value of the drug increases linearly with the concentration of the ligand. Specifically, it is given by

$$EC_{50} = \theta \cdot L_2 + L_{50;1} = L_{50;1}\left(\frac{L_2}{L_{50;2}} + 1\right). \tag{71}$$

This expression confirms that, as expected, in the absence of an endogenous ligand L_2, we obtain $EC_{50} = L_{50;1}$. In addition, it shows that when L_2 is small relative to $L_{50;2}$, the impact of L_2 is small.

The expressions for the complexes RL_1 and RL_2 can be used in the steady-state receptor balance equation

$$k_{\text{syn}} = k_{\text{deg}}\, R + k_{e(RL_1)} RL_1 + k_{e(RL_2)} RL_2 \tag{72}$$

to derive an expression for R in terms of the two ligand concentrations:

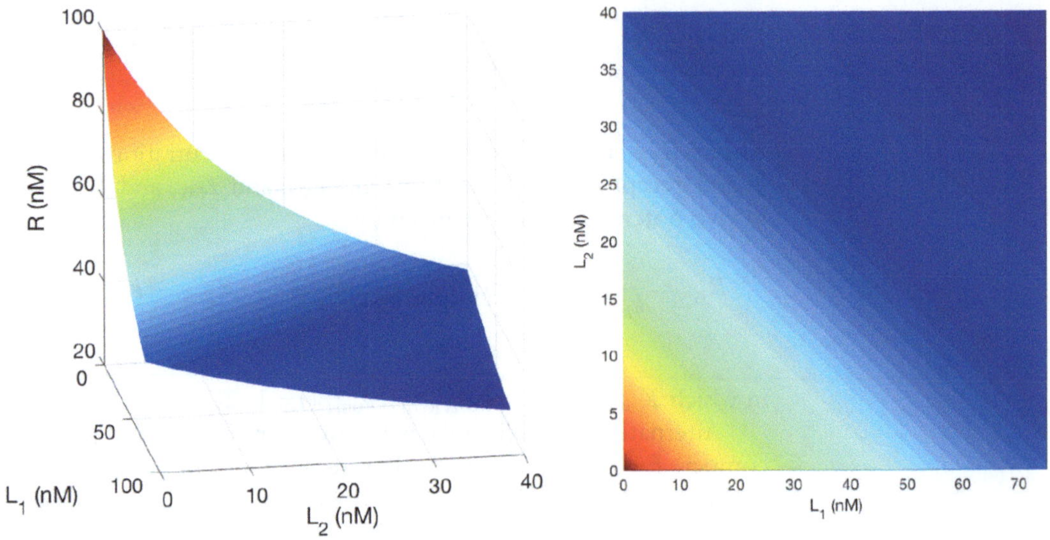

Fig. 9 Graphs of R versus L_1 and L_2 according to Eq. 73. Here $R_0 = 100$, $L_{50;1} = 50$, and $L_{50;2} = 25$, so that $\theta = 2$. Note that the level curves are straight lines with slope $L_1/L_2 = -2$

$$R = R_0\left(1 - \frac{L_1}{L_1 + \theta \cdot L_2 + L_{50;1}} - \frac{L_2}{L_2 + \theta^{-1} \cdot L_1 + L_{50;2}}\right). \tag{73}$$

Figure 9 shows graphically how R varies in the (L_1, L_2)-plane. See also Gibiansky [37], Koch et al [38], and [39] about drug–drug interactions.

The TMDD Model in PBPK Models: The model discussed and analyzed in the previous sections is the basic TMDD model in a single compartment. In recent years, the TMDD model, and often the reduced form based on the assumption that the system is in quasi-steady state, has been incorporated in more complex, physiologically based models, such as shown in Cao and Jusko [40], Shah and Betts [41], and Li et al. [42], and in connection with the so-called catch and release models (cf. [43] and [44]). Such models take account of the underlying physiological structure, incorporating different compartments in the model, such as the plasma compartment, tissue compartments, and compartments associated with different organs such as the liver and the lymphatic system.

Acknowledgement

It is a great pleasure to thank Johan Gabrielsson for many stimulating discussions about this chapter.

References

1. Levy G (1994) Pharmacologic target mediated drug disposition. Clin Pharmacol Ther 56:248–252
2. Sugiyama Y, Hanano M (1989) Receptor-mediated transport of peptide hormones and its importance in the overall hormone disposition in the body. Pharm. Res 6(3):192–202
3. Mager D, Jusko WJ (2001) General pharmacokinetic model for drugs exhibiting target-mediated drug disposition. J Pharmacokinet Phamacodyn 28(6):507–532
4. Krzyzanski W, Perez-Ruixo JJ, Vermeulen A (1999) Basic pharmacodynamic models for agents that alter production of natural cells. J. Pharmacokin Biopharm 27:467–489
5. Mager D, Krzyzanski W (2005) Quasi-equilibrium pharmacokinetic model for drugs exhibiting target-mediated drug disposition. Pharm Res 22(10):1589–1596
6. Mager, D (2006) Target-mediated drug disposition and dynamics. Biochem Pharmacol 72:1–10
7. Gibiansky L, Gibiansky E, Kakkar T, Ma P (2008) Approximations of the target-mediated drug disposition model and identifiability of model parameters. J Pharmacokinet Pharmacodyn 35(5):573–591
8. Krippendorff BF, Kuester K, Kloft C, Huisinga W (2009) Nonlinear pharmacokinetics of therapeutic proteins resulting from receptor mediated endocytosis. J Pharmacokinet Pharmacodyn 36:239–260
9. Peletier LA, Gabrielsson J (2009) Dynamics of target-mediated drug disposition. Eur J Pharm Sci 38:445–464
10. Betts AM, Clark TH, Yang J, Treadway JL, Li M, Giovanelli MA, Abdiche Y, Stone DM, Paralkar VM (2010) The application of target information and preclinical pharmacokinetic/pharmacodynamic modeling in predicting clinical doses of a Dickkopf-1 antibody for osteoporosis. J Pharm Exper Ther 333:2–13
11. Peletier LA, Gabrielsson J (2012) Dynamics of target-mediated drug disposition: characteristic profiles and parameter identification, J Pharmacokinet Pharmacodyn 39:429–451
12. Dua P, Hawkins E, Graaf PH van der (2015) A tutorial on target-mediated drug disposition (TMDD) models. CPT Pharmacometrics Syst Pharmacol 4:324–337. https://doi.org/10.1002/psp4.41
13. Luu KT, Kraynov E, Kuang B, Vicini P, Zhong W-Z (2013) Modeling, simulation, and translation framework for the preclinical development of monoclonal antibodies. AAPS J 15:551–558
14. Stein A, Peletier LA (2018) Predicting the onset of nonlinear pharmacokinetics. CPT Pharmacometrics Syst Pharmacol 7:670–677. https://doi.org/10.1002/psp4.12316
15. Gabrielsson J, Peletier LA, Hjorth S (2018) *In vivo* potency revisited - Keep the target in sight. Pharmacol Ther 184:177–188
16. Stein AM, Ramakrishna R (2017) AFIR: a dimensionless potency metric for characterizing the activity of monoclonal antibodies. CPT: Pharmacometrics Syst Pharmacol 6 (4):258–266. https://doi.org/10.1002/psp4.12169
17. Ahmed S, Ellis M, Li H, Pallucchini L, Stein AM (2019) Guiding dose selection of monoclonal antibodies using a new parameter (AFTIR) for characterizing ligand binding systems. J Pharmacokinet Pharmacodyn 46:287–304
18. Derendorf H, Schmidt S (2020) Rowland and Tozer's clinical pharmacokinetics and pharmacodynamics: concepts and applications, 5th edn. Wolters Kluwer, Philadelphia
19. Rosenbaum SE (2011) Basic pharmacokinetics and pharmacodynamics. Wiley, Hoboken
20. Bauer RJ, Dedrick RL, White ML, Murray MJ, Garovoy MR (1999) Population pharmacokinetics and pharmacodynamics of the anti-CD11a antibody hu1124 in human subjects with psoriasis. J Pharmacokinet Biopharm 27:397–420
21. Yan X, Mager DE, Krzyzanski, W (2010) Selection between Michaelis Menten and target-mediated drug disposition pharmacokinetic models. J Pharmacokinet Pharmacodyn 35:25–47
22. Ma P (2011) Theoretical considerations of target-mediated drug disposition models: simplifications and approximations. Pharm Res 29:866–882
23. Marathe A, Krzyzanski W, Mager DE (2009) Numerical validation and properties of a rapid binding approximation of a target-mediated drug disposition pharmacokinetic model. J Pharmacokinet Pharmacodyn 36:199–219. https://doi.org/10.1007/s10928-009-9118-8
24. Lin CC, Segel LA (1988) Mathematics applied to deterministic problems in the natural sciences. In: Classics in applied mathematics, vol 1. Society for Industrial and Applied Mathematics (SIAM), Philadelphia
25. Michaelis L, Menten, ML (1913) Die Kinetik der Invertinwirkung. Biochem J 49:333–369

26. Briggs GE, Haldane JBS (1913) A note on the kinetics of enzyme action. Biochem J 49:338–339
27. Brown AJ (1902) Enzyme action. J Chem Soc 81:373–388
28. Cornish-Bowden A (2013) The origins of enzyme kinetics. FEBS Lett 587:2725–2730
29. Gabrielsson J, Peletier LA (2018) Michaelis-Menten from an in vivo perspective: open versus closed systems. AAPS J 20(6):102. https://doi.org/10.1208/s12248-018-0256-z
30. Wagner JG (1971) A new generalised nonlinear pharmacokinetic model and its implications. In Wagner JG (ed) Biopharmaceutics and relevant pharmacokinetics. Drug Intelligence Publications, Hamilton, p 302–317
31. Wagner JG (1973) A modern view of pharmacokinetics. J Pharmacokinet Biopharmaceut 1 (5):363–401
32. Aston PJ, Derks G, Raji A, Agoram BM, Graaf PH van der (2011) Mathematical analysis of the pharmacokinetic-pharmacodynamic (PKPD) behaviour of monoclonal antibodies: predicting in vivo potency. J Theor Biol 281:113–121
33. Peletier LA, Gabrielsson J (2018) New equilibrium models of drug-receptor Interactions derived from target-mediated drug disposition. AAPS J 20:69. https://doi.org/10.1208/s12248-018-0221-x
34. Gibiansky L, Gibiansky E (2010) Target-mediated drug disposition model for drugs that bind to more than one target. J Pharmacokinet Pharmacodyn 37:323–346. https://doi.org/10.1007/s10928-010-9163-3
35. Benson N, Graaf PH van der, Peletier LA (2013) Cross-membrane signal transduction of receptor tyrosine kinases (RTKs): from systems biology to systems pharmacology. J Math Biol 66:719–742
36. Benson N, Graaf PH van der, Peletier LA (2016) Selecting optimal drug-intervention in a pathway involving receptor tyrosine kinases (RTKs). Nonlinear Anal 137:148–170
37. Gibiansky L, Gibiansky E (2017) Target-mediated drug disposition model for drugs with two binding sites that bind to a target with one binding site. J Pharmacokinet Pharmacodyn 44(5):463–475. https://doi.org/10.1007/s10928-017-9533-1
38. Koch G, Jusko WJ, Schropp J (2017) Target-mediated drug disposition with drug-drug interaction, Part I: single drug case in alternative formulations. J Pharmacokinet Phamacodyn 44(1):17–26
39. Koch G, Jusko WJ, Schropp J (2017) Target mediated drug disposition with drug-drug interaction, Part II: competitive and uncompetitive cases. J Pharmacokinet Phamacodyn 44(1):27–42
40. Cao Y, Jusko WJ (2012) Applications of minimal physiologically-based pharmacokinetic models. J Pharmacokinet Pharmacodyn. 39 (6):711–723. https://doi.org/10.1007/s10928-012-9280-2
41. Shah DK, Betts AM (2012) Towards a platform PBPK model to characterise the plasma and tissue disposition of monoclonal antibodies in preclinical species and human. J Pharmacokinet Pharmacodyn 39(1):67–86
42. Li L, Gardner I, Rose R, Jamei M (2014) Incorporating target shedding into a minimal PBPK-TMDD model for monoclonal antibodies. CPT Pharmacometrics Syst Pharmacol3: e96. https://doi.org/10.1038/psp.2013.73
43. Glassman PM, Balthasar JP (2016) Application of a catenary PBPK model to predict the disposition of "catch and release" anti-PCSK9 antibodies. Int J Pharm 505:69–78
44. Yuan D, Rode F, Cao Y (2018) A minimal physiologically based pharmacokinetic model with a nested endosome compartment for novel engineered antibodies. AAPS J 20 (3):48. https://doi.org/10.1208/s12248-017-0183-4 Refs. [45–47] are not cited in the text. Please provide the citation or delete them from the list.
45. Cao Y, Jusko WJ (2014) Incorporating target-mediated drug disposition in a minimal physiologically-based pharmacokinetic model for monoclonal antibodies. J Pharmacokinet Pharmacodyn 41:375–387
46. Gabrielsson J, Peletier LA (2017) Pharmacokinetic steady-states highlight interesting target-mediated disposition properties. AAPS J. 19 (3):772–786
47. Gibiansky L, Gibiansky E (2017) Target-mediated drug disposition model for drugs with two binding sites that bind to a target with one binding site. J Pharmacokinet Phamacodyn 44(5):463–475

Chapter 3

Beyond the Michaelis–Menten: Bayesian Inference for Enzyme Kinetic Analysis

Hyukpyo Hong, Boseung Choi, and Jae Kyoung Kim

Abstract

Although the Michaelis–Menten (MM) rate law has been widely used to estimate enzyme kinetic parameters, it works only under the condition of extremely low enzyme concentration. Furthermore, even when this condition is satisfied, parameter estimation is often imprecise due to the parameter identifiability issue. To overcome these limitations of the canonical approach to enzyme kinetics, we developed a Bayesian approach based on a modified form of the MM rate law, which is derived with the total quasi-steady state approximation. Here, we illustrate how to perform the Bayesian inference for the progress curve assay with our user-friendly computational R package. We also describe an optimal experimental design for the progress curve assay, with which enzyme kinetic parameters can be accurately and precisely estimated from minimal measurements of the progress curves.

Key words Enzyme kinetics, Progress curve assay, Bayesian inference, Michaelis–Menten rate law, Total quasi-steady state approximation

1 Introduction

The Michaelis–Menten (MM) rate law is the best-known and simplest approach to describe the rate of enzymatic conversion of a substrate (S) into a product (P) [1–4]:

$$\frac{dP}{dt} = \frac{k_{\text{cat}}\, E_{\text{T}} S}{S + K_{\text{M}}}, \quad (1)$$

where k_{cat}, K_{M}, and E_{T} are the catalytic constant, the Michaelis–Menten constant, and a total enzyme concentration, respectively.

A major assay estimating k_{cat} and K_{M} from the progress curves of the product is the initial velocity assay, which uses the initial product formation rates (i.e., $\frac{k_{\text{cat}}\, E_{\text{T}} S_{\text{T}}}{S_{\text{T}} + K_{\text{M}}}$) over a range of initial substrate concentrations (S_{T}) [5]. Another assay is the progress curve assay, where the whole progress curve is fitted to the differential equation of the MM rate law (Eq. 1) [5]. Although the progress curve assay is more technically challenging than the initial velocity

Quentin Vanhaelen (ed.), *Computational Methods for Estimating the Kinetic Parameters of Biological Systems*, Methods in Molecular Biology, vol. 2385, https://doi.org/10.1007/978-1-0716-1767-0_3,

assay (e.g., Lineweaver–Burk plots), it requires less data because the information of the whole progress curve is used [5, 6].

As these assays are based on the MM rate law (Eq. 1), they should be used only when the MM rate law is accurate, i.e., the enzyme concentration is low enough so that enzyme–substrate complex is negligible compared to the substrate [5, 7, 8]. Specifically, the total enzyme concentration (E_T) should be at least 10 times lower than the sum of the initial substrate concentration (S_T) and K_M [5, 7, 8]. To ensure this validity condition, in vitro experiments are performed with a large excess of substrate concentration over the enzyme concentration. However, this is not always plausible; e.g., sensitive QD-FRET-based probes require an extremely low concentration of substrate [9–11]. Importantly, in vivo enzyme concentrations are typically much higher than those used experimentally in vitro [5, 8, 12–14]. For instance, the hepatic CYP enzyme usually greatly exceeds the drug that is its metabolic substrate [15–17]. However, unfortunately, even in situations outside the range of validity of the MM rate law, it has been frequently misused due to its popularity and familiarity [4, 8]. This misuse of the MM rate law can cause serious errors in parameter estimation [6, 8, 17].

Such errors can be prevented by using a slightly modified form of the MM rate law, which is derived with the *total* quasi-steady state approximation (tQSSA):

$$\frac{dP}{dt} = \frac{k_{cat}}{2}\left[K_M + E_T + \widehat{S} - \sqrt{\left(K_M + E_T + \widehat{S}\right)^2 - 4E_T\widehat{S}}\right], \quad (2)$$

where $\widehat{S}$ is the total substrate concentration (i.e., free substrate + substrate and enzyme complex) [5, 8, 18–25]. In general, this tQSSA model is accurate for any ratio of the enzyme concentration to the substrate concentration, unlike the MM rate law, which is derived with the *standard* quasi-steady state approximation (sQSSA) [5, 8, 18–25]. Thus, unlike using the MM rate law, using the tQSSA model to estimate kinetic parameters does not cause bias even when the enzyme concentration is not low enough [6, 8]. By using the unbiased estimation afforded by the tQSSA model regardless of enzyme concentrations, we have been able to develop an optimal experimental design for the progress curve assay with which enzyme kinetic parameters can be precisely and accurately estimated from the minimal number of progress curves [6].

In this chapter, we describe the optimal progress curve assays based on the tQSSA model (Eq. 2). To do this, we introduce our Bayesian inference algorithm that estimates k_{cat} and K_M from the progress curve using either the MM rate law (i.e., sQSSA model) or the tQSSA model [6]. Importantly, we provide a user-friendly R package "EKMCMC" version 1.1.0, which we updated from our previous version 0.1.0 [6]. With this updated code, users can easily perform the Bayesian inference for progress curve assays after completing minimal tasks such as uploading progress curve data and specifying the initial concentrations of substrate and enzyme. By using this code, we illustrate that regardless of enzyme concentrations, progress curve assays based on the tQSSA model (Eq. 2), but not the MM rate law (Eq. 1), are accurate. Thus, from multiple progress curves measured under a wide range of enzyme concentrations, our code based on the tQSSA model can accurately and precisely estimate both k_{cat} and K_M.

2 Method

2.1 Overview of the Software: Installation

Previously, we developed the Bayesian inference algorithm based on either the MM rate law (Eq. 1) or the tQSSA model (Eq. 2) to estimate k_{cat} and K_M from progress curves [6]. In our algorithm, a Markov chain Monte Carlo (MCMC) method is used to obtain the posterior samples of k_{cat} and K_M to gain their estimates. In particular, when k_{cat} and K_M are estimated simultaneously, we used the Gibbs sampler method, an MCMC method that sequentially samples the two parameters from each conditional posterior distribution conditioned on the current value of the other variable. In this process, to draw the samples for K_M, the Metropolis–Hastings algorithm is used because its complicated formula for the conditional posterior distribution prevents direct sampling, unlike k_{cat}, which has a simple posterior distribution with a conjugate gamma prior [6].

In order to perform the Bayesian inference, the "EKMCMC" package [26] in R software [27] needs to be installed. Users can easily download and install the R software from the CRAN website. If RStudio [28] is installed with it, R programs can be used more easily. Throughout this chapter, all analyses were conducted with R version 3.6.1, so this or later version is recommended.

2.2 Step-by-Step Instructions for Computational Package for the Bayesian Inference

After installing the programs, the Bayesian inference for enzyme kinetics from progress curves can be performed as follows.

Step 1. In the R command line, run the code below to install and import the "EKMCMC" package.

```
> install.packages("EKMCMC")
> library("EKMCMC")
```

Step 2. Set the working directory and place the progress curve data named "timeseries_data.csv" in the current working directory.

Every two columns in the data file represent the time points when the concentrations of the products were measured and the corresponding measured concentrations. Thus, if the number of measured progress curves is d, then the number of columns is $2d$. Note that the duration of each progress curve measurement (i.e., the number of time points) can be different, and the time points in each progress curve do not need to be evenly spaced (e.g., time= 0, 1, 3, 4, 7 ...).

Step 3. Specify the units of time and concentration used in the "timeseries_data.csv" file by entering the units in the variables `t_unit_data` and `c_unit_data`, respectively. Any units can be used for time and concentration as long as they are used consistently throughout the code. In this example, the units of time and concentration are second and nM, respectively.

```
> t_unit_data <- "s"
> c_unit_data <- "nM"
```

Step 4. Import progress curve data from the "timeseries_data.csv" file with the command `read.csv()` as follows.

Below, the code `head(timeseries_data)` shows the first few rows of the data.

```
> timeseries_data <-read.csv("timeseries_data.csv", header =
F)
> head(timeseries_data)
 V1 V2 V3 V4 V5 V6 V7 V8
1 0 0.000 0 0.000 0 0.000 0 0.000
2 10 0.176 10 0.176 1 0.176 1 0.220
3 20 0.440 20 0.352 2 0.308 2 0.440
4 30 0.748 30 0.484 3 0.440 3 0.616
5 40 0.880 40 0.572 4 0.704 4 0.616
6 50 0.968 50 0.748 5 0.836 5 0.792
```

Step 5. Store the initial substrate and enzyme concentrations in the variables `enz_data` and `subs_data`, respectively, based on the experimental setting.

`enz_data` and `subs_data` have to be vectors of length d (i.e., the number of progress curves in `timeseries_data`). In this example, different initial enzyme concentrations (i.e., 4.4 nM and 440 nM) are used to measure the progressive curves while the same substrate concentration (4.4 nM) is used.

```
> enz_data <- c(4.4, 4.4, 440, 440)
> subs_data <- c(4.4, 4.4, 4.4, 4.4)
```

Step 6. Store the values of K_M or k_{cat} in the variables K_M_data or catal_data, respectively, if their true values are known.

The code estimates the parameters whose true values are not entered by the user. For instance, if the user enters only the true value of K_M, then the code estimates k_{cat}. If the user does not specify any of the parameter values (i.e., specify FALSE) as follows, the code estimates both of the parameters.

```
> K_M_data = FALSE
> catal_data = FALSE
```

Step 7. Run the function main_est() after specifying the parameters of main_est() as follows.

```
> result <- main_est(method=TRUE, timeseries = timeseries_da-
ta, enz = enz_data, subs = subs_data, K_M = K_M_data, catal =
catal_data, K_M_init = 10, std=10, tun=3.5, jump=10,
burn=1000, nrepeat=1000, catal_m=1, catal_v=100, K_M_m=1,
K_M_v =1e+4, volume = FALSE, t_unit = t_unit_data, c_unit =
c_unit_data)
```

method: This determines which model, the classical sQSSA model (Eq. 1; MM rate law) or the tQSSA model (Eq. 2), is used for the estimation. Specifically, the input for method is TRUE (FALSE); then the tQSSA (sQSSA) model is used.

K_M_init: The initial guess for K_M to perform the MCMC method for the estimation of K_M. If it is not specified, then our code uses the maximum value among the initial substrate concentrations (subs_data) as K_M_init. Note that the user does not need to specify the initial guess for k_{cat} as it is automatically generated by the code based on K_M_init.

std: The standard deviation of the proposal distribution for the Metropolis–Hastings step for K_M. If a user has any information about the scale of K_M, then the value can be assigned to std. Otherwise, one recommendation is to assign one of the initial substrate concentrations or the sample standard deviation of the concentrations of products. However, it should be adjusted to attain a reasonable acceptance ratio, between 0.3 and 0.5 (*see* **Notes 1** and **2** for tuning strategy) [29]. If std is FALSE, then the standard deviation is adaptively calculated by the code using the hessian of log posterior distribution. In this case, the proposal distribution can be tuned with the variable tun.

`tun`: The tuning parameter of the Metropolis–Hastings step for K_M when `std` is `FALSE`. The standard deviation of the proposal distribution is determined by `tun` $\times$ $\sqrt{\text{hessian}\,(\log\text{ posterior})^{-1}}$. It should be adjusted to attain a reasonable acceptance ratio, between 0.3 and 0.5 (*see* **Notes 1** and **2** for tuning strategy) [29].

`jump`: The reciprocal of the thinning rate of the MCMC iteration. If `jump` is 5, then 1/5 is the thinning rate. Its default value is 10 in our code, which can be adjusted depending on the autocorrelation function (*see* Figs. 1 and 2).

`burn`: The length of the burn-in period, which determines the number of initial MCMC samples discarded to ensure that the remaining MCMC samples are independent from their initial value. Its default value is 1000 in our code, which can be adjusted depending on the shape of the trace plot of the MCMC samples (*see* **Note 3**).

`nrepeat`: The number of effective MCMC iterations. Its default value is 1000. The total number of MCMC iterations is `burn`+ `nrepeat`×`jump`.

`catal_m`: The mean of the gamma prior distribution of k_{cat}.

`catal_v`: The variance of the gamma prior distribution of k_{cat}.

`K_M_m`: The mean of the gamma prior distribution of K_M.

`K_M_v`: The variance of the gamma prior distribution of K_M.

The above mean and variance of prior distributions need to be determined based on a user's prior knowledge about k_{cat} and K_M. If there is not enough prior knowledge, a noninformative prior should be considered by assigning an arbitrary value to the mean and a very large value to the variance. For the non-informative prior, if users do not specify values, `catal_m=1`, `catal_v=10^6`, `K_M_m=max(subs_data)`, and `K_M_v=max (subs_data)^2*1000` are automatically assigned.

`volume`: The volume of a system in which the experiments are performed. If the input is `FALSE`, then the code automatically decides its values so that the length of the data is comparable to the number of product molecules.

The execution time of the code depends on which parameters are estimated, the data size (i.e., the number of time points and the number of progress curve), and the total number of MCMC iterations (i.e., `burn`+`nrepeat`×`jump`). For the running example, the total number of iterations is 11,000, the number of progress curves is 4, and the number of time points for each progress curve is 101. In this case, the execution time is less than a second when only k_{cat} is estimated and is approximately 10 s when K_M (i.e., only K_M or both) is estimated in an environment of Intel(R) Core

(TM) i7-6700 CPU @ 3.40 GHz with 64 GB RAM on Windows 10 64-bits. Users can predict execution times by scaling the above result depending on their data and parameter settings.

3 Outputs of the Computational Package

3.1 Summary Statistics

After running the code, the summary statistics are generated in the console window as follows. `Posterior mean` is the sample mean of posterior samples for estimating a parameter. For instance, the posterior sample means 4.671×10^{-2} s^{-1} and 4.435×10^{1} nM are the estimates of k_{cat} and K_M, respectively. `Posterior sd`, `Credible interval(U)`, and `Credible interval(L)` are the sample standard deviation, the 97.5th percentile, and the 2.5th percentile of the posterior samples. As the posterior standard deviation and the credible interval (i.e., `U`−`L`) become smaller, parameter estimation becomes more precise (i.e., the parameter identifiability issue is less likely to occur). `Relative CV` is the ratio between the coefficient of variation (CV) of the posterior samples and the CV of the prior distribution used. The relative CV < 1 means that the posterior samples give more information than prior information (*see* **Note 4** for details). `Acceptance ratio` for K_M is the portion of accepted proposed samples when the Metropolis–Hastings algorithm is used to generate the posterior samples of K_M. Here, the acceptance ratio (4.221×10^{-1}) is close to the optimal number of 0.4, which supports that the current choice of proposal distribution determined by `std` or `tun` is appropriate (*see* **Note 1** for details about when the acceptance ratio is far from 0.4).

```
MCMC simulation summary of the catalytic constant (k_cat)
Posterior mean:          4.671e-02 (s^-1)
Posterior sd:            5.243e-03 (s^-1)
Credible interval(U):    5.754e-02 (s^-1)
Credible interval(L):    3.708e-02 (s^-1)
Relative CV:             1.122e-02

MCMC simulation summary of the Michaelis-Menten constant (K_M)
Posterior mean:          4.435e+01 (nM)
Posterior sd:            8.071e+00 (nM)
Credible interval(U):    6.18e+01 (nM)
Credible interval(L):    3.052e+01 (nM)
Relative CV:             1.82e-03
Acceptance ratio:        4.221e-01
```

3.2 Diagnostic Plots

The distribution of samples generated by the MCMC method is used to estimate the posterior distributions of the parameters as seen in the above summary statistics. To ensure the convergence of the samples generated by the MCMC method to the posterior distributions of the parameters, we need to check diagnostic plots (Figs. 1 and 2) [30], which are automatically generated after running the code. The trace plots for k_{cat} (Fig. 1, top) and K_M (Fig. 2, top) are the MCMC samples over time, which can be used to test how quickly the MCMC samples converge to the posterior distributions and become independent from the initial values. The exemplary trace plots show ideal rapid up-and-down fluctuation without long-term drifts, which indicates the convergence of the MCMC samples. If the shape of a trace plot is not ideal, we need to tune the standard deviation of the proposal distribution for the Metropolis–Hastings algorithm by using `tun` and `std` (*see* **Notes 1** and **2** for details). The autocorrelation function (ACF) plots for k_{cat} (Fig. 1, bottom left) and K_M (Fig. 2, bottom left) show whether the samples in the trace plots (Figs. 1 and 2, top) are positively autocorrelated (ACF ≈ 1), negatively autocorrelated (ACF ≈ -1), or independent of each other (ACF ≈ 0). For instance, the autocorrelations of both k_{cat} and K_M are nearly zero when the lag is larger than 2 (Figs. 1 and 2, bottom left). This indicates that the

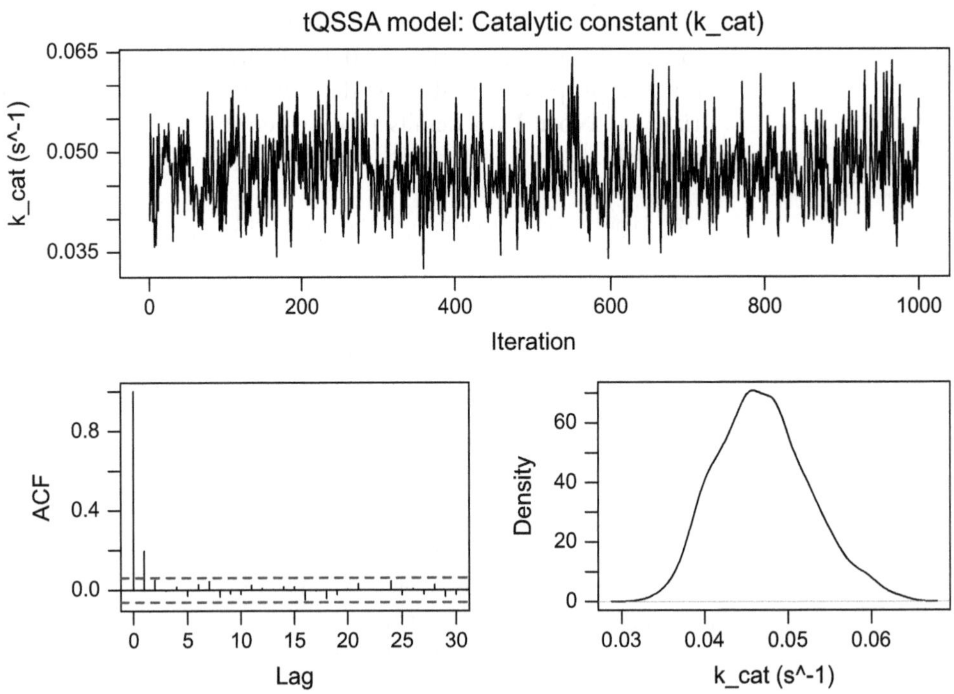

Fig. 1 Diagnostic plots for k_{cat} automatically generated by the code: trace plot (top), autocorrelation function plot (bottom left), and posterior density plot (bottom right). All plots were drawn with 1000 posterior samples after removing the first 1000 samples (burn-in period) and applying thinning rate 0.1

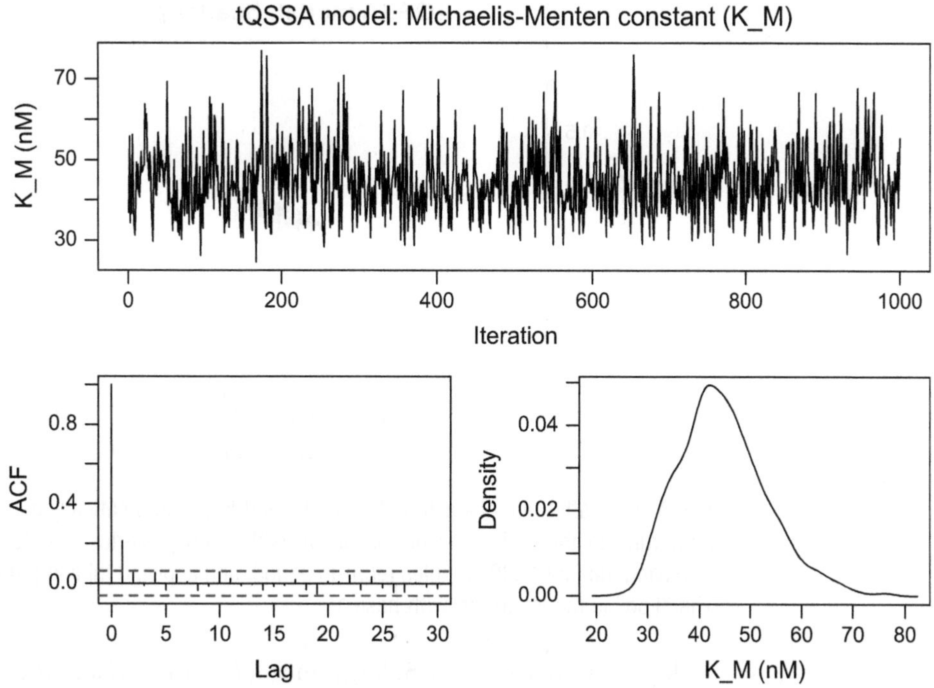

Fig. 2 Diagnostic plots for K_M automatically generated by the code: trace plot (top), autocorrelation function plot (bottom left), and posterior density plot (bottom right). All plots were drawn with 1000 posterior samples after removing the first 1000 samples (burn-in period) and applying thinning rate 0.1

samples in the trace plots whose sampling time differs by more than 2 are nearly independent, and thus the MCMC samples rapidly converge to the posterior distributions. If the ACF plots do not quickly drop to zero as the lag increases, the thinning rate should be decreased (i.e., increase jump). Finally, the posterior density plots for k_{cat} (Fig. 1, bottom right) and K_M (Fig. 2, bottom right) are unimodal and bell-shaped without extreme outliers, which again indicates the convergence of the MCMC samples.

When both k_{cat} and K_M are estimated, the scatter plot between the log-transformed posterior samples of the two parameters is provided (Fig. 3). While the scatter plot shows a weak positive correlation, according to the relative CV, acceptance ratio, and diagnostic plots, we can conclude that the MCMC simulation gives precise estimation (i.e., the identifiability problem does not occur).

The output variable `result` contains the posterior samples of the estimated parameters. To see the last few posterior samples, `tail(result)` can be used. Furthermore, to draw the posterior density plot (Figs. 1 and 2 bottom right) or the scatter plot (Fig. 3) with different ranges of axes, users can use `plot()` after adjusting `xlim` and `ylim`, which represent the ranges of the horizontal and vertical axes, respectively. For the scatter plot, the horizontal range

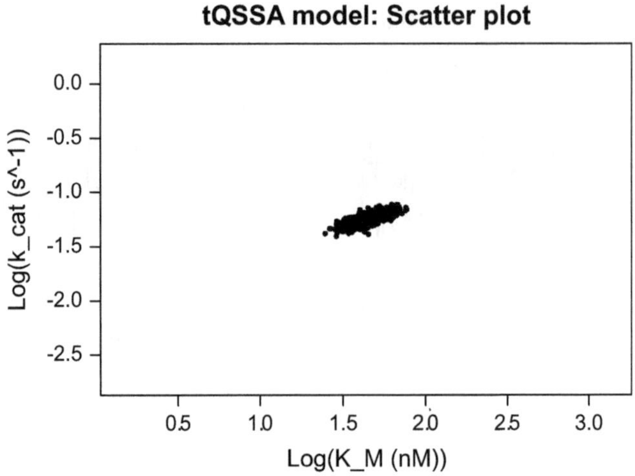

Fig. 3 The scatter plot of log-transformed K_M and k_{cat} is automatically generated after running the code. The plot is drawn with 1000 posterior samples after removing the first 1000 samples (burn-in period) and applying thinning rate 0.1. The "Log" is the logarithm with base 10

is $[\log_{10}(\text{mean}(K_M)) - 1.5, \log_{10}(\text{mean}(K_M)) + 1.5]$ and the vertical range is $[\log_{10}(\text{mean}(k_{cat})) - 1.5, \log_{10}(\text{mean}(k_{cat})) + 1.5]$ by their defaults. These may need to be adjusted to draw all points in a plot depending on data.

```
> tail(result)
     k_cat (s^-1) K_M (nM)
995    0.04434113 42.91697
996    0.04717631 52.49882
997    0.04523579 40.04123
998    0.05824598 55.12163
999    0.05037549 46.68125
1000   0.04258959 35.33761
> plot(density(result[,1]), xlim = c(0.02, 0.08))
> plot(density(result[,2]), xlim = c(10, 100))
> plot(log10(result[, 2]), log10(result[, 1]), type = "p", pch
= 19, cex = 0.5, xlim = c(1, 2), ylim = c(-2,-1))
```

4 Comparisons of Enzyme Kinetics Based on the sQSSA and the tQSSA Models

4.1 Estimation of k_{cat}

As the enzyme concentration increases, the simulated progress curves of the sQSSA model (i.e., MM rate law, Eq. 1) become inaccurate (Fig. 4). On the other hand, the tQSSA model (Eq. 2) successfully captures the full model based on mass action kinetics regardless of the enzyme concentrations (Fig. 4).

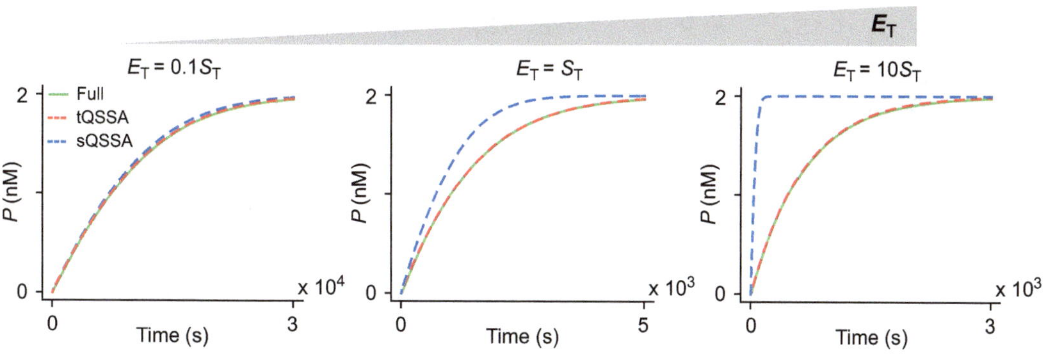

Fig. 4 Simulated progress curves of the original full model based on mass action kinetics, the sQSSA model (Eq. 1), and the tQSSA model (Eq. 2) under different enzyme concentrations. Here, k_{cat}=1.6 × $10^{-3}s^{-1}$, $S_T = K_M = 2$ nM, and $E_T = 0.1S_T$, S_T, or $10S_T$, where S_T is the initial substrate concentration. While the sQSSA model becomes inaccurate as enzyme concentration increases, the tQSSA model is accurate regardless of enzyme concentrations (i.e., matching the full model). The figures are reproduced from [6] with permission from *Scientific Reports*

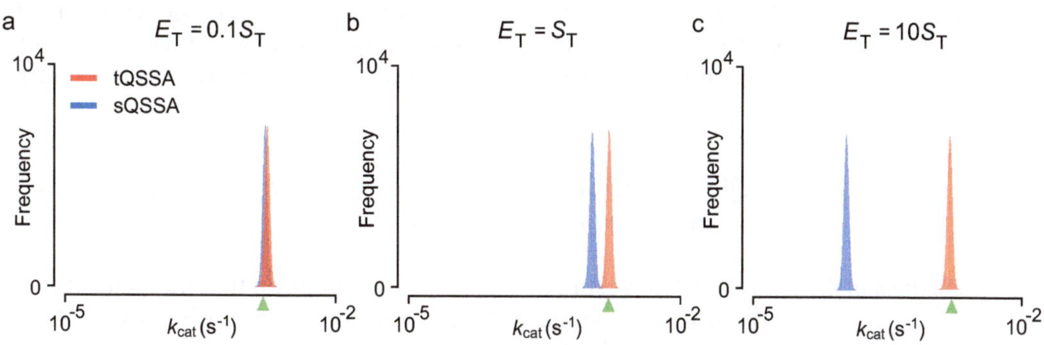

Fig. 5 The estimation of k_{cat} with either the sQSSA or the tQSSA model when K_M is fixed at its true value (2 nM). For each condition ($E_T = 0.1S_T$ (**a**), S_T (**b**) $10S_T$ (**c**)), 10^5 posterior samples obtained by applying our Bayesian inference algorithm to five noisy discrete progress curves are shown. While the estimates of k_{cat} with the sQSSA model become biased as enzyme concentration increases, those obtained with the tQSSA model are accurate regardless of conditions. Specifically, estimates of k_{cat} (mean ± SD) obtained with the sQSSA model are $1.69 \times 10^{-3} \pm 1.08 \times 10^{-4}$ (**a**), $1.06 \times 10^{-3} \pm 6.87 \times 10^{-5}$(**b**), and $1.06 \times 10^{-4} \pm 7.03 \times 10^{-6}$ (**c**), and those obtained with the tQSSA model are $1.77 \times 10^{-3} \pm 1.13 \times 10^{-4}$ (**a**), $1.62 \times 10^{-3} \pm 1.05 \times 10^{-4}$ (**b**), and $1.51 \times 10^{-3} \pm 1.00 \times 10^{-4}$ (**c**). Here, the green triangles indicate the true value of k_{cat} (1.6 × $10^{-3}s^{-1}$), and the horizontal axis has a log scale but the vertical axis has a linear scale. The figures are reproduced from [6] with permission from *Scientific Reports*

As model accuracy is required for accurate parameter estimation without bias, we can expect that the sQSSA model fails to estimate the enzyme kinetic parameters when enzyme concentration is high. Indeed, even though the true value of K_M is assumed to be known, posterior samples of k_{cat} generated with the code based on the sQSSA model fail to capture the true value of k_{cat} when the enzyme concentration is high (Fig. 5b, c). On the other hand, posterior samples obtained with our code based on the

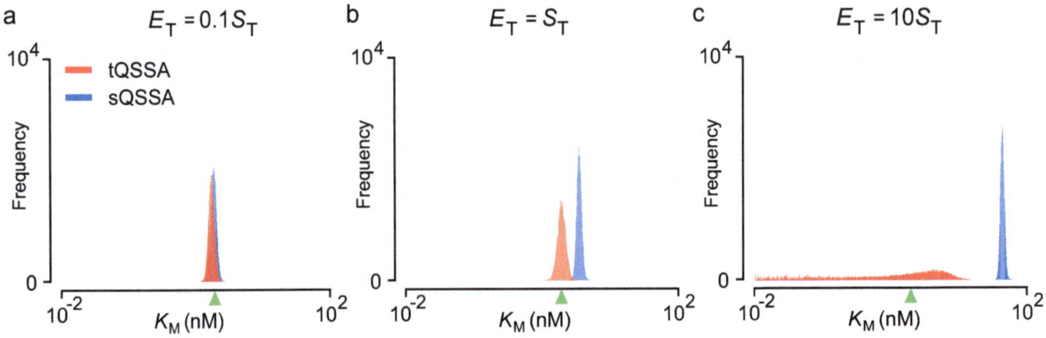

Fig. 6 The estimation of K_M with either the sQSSA or the tQSSA model when k_{cat} is fixed at its true value ($1.6 \times 10^{-3} s^{-1}$). For each case, 10^5 posterior samples obtained by applying our Bayesian inference algorithm to five noisy discrete progress curves are shown. While the estimates of K_M with the sQSSA model become biased as enzyme concentration increases, those obtained with the tQSSA model are less biased for all conditions. Specifically, estimates of K_M (mean $\pm$ SD) obtained with the sQSSA model are 1.88 ± 0.17 (**a**), 3.51 ± 0.29 (**b**), and 43.7 ± 3.00 (**c**), and those obtained with the tQSSA model are 1.73 ± 0.17 (**a**), 1.95 ± 0.27 (**b**), and 1.66 ± 2.38 (**c**). Note that at high enzyme concentrations (**c**), the SD of the estimates of K_M obtained with the tQSSA model is large (i.e., 2.38), which indicates that K_M is not identifiable (i.e., estimation is imprecise). Here, the green triangles indicate the true value of K_M (2 nM), and the horizontal axis has a log scale but the vertical axis has a linear scale. The figures are reproduced from [6] with permission from *Scientific Reports*

tQSSA model accurately capture the true value of k_{cat} regardless of the enzyme concentrations (Fig. 5).

4.2 Estimation of K_M

When K_M is estimated under the assumption that the value of k_{cat} is known, the estimate using posterior samples obtained with the sQSSA model becomes seriously biased as the enzyme concentration increases (Fig. 6). On the other hand, such bias does not occur when the tQSSA model is used (Fig. 6). However, the distribution of posterior samples of K_M become wider as the enzyme concentration increases, indicating that K_M is not identifiable with the tQSSA model at high enzyme concentrations (i.e., estimation is not precise). Such imprecise estimation occurs because if $K_M \ll S_T + E_T$, K_M is negligible in the tQSSA model (Eq. 2) (*see* ref. 6 for details of identifiability of K_M estimates with the tQSSA model).

4.3 Estimation of k_{cat} and K_M

When k_{cat} and K_M are estimated together, which is the typical goal of enzyme kinetics, both models unfortunately show imprecise estimations. Specifically, the posterior samples of k_{cat} and K_M obtained with the sQSSA model are strongly correlated, indicating that only the ratio between k_{cat} and K_M is identifiable (Fig. 7). This is because the dynamics of the sQSSA model (Eq. 1) are mainly determined by the ratio between k_{cat} and K_M. Similarly, when enzyme concentration is low, the posterior samples of k_{cat} and K_M obtained with the tQSSA model are strongly correlated (Fig. 7a). On the other hand, as enzyme concentration increases, the scatter

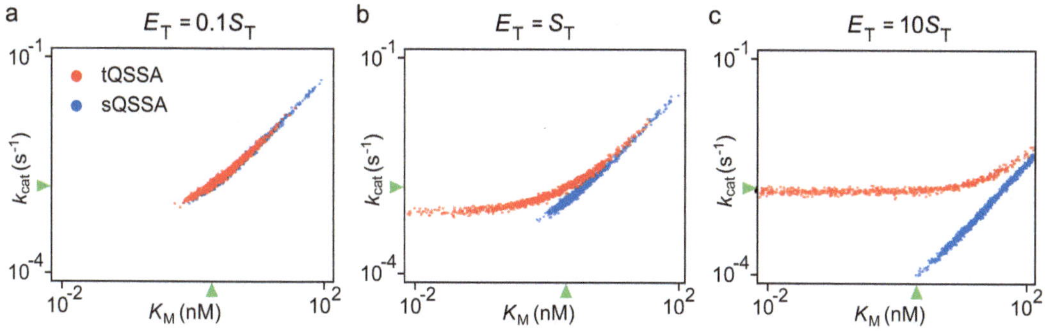

Fig. 7 The scatter plots of posterior samples of k_{cat} and K_M when they are simultaneously estimated with either the sQSSA or the tQSSA model. The scatter plots obtained with the sQSSA model show strong correlations between k_{cat} and K_M for all enzyme concentrations. On the other hand, the positively correlated scatter plots obtained with the tQSSA model become horizontal as enzyme concentration increases. The horizontal scatter plot indicates that at high enzyme concentrations, k_{cat}, but not K_M, is identifiable with the tQSSA model. Here, the green triangles represent the true values of the parameters, and both axes have log scales. The figures are reproduced from [6] with permission from *Scientific Reports*

plots of the posterior samples obtained with the tQSSA model become horizontal (Fig. 7b, c), which indicates that K_M is not identifiable (*see* **Note 5**). This is because the tQSSA model does not depend on K_M at high enzyme concentrations (*see* ref. 6 for details). Importantly, at high enzyme concentration, although K_M is not identifiable, k_{cat} can still be accurately and precisely estimated (Fig. 7c).

4.4 An Optimal Progress Curve Assay

The estimation of both k_{cat} and K_M using a single progress curve suffers from the lack of identifiability (Fig. 7). Furthermore, the posterior samples obtained with the sQSSA model are biased (i.e., as enzyme concentration increases, the scatter plot moves toward the right and the bottom in Fig. 7). However, the posterior samples obtained with the tQSSA model do not show bias, regardless of enzyme concentration. Thus, unlike the sQSSA model, the tQSSA model can be used to estimate parameters from progress curves measured with varying enzyme concentrations (i.e., typical progress curve assay conditions) [9, 11] without risk of bias. This is the great advantage of using the tQSSA model because progress curves measured with low and high enzyme concentrations provide different information for parameter estimation, which is reflected in different types of scatter plots (i.e., correlated (Fig. 7a) and horizontal scatter plots (Fig. 7c)). In particular, with the horizontal scatter plot (Fig. 7c), k_{cat} can be accurately and precisely estimated. With this information, from the correlated scatter plot (Fig. 7a), K_M can be accurately and precisely estimated. This allows us to develop an optimal design of progress curve assay experiments based on the tQSSA model for accurate and precise parameter estimation.

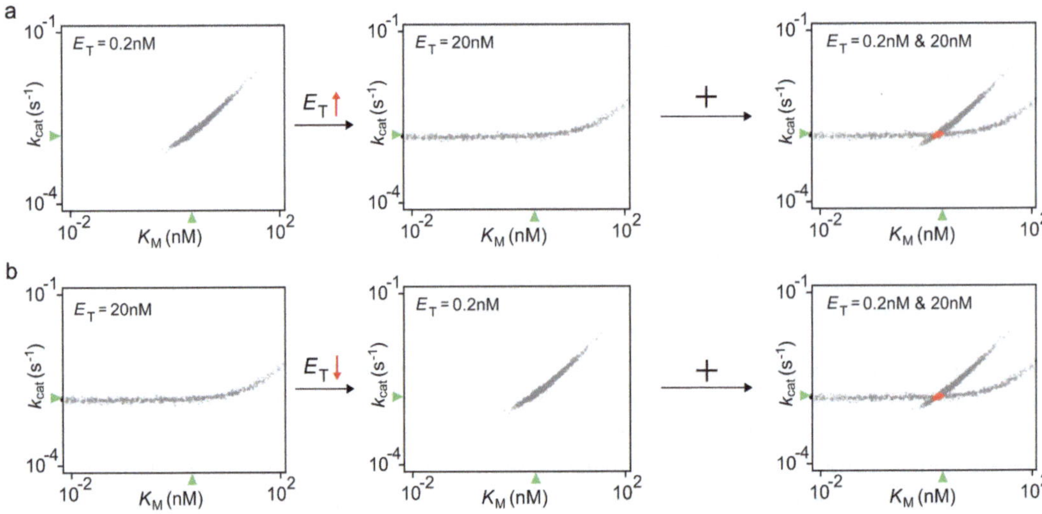

Fig. 8 Optimal experimental design for accurate and precise progress curve assays. (**a**) When the scatter plot of posterior samples from the first experiment shows a strong positive correlation (left), the enzyme concentration needs to be increased until the newly obtained scatter plot becomes horizontal (middle). Then, applying our code to the combination of the progress curves measured under different enzyme concentrations leads to the accurate and precise estimation of both k_{cat} and K_M (right; red zone). (**b**) When the scatter plot from the first experiment is horizontal (left), the enzyme concentration needs to be decreased until the scatter plot shows a strong positive correlation (middle). Then, with the progress curves obtained with different enzyme conditions, our code will provide the accurate and precise estimation of both k_{cat} and K_M (right; red zone). Note that in the right figures, the gray dots are copied from the left and middle figures. Here, the green triangles represent the true values of parameters. The figures are reproduced from [6] with permission from *Scientific Reports*

Step 1. With a certain choice of enzyme concentration, measure a progress curve.

Step 2. With the progress curve, estimate both k_{cat} and K_M with our code based on the tQSSA model and then draw the scatter plot of the posterior samples as described in the instructions above.

Step 3. If the scatter plot shows a strong correlation (Fig. 8a, left) or becomes horizontal (Fig. 8b, left), increase or decrease the enzyme concentration, respectively. Then, with the new enzyme concentration, measure a new progress curve.

Step 4. With the newly obtained progress curve, estimate both k_{cat} and K_M with our code based on the tQSSA model. If the type of the scatter plot differs from that obtained with the initial progress curve (Fig. 8a, b, middle), move to **step 5**. If not, go back to **step 3**.

Step 5. Apply our code to progress curves obtained under different conditions together (i.e., put all progress curves into "time-series_data.csv"). This will lead to accurate and precise estimation of both k_{cat} and K_M (Fig. 8a, b, right; red zones).

For simplicity, we illustrate a case when enzyme concentration is varied to get the different types of scatter plots in this chapter. However, substrate concentration instead of enzyme concentration can also be varied to get different types of scatter plots in **step 3** (*see* ref. 6 for details).

5 Notes

1. For the Metropolis–Hastings step for K_M, the acceptance ratio should be close to the optimal value of 0.4 [29]. When the acceptance ratio is too low (0.01), the trace plot of its posterior samples does not show up-and-down fluctuation (Fig. 9a). On the other hand, when the acceptance ratio is too high (0.98), the trace plot shows a long-term drift (Fig. 9b). In both cases, the standard deviation of the proposal distribution for K_M needs to be adjusted to get the optimal acceptance ratio between 0.3 and 0.5 (Fig. 2, top). Specifically, users need to decrease (increase) `std` or `tun` to increase (decrease) the acceptance ratio.
2. When `std = FALSE`, the standard deviation of a proposal distribution is adaptively determined via the hessian of log

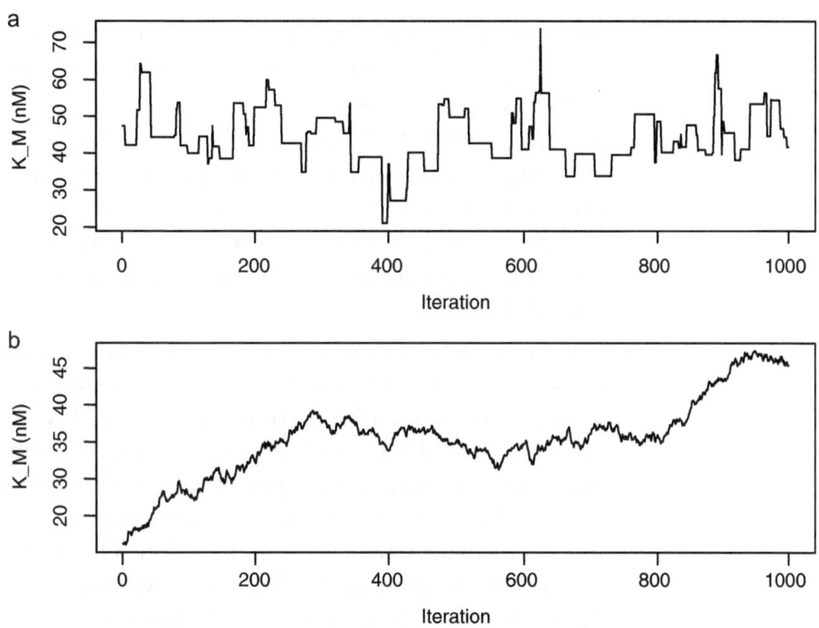

Fig. 9 Problematic trace plots of posterior samples of K_M when the standard deviation of its proposal distribution for the Metropolis–Hastings algorithm is chosen inappropriately. (**a**) A trace plot with a low acceptance ratio (0.01) due to the choice of a large standard deviation of the proposal distribution (10^{10}). (**b**) A trace plot with a high acceptance ratio (0. 98) due to the choice of a small standard deviation of the proposal distribution (10^6)

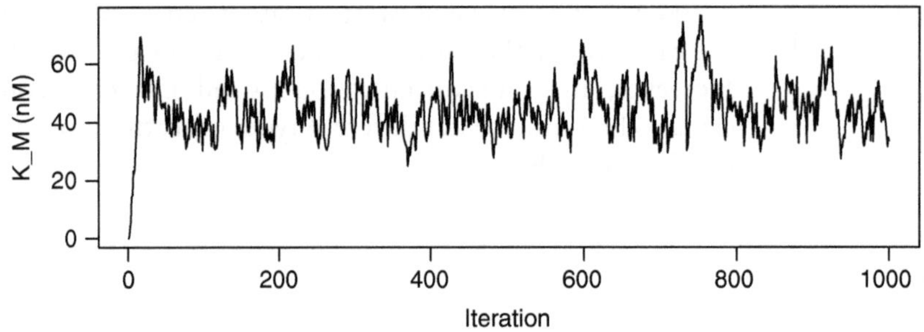

Fig. 10 A problematic trace plot of posterior samples of K_M when the burn-in period is too short

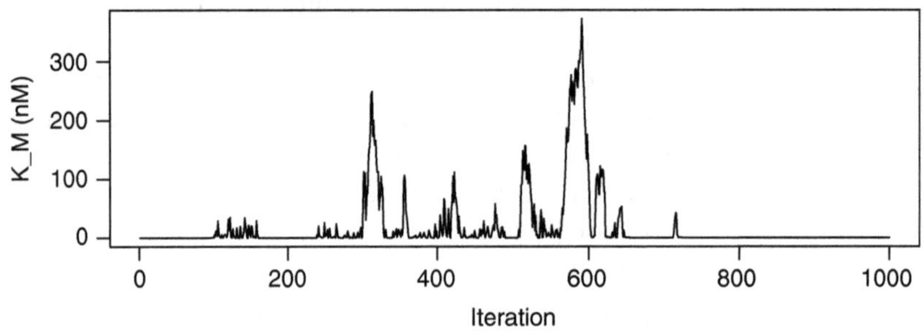

Fig. 11 A problematic trace plot of posterior samples of K_M when an identifiability problem occurs

posterior distribution by our code. However, error can occur if the hessian of log posterior is not calculated. In this case, users should specify the value of `std`.

3. The first few MCMC samples highly depend on the initial value (Fig. 10). Thus, "burn-in" is needed, i.e., the first few samples should be discarded to obtain samples in the stationary state. Specifically, if there exists a drift at the beginning of the trace plot, the region containing the drift should be discarded (e.g., 0 to 200).

4. The scale of a standard deviation depends on the scale of means. Thus, the CV (standard deviation/mean) can be used as the scaled standard deviation. The relative CV, which is the ratio between the CV of the posterior samples and prior CV, can be used to measure the performance of the Bayesian inference. If the relative CV is less than 1, the posterior samples give more information than prior information. That is, if the relative CV is not less than 1, the Bayesian inference is mainly determined by prior information rather than observed data. Thus, when a noninformative prior is used, the relative CV is usually less than 1. On the other hand, when an informative prior is used, we should make sure that the relative CV is less than 1.

5. Even when the acceptance ratio is optimal, the trace plot can show long-term drifts if a parameter identifiability problem occurs (Fig. 11). This identifiability problem can be resolved by measuring additional progress curves after changing experimental conditions (i.e., change the enzyme or substrate concentration) to get different types of $K_M - k_{cat}$ scatter plots (Fig. 8).

Acknowledgments

This work was supported by Global Ph. D. Fellowship Program 2019H1A2A1075303 (H.H.), (NRF-2017R1D1A3B03031008 (B.C.), NRF-2020R1F1A1A01066082 (B.C.), National Research Foundation of Korea grants NRF-2016 RICIB 3008468 (J.K.K.), the Institute for Basic Science IBS-R029-C3 (J.K.K.), and the TJ Park Science Fellowship of the POSCO TJ Park Foundation (J.K. K.).

References

1. Henri V (1903) Lois générales de l'action des diastases. Librairie Scientifique A, Hermann
2. Menten L, Michaelis M (1913) Die kinetik der invertinwirkung. Biochem Z 49(333–369):5
3. Briggs GE, Haldane JBS (1925) A note on the kinetics of enzyme action. Biochem J 19 (2):338
4. Gunawardena J (2014) Time-scale separation–Michaelis and Menten's old idea, still bearing fruit. FEBS J 281(2):473–488
5. Schnell S, Maini P (2003) A century of enzyme kinetics: reliability of the KM and vmax estimates. Comment Theor Biol 8:169–187
6. Choi B, Rempala GA, Kim JK (2017) Beyond the Michaelis-Menten equation: accurate and efficient estimation of enzyme kinetic parameters. Sci Rep 7(1):1–11
7. Segel LA, Slemrod M (1989) The Quasi-Steady-state assumption: a case study in perturbation. SIAM Rev 31(3):446–477
8. Kim JK, Tyson JJ (2020) Misuse of the Michaelis-Menten rate law for protein interaction networks and its remedy. PLoS Comput Biol 16(10):e1008258
9. Algar WR, Malanoski AP, Susumu K, Stewart MH, Hildebrandt N, Medintz IL (2012) Multiplexed tracking of protease activity using a single color of quantum dot vector and a time-gated Forster resonance energy transfer relay. Anal Chem 84(22):10136–10146
10. Sapsford KE, Farrell D, Sun S, Rasooly A, Mattoussi H, Medintz IL (2009) Monitoring of enzymatic proteolysis on a electroluminescent-CCD microchip platform using quantum dot-peptide substrates. Sensors Actuators B Chem 139(1):13–21
11. Algar WR, Malonoski A, Deschamps JR, Blanco-Canosa JB, Susumu K, Stewart MH, Johnson BJ, Dawson PE, Medintz IL (2012) Proteolytic activity at quantum dot-conjugates: Kinetic analysis reveals enhanced enzyme activity and localized interfacial "hopping". Nano Lett 12(7):3793–3802
12. Albe KR, Butler MH, Wright BE (1990) Cellular concentrations of enzymes and their substrates. J Theor Biol 143(2):163–195
13. Fujioka A, Terai K, Itoh RE, Aoki K, Nakamura T, Kuroda S, Nishida E, Matsuda M (2006) Dynamics of the Ras/ERK MAPK cascade as monitored by fluorescent probes. J Biol Chem 281(13):8917–8926
14. Blüthgen N, Bruggeman FJ, Legewie S, Herzel H, Westerhoff HV, Kholodenko BN (2006) Effects of sequestration on signal transduction cascades. FEBS J 273(5):895–906
15. Houston JB, Kenworthy KE (2000) In vitro-in vivo scaling of CYP kinetic data not consistent with the classical Michaelis-Menten model. Drug Metab Dispos 28(3):246–254
16. Wienkers LC, Heath TG (2005) Predicting in vivo drug interactions from in vitro drug

discovery data. Nat Rev Drug Discov 4 (10):825

17. H-M B, H-Y Y, Kim SK, Kim JK (2020) Beyond the Michaelis-Menten: accurate prediction of in vivo hepatic clearance for drugs with low KM. Clin Transl Sci 13:1–9
18. Cha S (1970) Kinetic behavior at high enzyme concentrations magnitude of errors of Michaelis-Menten and other approximations. J Biol Chem 245(18):4814–4818
19. Laidler KJ (1955) Theory of the transient phase in kinetics, with special reference to enzyme systems. Can J Chem 33 (10):1614–1624
20. Tzafriri AR (2003) Michaelis-Menten kinetics at high enzyme concentrations. Bull Math Biol 65(6):1111–1129
21. Borghans JA, De Boer RJ, Segel LA (1996) Extending the quasi-steady state approximation by changing variables. Bull Math Biol 58 (1):43–63
22. Schnell S, Maini PK (2002) Enzyme kinetics far from the standard quasi-steady-state and equilibrium approximations. Math Comput Model 35(1–2):137–144
23. Bersani AM, Bersani E, Dell'Acqua G, Pedersen MG (2015) New trends and perspectives in nonlinear intracellular dynamics: one century from Michaelis–Menten paper. Contin Mech Thermodyn 27(4–5):659–684
24. Tzafriri AR, Edelman ER (2007) Quasi-steady-state kinetics at enzyme and substrate concentrations in excess of the Michaelis–Menten constant. J Theor Biol 245(4):737–748
25. Lim HC (1973) On kinetic behavior at high enzyme concentrations. AICHE J 19 (3):659–661
26. Choi B, Kim JK, Rempala GA (2020) EKMCMC: MCMC procedures for estimating enzyme kinetics constants. R Package version 1.1.0
27. R Core Team (2019) R: A language and environment for statistical computing. R Foundation for Statistical Computing, Vienna, Austria. http://www.R-project.org/
28. RStudio Team (2020) RStudio: integrated development for R. RStudio, PBC, Boston, MA. http://www.rstudio.com/
29. Gelman A, Carlin JB, Stern HS, Dunson DB, Vehtari A, Rubin DB (2013) Bayesian data analysis. Chapman & Hall/CRC texts in statistical science. CRC Press, Boca Raton
30. Roy V (2020) Convergence diagnostics for Markov chain Monte Carlo. Annu Rev Stat Appl 7(1):387–412

Chapter 4

Multi-Objective Optimization Tuning Framework for Kinetic Parameter Selection and Estimation

Yadira Boada, Jesús Picó, and Alejandro Vignoni

Abstract

Semi-mechanistic kinetic (i.e., dynamic) models based on first principles are particularly relevant in biology, as they can explain and predict functional behavior that arises from varying concentrations of the cellular components over time. Here, we describe a computational tuning framework to facilitate both the selection of kinetic parameters for these models and its estimation from experimental data. On the one hand, the tuning framework uses multi-objective optimization to generate a model-based set of guidelines for the selection of the kinetic parameters. These parameter values are the required ones to provide a biological system with desired behavior, while fulfilling the design criteria encoded in the optimization problem itself. On the other hand, this framework can also be used to estimate the parameter values of biological systems from experimental data, once the optimization objectives had been defined appropriately. The methodology gives accurate identification results, as it provides clear orientation on the effect of the parameter values over the system's behavior even under different experimental scenarios. It is particularly useful for easily combining time-course-averaged data and steady-state distribution data. This protocol also addresses aspects related to the appropriate description of the kinetic models and the settings of the software tools. Therefore, it supplies for hands-on testing to evaluate the validity of the underlying technical assumptions of the biological kinetic models.

Key words Kinetic models, Parameter selection, Parameter estimation, Multi-objective optimization

1 Introduction

In the era of artificial intelligence (AI), traditional sciences such as biology, medicine, or chemistry are looking more often at these computational-based methods as intuitive tools to understand a system's behavior. For biological networks, a specific behavior can be attributed to two main factors: (i) the dynamical nature of the components and interactions that embody a particular function and (ii) the evolution of the underlying system structure including the

This work is partially supported by grants MINECO/AEI, EU DPI2017-82896- C2-1-R and MICINN/AEI, EU PID2020-117271RB-C21.

Quentin Vanhaelen (ed.), *Computational Methods for Estimating the Kinetic Parameters of Biological Systems*,
Methods in Molecular Biology, vol. 2385, https://doi.org/10.1007/978-1-0716-1767-0_4,

dynamics of its components [13, 14]. Several efforts have been made to model properly these properties: mechanistic modeling that completely describes a network based on first principles [1, 10] or data-driven modeling that employs an *ad-hoc* model adapted to measured data obtained from experiments on the network [18, 26]. But there is still a lack of a framework that can combine the main features of both mechanistic and data-driven models to accurately predict the evolution of the biological network.

Semi-mechanistic kinetic models separate the model into a first principle part including dynamic equations and an experimental data part [24]. These types of models take advantage of the a priori system knowledge and the possibility of having limited "informative" measurements from the system or its components. The first principle equations can be modeled by following either a deterministic or stochastic approach. So, this will mainly depend on the purpose of the model or the nature of the experimental measurements. However, making use of stochastic models and the additional information conveyed by noisy experimental data may allow an improvement of model parameters estimation [10, 23]. This is because the biochemical reactions that take place in a biological network are inherently random. In any case, both deterministic and stochastic models are becoming particularly relevant in biology and synthetic biology since they accurately predict the functional performing of a network while demanding significantly less measurements than the data-driven and AI models.

Multi-objective optimization is a well-known and efficient method in computer science. Several real-life problems have multiple conflicting objective functions that need to be addressed simultaneously to fulfill the desired goal [20]. Multi-objective optimization frameworks have proven to have great impact in designing sophisticated learning systems, especially building robust biological systems [19]. Robustness and function reliability are precisely important features seeked by a new research area known as synthetic biology. Synthetic biology is defined as the engineering of biology, where synthetic gene circuits are deliberately (re)-designed and built from novel biological components to perform new functions for useful purposes [17]. It emphasizes engineering principles and methodology in designing, constructing, and characterizing biological systems to be used in several biotech applications [11].

Here, we introduce a framework called Multi-Objective Optimization Tuning (MOOT) that aims to facilitate the selection of kinetic model parameters and the parameter estimation of mechanistic models using experimental data. On the one hand, the MOOT method relies on multi-objective optimization to generate a set of guidelines for selecting the kinetic model parameters of a network [3, 4, 8]. These qualitative guidelines define the parameter values required to provide a biological system with a desired behavior. Thus, the mathematical formulation of the objectives encodes both the design criteria and the desired purpose in the optimization problem itself. Among the optimal solutions contained in the Pareto front [20], not only does the MOOT framework allow a robust computational design but also lead to a well-constructed and characterized biological networks. On the other hand, MOOT can also be used to estimate the model parameter values from experimental data, once the optimization objectives had been appropriately defined. The methodology has demonstrated good identification results as it provides clear orientation on how the parameter values affect the system's behavior [5, 7, 10, 25]. MOOT is suitable for model parameter estimation even under different scenarios that emulate changes made in the variables of the system or the conditions of certain experiments [5, 25]. It is also particularly useful for easily combining time-series data and steady-state distribution data [10].

Finally, the MOOT protocol also addresses aspects related to the proper description of the kinetic models and the settings of the software tools. Therefore, it supplies for hands-on testing to evaluate the validity of the underlying dynamic models and its technical assumptions against experimental data.

2 Materials

For the MOOT protocol, the first and indispensable material to work with is the mathematical model. The types of models that can be used include both ordinary and stochastic differential equations (ODEs and SDEs). To clarify the protocol steps, we used a stochastic model (to account for molecular noise and stochasticity) of a genetic circuit depicted in Fig. 1 as an example. This model is a set of stochastic differential equations with intracellular and extracellular species modeling gene expression of a protein of interest (PoI) across the *N* cells from a population. Quorum sensing was also modeled since it allows individual bacteria to coordinate protein expression of the protein and it is one of the most common mechanisms of communication in nature. One important fact is that it can be used only with the deterministic part, in case we are looking for macroscopic results of a biological model. If we also include the stochastic part, we will work with the whole model accounting for gene expression noise. So, depending on our needs, we can use one or the other. The gene circuit in Fig. 1, hereafter denoted as the QS/Fb circuit, integrates a cell-to-cell communication mechanism and an intracellular feedback loop [6, 10]. As a consequence, there is a biochemical species diffusing from inside to outside the cell as communication signal, so the individual-cell level and the population one are interconnected. The circuit aims to achieve a desired mean value of protein

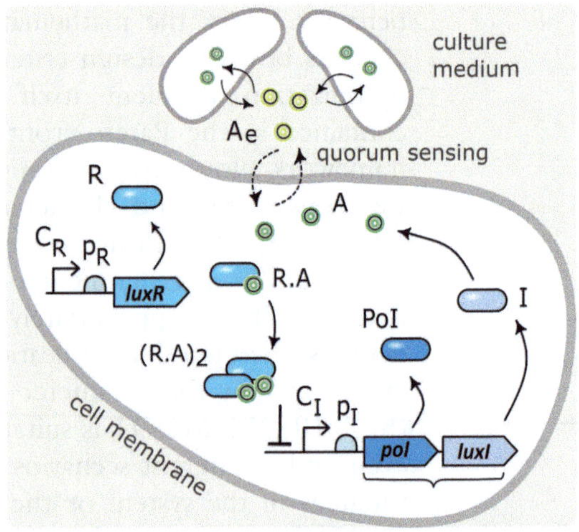

Fig. 1 QS/Fb circuit. The gene circuit aims to regulate the average expression of a protein of interest (PoI) while minimizing the noise strength. To this end, it relies on combining a cell-to-cell communication mechanism via quorum sensing (QS), and an intracellular negative feedback loop (Fb). The Fb subsystem regulates the expression of the protein of interest inside each cell and reduces its noise strength. The QS subsystem induces consensus among cells exchanging diffusible molecules to achieve homogeneous expression across the cell population

expression while minimizing its variability in time across the population of the N cells. Therefore, the circuit inherent randomness and performance will require stochastic analysis. For further details on this model, *see* ref. [21].

2.1 Getting the Model in Proper Form

1. **Define a vector containing the number of molecules of the biochemical species for the population of cells.** The circuit dynamics will later be expressed using this vector containing the number of molecules of the relevant biochemical species as the model state variables (*see* **Note 1**). Set the number N of cells to be simulated. This protocol assumes N as a constant throughout the computational simulation. Consider all N cells have the same set of relevant intracellular biochemical species. For a system with c common intracellular species for all N cells, and e extracellular species, define the column vector $\mathbf{n} = [\mathbf{n}^i, \ldots, \mathbf{n}^N, n_{c+1}, \ldots, n_{c+e}]^T$ containing all vectors $\mathbf{n}^i = [n_1, \ldots, n_c]^i$ with the number of molecules for the c intracellular species in the i-th cell, and the variables $n_{c+1} \ldots n_{c+e}$ containing the number of molecules of each extracellular species (*see* **Example 1**).

Example 1: For the QS/Fb circuit, the relevant species are the intracellular protein of interest PI (n_1), the constitutively expressed protein R (n_2), the transcription factor $(R{\cdot}A)_2$ (n_3), the small signaling molecule A (n_4) that can freely diffuse across the cellular membrane, and the monomer R·A (n_6). n_6 can be obtained algebraically as a function of the others, $n_6 = g(n_2, n_3, n_4)$ (*see* **Point 4** below to handle species algebraically determined). Additionally, n_5 (A_e) represents the amount of extracellular molecules of the diffusible species A. With this set of species, $\mathbf{n} = [\mathbf{n}^i, \ldots, \mathbf{n}^N, n_5]^T$, with $\mathbf{n}^i = [n_1^i, n_2^i, n_3^i, n_4^i, n_6^i]$ for the five intracellular species in each *i*-th cell.

2. **Define the vector of reaction propensities.** For a system with *N* cells where:
   - r_c reactions are common to all cells and affect both the dynamics of intracellular species and those of the extracellular ones
   - r_e reactions only affect the dynamics of extracellular species,

   define the column vector $\mathbf{a(n)}$ (*see* Eq. 1 containing all vectors $\mathbf{a(n)}^i$ of propensities for the r_c intracellular reactions in the *i*-th cell, and the reaction propensities $a(n)_{r_c+1} \ldots a(n)_{r_c+r_e}$ for the extracellular reactions. Refer to **Note 2** on how to model the reaction propensities.

$$\mathbf{a(n)} = \left[\begin{array}{c} \mathbf{a(n)}^1 \\ \mathbf{a(n)}^2 \\ \vdots \\ \mathbf{a(n)}^N \\ \hline a_{r_c+1}(\mathbf{n}) \\ \vdots \\ a_{r_c+r_e}(\mathbf{n}) \end{array}\right], \quad \mathbf{a(n)}^i = \begin{bmatrix} a_1(\mathbf{n}) \\ \vdots \\ a_{r_c}(\mathbf{n}) \end{bmatrix}. \tag{1}$$

Example 2: For the QS/Fb circuit, consider the set of reactions among the species depicted below. This set includes some pseudo-reactions: the first reaction with the lumped Hill-like functional propensity $f_1(n_3^i) = \frac{\mathrm{C_I p_I}}{\mathrm{dm_I}} \frac{\mathrm{kd_{lux}} + \alpha_I n_3^i}{\mathrm{kd_{lux}} + n_3^i}$ resulting from a previous model-order reduction (where α_I is the basal expression), and the 9th reaction accounting for the diffusion process. The corresponding vector of propensities is shown on the right. For every cell, the last propensity $\mathrm{DV_c} n_5$ depends on the extracellular species A_e (n_5), but it is included as an intracellular reaction as it affects the dynamics of the intracellular species A (n_4^i). On the contrary, the propensity function $\mathrm{d_{A_e}} n_5$ only affects directly the dynamics of the extracellular species n_5. Therefore, it is included as an extracellular reaction in the vector of propensities. Refer to **Example 12** for the software code implementation.

$$
\begin{aligned}
(\mathrm{R}\cdot\mathrm{A})_2{}^{\mathrm{i}} &\xrightarrow{f_1(n_3^i)} \mathrm{PI}^{\mathrm{i}} + (\mathrm{R}\cdot\mathrm{A})_2{}^{\mathrm{i}}\\
\mathrm{PI}^{\mathrm{i}} &\xrightarrow{\mathrm{d_I}} \emptyset\\
&\xrightarrow{\mathrm{c_{LR}}} \mathrm{R}^{\mathrm{i}}\\
\mathrm{R}^{\mathrm{i}} + \mathrm{A}^{\mathrm{i}} &\underset{\mathrm{k_1^-}}{\overset{\mathrm{k_1^+}}{\rightleftharpoons}} \mathrm{R}\cdot\mathrm{A}^{\mathrm{i}}\\
\mathrm{R}^{\mathrm{i}} &\xrightarrow{\mathrm{d_R}} \emptyset\\
\mathrm{R}\cdot\mathrm{A}^{\mathrm{i}} + \mathrm{R}\cdot\mathrm{A}^{\mathrm{i}} &\underset{\mathrm{k_2^-}}{\overset{\mathrm{k_2^+}}{\rightleftharpoons}} (\mathrm{R}\cdot\mathrm{A})_2{}^{\mathrm{i}}\\
(\mathrm{R}\cdot\mathrm{A})_2{}^{\mathrm{i}} &\xrightarrow{\mathrm{d_{RA}}} \emptyset\\
\mathrm{PI}^{\mathrm{i}} &\xrightarrow{\mathrm{k_A}} \mathrm{PI}^{\mathrm{i}} + \mathrm{A}^{\mathrm{i}}\\
\mathrm{A}^{\mathrm{i}} &\underset{\mathrm{DV_c}}{\overset{\mathrm{D}}{\rightleftharpoons}} \mathrm{A_e}\\
\mathrm{A}^{\mathrm{i}} &\xrightarrow{\mathrm{d_A}} \emptyset\\
\mathrm{A_e} &\xrightarrow{\mathrm{d_{Ae}}} \emptyset
\end{aligned}
$$

$$
\mathbf{a}(\mathbf{n}) = \begin{bmatrix} \mathbf{a}(\mathbf{n})^1 \\ \mathbf{a}(\mathbf{n})^2 \\ \vdots \\ \mathbf{a}(\mathbf{n})^{\mathrm{N}} \\ \hline \\ \mathrm{d}_{\mathrm{A_e}} n_5 \end{bmatrix}, \quad \mathbf{a}(\mathbf{n})^i = \begin{bmatrix} f_1(n_3^i) \\ \mathrm{d_I} n_1^i \\ \mathrm{C_{LR}} \\ \mathrm{k}_1^- n_6^i \\ \mathrm{k}_1^+ n_2^i n_4^i \\ \mathrm{d_R} n_2^i \\ \mathrm{k}_2^+ (n_6^i)^2 \\ \mathrm{k}_2^- n_3^i \\ \mathrm{d}_{\mathrm{RA_2}} n_3^i \\ \mathrm{k_A} n_1^i \\ \mathrm{d_A} n_4^i \\ \mathrm{D} n_4^i \\ \mathrm{DV_c} n_5 \end{bmatrix}
$$

3. **Define the extended stoichiometry matrix.** For a system with N cells where r_c common intracellular reactions and r_e extracellular reactions take place among c common intracellular

non-algebraic species—i.e., species that cannot be obtained algebraically as a function of others (*see* **Point 4** below to consider algebraic species)—for all N cells and e extracellular non-algebraic species, the extended stoichiometry matrix S has a blocks structure:

$$\mathbf{S} = \left[\begin{array}{c|c} \mathbf{I_N} \otimes \mathbf{S_{cc}} & \mathbf{0_{cN \times r_e}} \\ \hline \mathbf{1_{1 \times N}} \otimes \mathbf{S_{ec}} & \mathbf{S_{ee}} \end{array} \right] \tag{2}$$

where:

- $\mathbf{S_{cc}}$ is a $c \times r_c$ matrix formed by the stoichiometric coefficients for the c intracellular non-algebraic species accounting only for the r_c intracellular reactions.
- $\mathbf{0_{cN \times r_e}}$ is a $c \cdot N \times r_e$ null matrix.
- $\mathbf{S_{ec}}$ is a $e \times r_c$ matrix formed by the stoichiometric coefficients for the e extracellular non-algebraic species accounting for the interactions with the intracellular ones via the r_c intracellular reactions affecting them.
- $\mathbf{S_{ee}}$ is a $e \times r_e$ matrix formed by the stoichiometric coefficients for the e extracellular non-algebraic species accounting only for the r_e extracellular reactions.
- $\mathbf{I_N}$ is the $N \times N$ identity matrix.
- $\mathbf{1_{1 \times N}}$ is a $1 \times N$ row vector of ones.

and $\otimes$ is the Kronecker product (*see* **Note 3**).

Example 3: For the QS/Fb circuit, using the reactions defined in **Example 2**, we have:

$$\begin{aligned} \mathbf{S_{cc}} &= \begin{bmatrix} 1 & -1 & 0 & 0 & 0 & 0 & 0 & 0 & 0 & 0 & 0 & 0 & 0 \\ 0 & 0 & 1 & 1 & -1 & -1 & 0 & 0 & 0 & 0 & 0 & 0 & 0 \\ 0 & 0 & 0 & 0 & 0 & 0 & 1 & -1 & -1 & 0 & 0 & 0 & 0 \\ 0 & 0 & 0 & 1 & -1 & 0 & 0 & 0 & 0 & 1 & -1 & -1 & 1 \end{bmatrix} \\ \mathbf{S_{ec}} &= [0 \; 0 \; 0 \; 0 \; 0 \; 0 \; 0 \; 0 \; 0 \; 0 \; 0 \; 1 \; -1] \\ \mathbf{S_{ee}} &= [-1]. \end{aligned} \tag{3}$$

If we consider $N=2$ cells, the extended stoichiometry matrix takes the form:

$$\mathbf{S} = \begin{bmatrix} [1 \cdot \mathbf{S_{cc}}] & [0 \cdot \mathbf{S_{cc}}] & [\mathbf{0_{4\times 1}}] \\ [0 \cdot \mathbf{S_{cc}}] & [1 \cdot \mathbf{S_{cc}}] & [\mathbf{0_{4\times 1}}] \\ [1 \cdot \mathbf{S_{ec}}] & [1 \cdot \mathbf{S_{ec}}] & [\mathbf{S_{ee}}] \end{bmatrix}. \tag{4}$$

Refer to **Example 14** for the software code implementation.

4. **Consider algebraically determined species.** Algebraic relationships among the species in the circuit may arise as a product of model-order reduction. Thus, for each i-th cell, the vector of propensities $\mathbf{a}(\mathbf{n})^i$ may depend on species $n^i_{a_1} \ldots n^i_{a_a}$ that in turn can be obtained as an algebraic function of the species in the system. These algebraic species need not to be considered in the vector of the species defined in **Point 1** above. The algebraic functions $n^i_{a_j} = g^i_j(\mathbf{n}),\ j = 1 \ldots a$, will be used as constraints during the simulation of the system dynamics (*see* **Point 5** below). For a system with c_a common intracellular algebraic species for all N cells and e_a extracellular algebraic species:
   (a) Define the column vector $\mathbf{n_a} = [\mathbf{n_a}^i, \ldots, \mathbf{n_a}^{\mathrm{N}}, n_{c_a+1}, \ldots, n_{c_a+e_a}]^T$ containing all vectors $\mathbf{n_a}^i = [n_{a_1}, \ldots, n_{a_{c_a}}]^i$ for the number of molecules of the c_a intracellular algebraic species in the i-th cell, and the variables $n_{c_a+1}, \ldots, n_{c_a+e_a}$ containing the number of molecules of each algebraic extracellular species (*see* **Example 4**).
   (b) Define the column vector $\mathbf{g}(\mathbf{n}) = [\mathbf{g_a}^i(\mathbf{n}), \ldots, \mathbf{g_a}^{\mathrm{N}}(\mathbf{n}), g_{c_a+1}(\mathbf{n}), \ldots, g_{c_a+e_a}(\mathbf{n})]^T$ containing all vectors of the algebraic functions $\mathbf{g_a}^i(\mathbf{n}) = [g_{a_1}(\mathbf{n}), \ldots, g_{a_{c_a}}(\mathbf{n})]^i$ for the c_a intracellular algebraic species in the i-th cell, and the functions $g_{c_a+1}(\mathbf{n}) \ldots g_{c_a+e_a}(\mathbf{n})$ for the algebraic extracellular species (*see* **Example 4**).

   Example 4: For the QS/Fb circuit, notice the fourth component in the i-th cell vector of propensities $\mathbf{a}(\mathbf{n})^i$ depends on the species n^i_6 (*see* **Example 1**). This one can be obtained algebraically as a function of the others, $n^i_6 = g^i_6(n^i_2, n^i_3, n^i_4)$, so there was no need to explicitly consider it as a component of the vector containing the number molecules of the biochemical species. To simulate N cells, we set the algebraic constraint $\mathbf{n_6} = \mathbf{g}(\mathbf{n})$ with $\mathbf{n_6} = [n^1_6, \ldots n^{\mathrm{N}}_6]^T$, and $\mathbf{g_6}(\mathbf{n}) = [g^1_6(n^1_2, n^1_3, n^1_4), \ldots, g^{\mathrm{N}}_6(n^N_2, n^N_3, n^N_4)]^T$.

5. **Define the general structure of the dynamic model.** *See* **Note 1** for the technical background. Split the column vector $\mathbf{n}$ such that $\mathbf{n} = [\mathbf{n_{na}}, \mathbf{n_a}]$ where $\mathbf{n_{na}}$ contains the number of molecules for the set of non-algebraic species and $\mathbf{n_a}$ for the algebraic

ones. The temporal evolution of the number of molecules for each biochemical species of interest will be expressed as

$$\begin{aligned} \frac{d\mathbf{n_{na}}(t)}{dt} &= \mathbf{S} \cdot \mathbf{a}(\mathbf{n}(\mathbf{t})) \\ \mathbf{n_a}(t) &= \mathbf{g}(\mathbf{n}(t)) \end{aligned} \tag{5}$$

for the deterministic model and

$$\begin{aligned} \mathbf{n_{na}}(t+\delta t) &= \mathbf{n_{na}}(t) + \mathbf{S} \cdot \mathbf{a}(\mathbf{n}(\mathbf{t}))\delta t + \mathbf{S} \cdot \mathcal{N}_t \cdot \sqrt{\mathbf{a}(\mathbf{n}(\mathbf{t}))}\sqrt{\delta t} \\ \mathbf{n_a}(t+\delta t) &= \mathbf{g}(\mathbf{n}(t+\delta t)) \end{aligned} \tag{6}$$

for the stochastic model.

***Example** 5*: For the QS/Fb circuit, define $\mathbf{n_{na}} = [\mathbf{n_{na}}^{i}, \ldots, \mathbf{n_{na}}^{N}, n_5]^T$, with $\mathbf{n_{na}}^{i} = [n_1^i, n_2^i, n_3^i, n_4^i]$ (*see* **Example 1**) and $\mathbf{n_a} = \mathbf{n_6}$ as defined in **Example 4**. The propensities vector $\mathbf{a(n(t))}$ and the extended stoichiometry matrix **S** were defined in **Examples 2** and **3**, respectively.

2.2 Software

Optimization and visualization of solutions can be performed with any software designed for it. As an example here, we use two toolboxes developed in MATLAB, **spMODEx**[1] as multi-objective optimization algorithm [22], and **LevelDiagrams**[2] for visualization and to help in the decision-making process [2]:

1. **spMODEx** Includes five files from which the following three have relevance here:
   - `MOEAparam.m`: The configuration file tells the algorithm how many objectives, decision variables, value ranges, constraints there are.
   - `CostFunction.m`: The definition of the cost function. It needs to include a piece of code to simulate the model and to calculate the objectives. It is called by the algorithm in each iteration to evaluate the different candidates for the solution.
   - `spMODEx.m`: The algorithm itself is necessary to run the optimization using the previous two files.
2. **LevelDiagrams** Includes nine files from which the following three have relevance here:
   - `conceptCreate.m` That is used to initialize the data.
   - `basicNorm.m` Calculates different norms from the solutions for visualization.

[1] Tool available in http://www.mathworks.com/matlabcentral/fileexchange/65145.

[2] Tool available in http://www.mathworks.com/matlabcentral/fileexchange/62224.

– `ldDraw.m` Draws the LevelDiagram using the output of the previous files.

3 Methods

The framework setup is practically the same independently if our goal is to select parameters for a model, or to estimate values of the model parameters from experimental data. Thus, we depict the general method and particularize each part either for selection or estimation correspondingly. The Multi-Objective Optimization Tuning (MOOT) framework consists of three parts (See Fig. 2): multi-objective problem definition (MOP), multi-objective optimization (MOO), and multi-criteria decision-making (MCDM). Particularly, the first step is the one that differs depending on the purpose. For the parameter selection case, the MOP includes a definition of the desired behavior. However, the parameter estimation case involves the definition of the estimation error, that is, the error between the model prediction and the experimental data.

3.1 Define the Multi-Objective Problem. Objectives

Parameter selection.

The desired behavior has to be translated into design objectives. Any behavior that can be written in terms of a cost index can be used. Some examples are: steady-state or equilibrium values, time required for the response to reach a certain value, shape or amplitude of the response, presence or not of oscillations, height and/or width of a peak response [4], error with respect to a desired response. A particular class of cost functions is the class of control performance indexes (with respect to the desired steady state) such as Integral Absolute Error (IAE) or Integral Time Absolute Error (ITAE). Finally, in a stochastic setup, a desired behavior can be certain steady-state mean value with its corresponding noise strength or variance [8].

Example 6: For the QS/Fb example, the desired behavior is to obtain the minimum noise strength $\eta_{n_1}^2$ for a given mean value μ_{n_1} of the protein of interest (PI), which is represented by the intracellular species n_1 (*see* ref. [10] for details in the stochastic aspects) where the noise strength is defined as

$$\eta_{n_1}^2 = \frac{\sigma_{n_1}^2}{\mu_{n_1}^2}.$$

The indexes μ_{n_1} and $\eta_{n_1}^2$ are interrelated magnitudes that are also in conflict. When one tries to minimize them by finding a single set of parameters, a trade-off must be reached. Thus, we can solve our

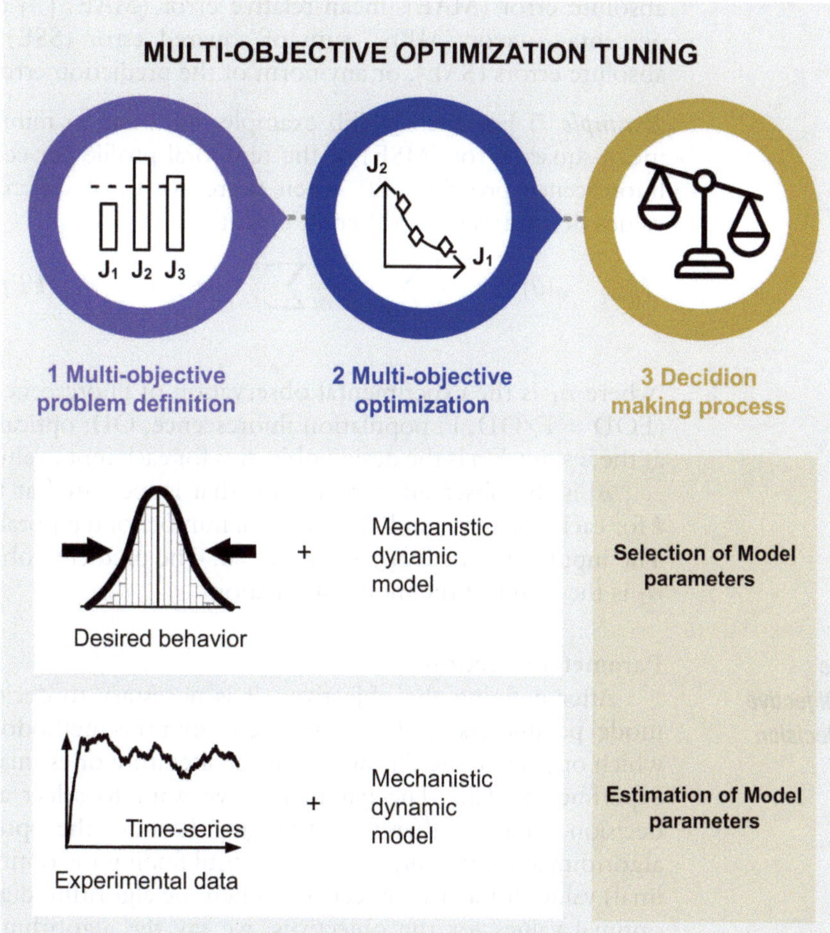

Fig. 2 Multi-Objective Optimization Tuning framework. Successful implementation of the MOOT includes: (1) The multi-objective problem definition: defining the circuit behavioral specifications or the errors between the model simulation and the experimental data in a proper way. (2) The optimization process: tuning or estimating the parameters using multi-objective global optimization. (3) The multi-criteria decision-making process: obtaining the most suitable values for the parameters or finding the proper estimates for the parameters among the obtained solutions

problem by minimizing both μ_{n_1} and $\eta^2_{n_1}$ as two competing objectives. Thus, the set of objectives J to optimize are

$$J_1(\theta) = \mu_{n_1}, \;\; J_2(\theta) = \eta^2_{n_1}, \tag{7}$$

where θ is the set of parameters we want to select with this methodology.

Parameter estimation.

In this case, the estimation error has to be defined as a design objective. Estimation errors commonly used include: Root-mean-square error (RMSE), mean-square error (MSE) [5], mean-absolute error (MAE), mean-relative error (MAE) [7] or mean-percentage error (MPE), sum of squared error (SSE), sum of absolute errors (SAE), or any norm of the prediction error [10].

Example *7*: For the QS/Fb example, we want to minimize the mean-square error (MSE) of the temporal profile per cell for the fluorescence protein GFP, when there are three different input values defined as three different objectives:

$$J_{[a=1,\ldots,3]}(\theta) = \frac{1}{n}\sum_{q=1}^{n}\frac{1}{m}\sum_{k=1}^{m}\left(\hat{n}_{1_{aq}}(k) - n_{1_{aq}}(kT)\right)^2, \tag{8}$$

where $\hat{n}_1$ is the experimental observation of fluorescence per cell (FOD = F/OD, F: population fluorescence, OD: optical density) at the instant k, a is the design objective for each input value, $q = [1, \ldots, n]$ is the observation repetition that is measured at the point k for each objective, and m is the total number of temporal samples. The input stimulus is applied at $t_0 = 0$. The predicted observation n_1 is the result of the model simulation.

3.2 Define the Multi-Objective Problem. Decision Variables

Parameter selection.

After defining the objectives, it is necessary to decide which model parameters need to be selected using this methodology, and which ones are either known from the literature or estimated from experimental data. The parameters we want to select are called decision variables; therefore, they are the ones the optimization algorithm automatically will change until finding the optimal (minimal) values for all the objectives. When the algorithm digs out the optimal values for the objectives, we say the algorithm found a solution. One solution includes the optimal values of the objectives and the corresponding values of the parameters that we want to select (decision variables). They are named Pareto front and Pareto set, respectively. For this case, it is necessary to check if the parameters we want to select (decision variables) have influence on the objectives as this is crucial for the optimization algorithm. Also, another criterion to choose the decision variables is parameters that can be modified physically, that is in the lab implementation of the system. Subsequently, it is necessary to define the initial range of each decision variable. The algorithm searches for values in these ranges.

Example 8: For our parameter selection in the QS/Fb example, we have chosen the following parameters and ranges:

Parameter to optimize	Description	Range of values
α_I	*luxI* basal expression	[0.01 0.2]
c_R	*luxR* transcription rate	[4 60] molecules · min^{-1}
p_I	LuxI translation rate	[0.2 10] min^{-1}
d_R	LuxR degradation rate	[0.02 0.2] min^{-1}
kd_{lux}	$(R{\cdot}A)_2$ dissociation constant	[10 2000] molecules

Parameter estimation. For the estimation case, we should choose the parameters to be estimated and their corresponding initial intervals (range) as they are consistent with the experiments.

Example 9: For the parameter estimation in the QS/Fb example, we looked for the parameters listed in the table below. The initial ranges for each parameter were defined according to their corresponding experimental data. *See* **Note 4** for how to deal with range selection.

Parameter	Description	Initial Range
p_I	Translation rate of $mRNA_{LuxI}$	[0.1 10] min^{-1}
d_I	Degradation rate of LuxI	[0.005 0.1] min^{-1}
kd_{lux}	Dissociation constant of $(R\cdot A)_2$ to promoter	[1 1000] molecules
p_R	Translation rate of $mRNA_{LuxR}$	[0.1 10] min^{-1}

Finally, the optimization problem can be defined as

$$\begin{aligned} \min_{\theta} J(\theta) &= [J_1(\theta), \ldots, J_n(\theta)] \\ \text{subjectto}: &\quad \text{Dynamical model.} \end{aligned} \tag{9}$$

The outcome of the multi-objective optimization is a set of solutions often called the Pareto front. The Pareto solutions (J_P) comprise the optimal values for each objective, and the corresponding Pareto set (θ_P) are the values of the decision variables giving those optimal objective values.

3.3 Multi-Objective Optimization

The multi-objective optimization process finds the best parameters θ_P^* that produce the best Pareto front approximation J_P^*. For problems with a large number of decision variables, as our case, it is more efficient to use an appropriate multi-objective optimization algorithm to approximate this solution.

In this chapter, we use a multi-objective evolutionary algorithm based on differential evolution that uses a spherical pruning to approximate the Pareto front. The implementation used in this chapter is the sp-MODEx[3] algorithm. The next step is to set up the multi-objective optimization. It includes three steps: parameter file configuration, cost function definition, and optimization. For this stage, we will use the software described in Materials section; however, any multi-optimization algorithm can be used.

- **Parameter file configuration: `MOEAparam.m`**

  Here, we define the number of objectives and decision variables, their ranges, the pertinence of the objectives (the ranges for desirable and undesirable values), and the cost function to use. See Example 10 for details.
- **Cost function file: `CostFunction.m`**

  Here, we define cost function. Normally, it includes a piece of code for running the simulation of our model and the comparison with either the desired behavior (parameter selection) or the experimental data (parameter estimation). It returns the values of the objectives for a given value of the parameters. See Example 11 for details.
- **Optimization**

  Finally, run the optimization.

The code corresponding to the parameter file configuration `MOEAparam.m` where objectives number, decision variables number are set is:

```
%% Variables regarding the multi-objective problem

spMODExDat.NOBJ = 2;                    % Number of decision
objectives.
spMODExDat.NRES = 0;                 % Number of constraints plus
                                      % non-decision objectives
spMODExDat.NVAR = 10;                   % Number of decision
variables
```

And the code to define the ranges of the decision variables is:

[3] Tool available in http://www.mathworks.com/matlabcentral/fileexchange/65145.

```
spMODExDat.FieldD    = ...          % Optimization bounds
[zeros(10,1), ones(10,1)];
```

Example 10 demonstrates how to translate the bounds defined in **step 3.2** into their corresponding code.

Example 10: For our parameter selection in the QS/Fb example, the ranges can be written as follows:

```
spMODExDat.FieldD    = [0.01 0.2;...     % alpha_luxI
                         4 60; ...        % cR [molecules/min]
                         0.2 10;...       % pI   [1/min]
                         0.02 0.2;...     % dR [1/min]
                         10 2000];        % kdlux [molecules]
```

where `FielD` is a two column, five row matrix with the lower bound in the first column and the higher bound in the second one, each parameter in one row.

The code to define the pertinence of the objectives in terms of desirable and undesirable values is:

```
% Physical Matrix Definition.
% HD Highly Desirable
% D  Desirable
% T  Tolerable
% U  Undesirable
% HU Highly Undesirable
%                                 HD  -  D  -  T  -  U  - HU
spMODExDat.PhyMatrix{1} = [0.0  0.20  0.70  0.8  2.0   10;
                           0.0  0.60  0.90  1.2  2.0   10];

```

where `PhyMatrix{1}` is a matrix six column, with as many rows as objectives we have (in this case two). Each column defines the limits of the intervals as follows: the first and second rows are the lower and higher limits of the highly desirable zone; the second and third rows are the lower and higher limits of the desirable zone; the third and fourth rows are the lower and higher limits of the tolerable zone; and so on and so forth. The optimization algorithm will try to return solutions that are from tolerable to highly desirable, and will discard the others.

Next, it is necessary to define the problem name in `MOEA-param.m`:

```
spMODExDat.CostProblem   = 'DTLZ2'; % Problem Instance in CostFunction
```

where instead of DTLZ2, it should say your problem name. This same name (`YourProblem`) has to go in the `CostFunction.m` file:

```
37 elseif strcmp(Dat.CostProblem, 'YourProblem')
38     % Here comes the call for a cost function of your own multi objective
39     % problem.
40 end
```

After line 37, you need to include the code that executes your simulation. It requires to follow the template, this is, to accept the present values for the decision variables in `X`, and it has to return the value of the objectives `J` for those `X`.

Example 11: For our parameter selection in the QS/Fb example, the cost function includes the following lines:

```
 %Loop for all the different sets of values in X
for xpop=1:size(X,1)
 % Decision variables
alphaI = X(xpop,1);   % leakage of the repressor Plux
 % LuxR
cR = X(xpop,2);      % constitutive expression rate of LuxR [molecules/min]
 % LuxI
pI  =  X(xpop,3);         % translation rate of LuxI #mRNA [1/min].
dR = X(xpop,4);       % degradation rate of LuxR [1/min]
kdLux  =  X(xpop,5);      % dissociation cte (LuxR.A)2 to promoter [molecules]
...
```

Then, it goes the code that runs the model simulation. Next using these values from the simulation results, the code computes the values of each objective and returns them as follows:

```
        J1 = mx1;          % Mean of LuxI
        J2 = (vx1/mx1).^2;  % Noise of LuxI
        J(xpop,:)=[J1, J2];
end
```

Notice if the objectives vary over a wide range of magnitude, some computing problems could appear. For instance, the sp-MODE algorithm can give more importance to one objective than another; therefore, *see***Note 5** for details on how to deal with this problem.

Finally, the optimization can be executed with the following lines:

```
MOEAparam;
OUT=spMODEx(spMODExDat)
```

The output of the algorithm (`OUT`) is a structure with the solutions obtained including the values of the decision variables or Pareto set (`OUT.PSet`, a matrix with each decision variable in the columns and solution in the rows) and the values of the objectives (`OUT.PFront`, a matrix with each objective in the columns and solutions in the rows) corresponding to those decision variables. These solutions are ready to be analyzed and plotted so we can select the ones that are suitable for our purposes.

3.4 Multi-Criteria Decision-Making

After the optimization process, the selection of the preferable solution according to designer's criteria takes place in an *a posteriori* multi-criteria analysis of the Pareto front and Pareto set approximation. The MOOT framework uses tools that simplify the visualization and the analysis of the trade-off among competing objectives. Such visualization and analysis are a non-trivial task especially when the number of objectives and/or the number of decision variables in the Pareto set are larger than three, like in our case. Here, we show the necessary steps to perform a multi-criteria decision process (MCDM) using the parameter selection case as the main example, the **LevelDiagram**[4] toolbox and a custom script available in our Git repository.[5]

[4] Toolbox available in http://www.mathworks.com/matlabcentral/fileexchange/62224.

[5] Git repository https://github.com/sb2cl/MOOT_Selection_and_Estimation/.

The basic manipulations to obtain the LevelDiagram of our solutions include:

– Create the concept variables to be used by the toolbox from the optimization results.

```
% Create variables related with the concept:
% concept1 and concept1_data
conceptCreate(OUT.PFront,OUT.PSet, 'concept1')
```

– Obtain bounds from the concept variable and calculate norms on the objectives.

```
% Bounds for Pareto front normalization
% bounds =[upperbounds; lowerbounds]
bounds=[concept1.maxpf;concept1.minpf];

% Calculate the value of the norm (2-norm) for y-axis synchronization
% and create the associated variable concept1_ld1
basicNorm( 'ld1', 'concept1',bounds,2);
```

– Draw the LevelDiagram.

```
% Draw the Level Diagram ld1 with a unique concept1
ldDraw( 'ld1', 'concept1')
```

The LevelDiagram toolbox is flexible and allows the designer to decide the synchronization value that is the one shown in the y-axis of all plots, *see* **Note 6** for further details. See Example 12 for Pareto front visualization, and Example 13 for Pareto set visualization in both the parameter selection cases.

Example 12: For the parameter selection in the QS/Fb example, the LevelDiagram showing the mean μ and the noise strength η^2 of the protein of interest as the optimization objectives is illustrated below.

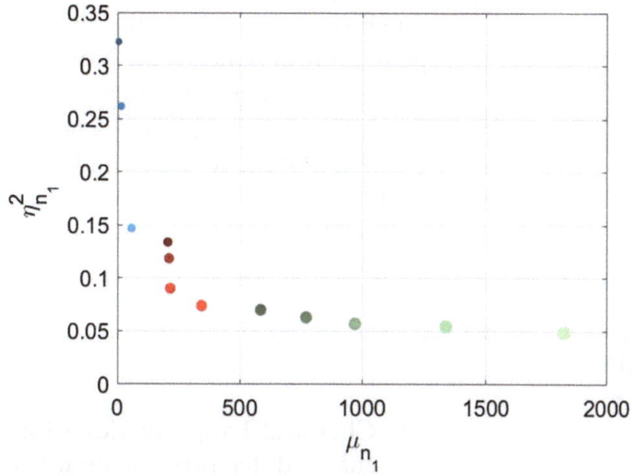

Pareto front visualization using LevelDiagrams. 2-dimensional Pareto front was successfully plotted in the conventional way, that is, one objective in the x-axis and the other one in the y-axis. The objective in the y-axis is then used to synchronize the remaining Pareto set plots. Here, we also order the size of the points with increasing values of the objective J_1. *See* **Note 7** for details on how to customize the LevelDiagram changing the size of the points.

Example 13: The LevelDiagram showing the decision variables (the Pareto set) for our parameter selection of the QS/Fb example is depicted below.

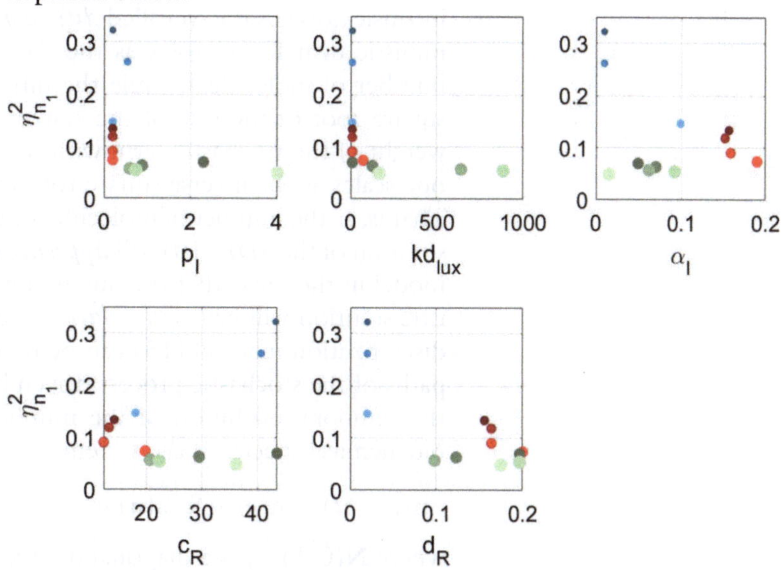

Pareto set visualization using LevelDiagrams. Successful visualization of the Pareto set including the five decision variables defined in the MOP. For each subplot, the x-axis represents the value of the decision variable and the synchronization value is shown in the y-axis. In this case, we chose the noise (J_2) as the synchronization value since the QS/Fb example aims to decrease gene expression noise. *See* **Note 6** for details on how to customize the LevelDiagram to change the y-axis.

Advanced manipulations of the LevelDiagram toolbox include clustering of the solutions and changing the color of the point. *See* **Note 8** related to this topic. Further reading about the multi-criteria decision process and visualization for parameter estimation is available in [5, 7, 9, 10].

4 Notes

1. **Chemical Langevin Equation.** The general form of the stochastic differential chemical Langevin equation is

$$d\mathbf{n}(t) = \mathbf{S} \cdot \mathbf{a}(\mathbf{n}) \cdot dt + \mathbf{S} \cdot \sqrt{\mathbf{a}(\mathbf{n})} \cdot d\mathbf{W}, \tag{10}$$

where $\boldsymbol{n}(t)$ is a vector containing the number of molecules of each species in the model, $\mathbf{S}_{K\times J}$ is the stoichiometry matrix, where K is the number of species and J is the number of biochemical reactions, $\mathbf{a(n)}_{J\times 1}$ is the vector of propensities containing the reaction kinetics, and $d\,W$ are scalar independent Brownian motions associated to each reaction [16]. The first term on the right-hand side of the equation corresponds to the deterministic kinetics, also called the macroscopic *drift term* within this stochastic context. The second term, accounting for intrinsic noise, is the so-called *diffusion term*. Notice the deterministic drift term grows as the size of the system (i.e., the number of molecules), while the diffusion term grows as the square root of the size of the system. Therefore, the relative weight of the stochastic term with respect to the deterministic one scales as the inverse square root of the size of the system. That is, as the number of molecules of the species increases, the solution of the SDE (10) will *approach* that of the deterministic model in the sense that the fluctuations around the deterministic solution will have less *relative* size. The Euler–Maruyama discretization method [15] can be used for generating sample paths of the stochastic process driven by the CLE. It describes the temporal evolution of the number of molecules of each biochemical species in the system as

$$\mathbf{n}(t + \delta t) = \mathbf{n}(t) + \mathbf{S} \cdot \mathbf{a}(\mathbf{n})\delta t + \mathbf{S} \cdot \mathcal{N} \cdot \sqrt{\mathbf{a}(\mathbf{n})}\sqrt{\delta t}, \tag{11}$$

where $\mathbf{N}(0,1)_{J\times J}$ is a diagonal matrix containing J statistically independent normal random variables, and δt is the discretization time step.

2. **Reaction propensities.** The reaction propensities $\mathbf{a(n)}$ in Eq. 10 can be obtained by applying the mass action kinetics formalism [12]. The law of mass action states that the rate of a chemical reaction is proportional to the product of the reactant concentrations raised to a given power given by the stoichiometry of the reaction. If one of the required products is lacking, the reaction will not take place. The reaction proceeds faster as the concentration of the required substrates increases. Doing it proportionally to the product of the reactant concentrations—also called *substrates*—basically accounts for the probability of encounter (collision) among the reactants. Thus, the rationale behind mass action kinetics is that the rate at which a reaction proceeds is proportional to the probability that the required reactants encounter. This probability, in turn, is proportional to the product of their concentrations. Consider a system with m species and a reaction R_j relating them:
   R_j: ,
   where p_{ji}, q_{ji} are the consumption and production stoichiometry coefficients for the i-th species, and $\mathrm{k}_j^+, \mathrm{k}_j^-$ the specific reaction rates of the forward and reverse reactions, respectively. Define the net stoichiometry coefficient $r_{ji} = q_{ji} - p_{ji}$. According to the mass action formalism, reaction R_j will contribute to the deterministic dynamics of the i-th species, n_i, as

$$\dot{n}_i(t) = r_{ji}\mathrm{k}_j^+ \prod_{s=1}^{m} n_s^{p_{js}} - r_{ji}\mathrm{k}_j^- \prod_{s=1}^{m} n_s^{q_{js}} + \ldots,$$

   where notice other reactions will contribute with analogous terms. The functions $a_j^+(\mathbf{n}) = \mathrm{k}_j^+ \prod_{s=1}^{m} n_s^{p_{js}}$ and $a_j^-(\mathbf{n}) = \mathrm{k}_j^- \prod_{s=1}^{m} n_s^{q_{js}}$ are the propensity functions corresponding to the forward and reverse j-th reactions.

3. **Kronecker product.** If $\mathbf{A}$ is an $m \times n$ matrix and $\mathbf{B}$ is a $p \times q$ matrix, the Kronecker product $\mathbf{A} \otimes \mathbf{B}$ is the $mp \times nq$ matrix:

$$\mathbf{A} \otimes \mathbf{B} = \begin{bmatrix} a_{11}\mathbf{B} & \ldots & a_{1n}\mathbf{B} \\ \vdots & \ddots & \vdots \\ a_{m1}\mathbf{B} & \ldots & a_{mn}\mathbf{B} \end{bmatrix}.$$

4. **Selection of the decision variables ranges.** In order to appropriately choose the ranges for the decision variables, sometimes it is recommended to perform a preliminary optimization to observe the behavior of the solution. If solutions stack at one end of the range for any of the decision variables, it means the optimum value could lie outside the initial proposed range. Expand the range and optimize again until solutions do not pile up in the range limits. Notice that the limit could be imposed by a physical or biological situation. In this case, there is nothing to do.

5. **Normalization of objectives.** Sometimes the objectives have different orders of magnitude, for example, one objective ranges from 0 to 10, and another from 5 to 5000. In this case, it is convenient to normalize the objectives in order to not give (without knowing) more relative importance to the one with larger range. In general, the inconvenient is that a priori the range of the objectives it is not know. Hence, it is recommended to perform a preliminary run of the optimization to see the range of the objectives. Then, it is possible to modify the calculation of the objectives including a normalization.

6. **Changing the y-axis in the LevelDiagram visualization.** The normal representation in LevelDiagram toolbox is to show the same type of norm for each objective (e.g., the Euclidean norm, or 2-norm) as the synchronization value, to appear in the y-axis of all the plots. However, it is possible to change that value to any other that suits our purpose. In the special case of having an objective of paramount importance (all the objectives are important in a multi-objective problem, but sometimes an objective plays a distinctive role), or in the case of a 2-dimensional problem (like in the example, mean and noise), this objective can be used as a synchronization value. To set the second objective as a synchronization value is as easy as adding the following line before drawing the LevelDiagram with `ldDraw` command.

```
%Change the x-axis of all the plots to be the second objective
concept1_ld1 = OUT.PFront(:,2)
```

Here, `OUT.PFront(:,2)` points to all the solutions, second objective.

7. **Changing the size of the point according to an objective or decision variable.** Further customization can be done to help in the decision-making process. To change the size of the points according to increasing values of an objective or decision variable can be achieved setting up the following code:

```
%% Customization of the diagrams
%% The marker shape can be selected between 's','o',etc.
ldChangeMarker(ld1, 'o', 'concept1')
```

```

% Creating an array with sizes for each point
s=(20:5:20+5*(concept1.nind-1))';

% Ordering sizes according to the values of J1
[~,idx]=sort(concept1_data(:,1));
s2(idx,1)=s;
ldChangeSize(ld1,s2, 'concept1')
```

8. **Clustering optimization solutions.** Visualization aids such as LevelDiagrams help the designer in the decision-making process. However, having many objectives and decision variables sometimes renders the process more complicated. Clustering the solutions and then coloring them accordingly might alleviate this problem. First, we need to normalize the data we want to consider for the clustering (in this example, all the objectives and all the decision variables). This can be achieved with the following code:

```
%% Clustering
%Constructiong the Data matrix with Objectives and Desi-
cion Variables
pf = OUT.PFront;
PS = OUT.PSet;
Points = size(PS,1);
PSn = [];
for m=1:Points
     PSn=[PSn; PS(m,:)./max(PS)];
end
PFn = [];
for m=1:Points
     PFn=[PFn; pf(m,:)./max(pf)];
end
Data =[PSn PFn];
```

Then, we can continue by computing the distance between the elements and the clustering tree:

```
%Calculation of the distance and clustering tree
eucD = pdist(Data,   'euclidean'); %
```

```
clustTreeEuc = linkage(eucD, 'ward'); %
cophenet(clustTreeEuc,eucD);

 % Number of desired clusters
Desired_Clusters = 3;
Clusters_out = cluster(clustTreeEuc, 'maxclust',Desired_Clusters);
```

Next, assigning colors to each solution according to the clusters is needed. The following code implements this end:

```
 % Using Clusters to set the colors of the points
maxC = max(Clusters_out);
Color_Map = make_color_map(maxC);

 % Seting the color of each point
Color_Clusters = zeros(length(Clusters_out),3);
for i=1:maxC
    Color_Clusters(Clusters_out==i,:) = ones(size( Color_Clusters(Clusters_out==i,:),1),1)*Color_Map(i,:);
end
```

And finally, we can change the color of the points in the LevelDiagram:

```
 %Change the color of the points in the plots
ldChangeColor(ld1,Color_Clusters, 'concept1');
```

Examples 11 and 12 were obtained using this previous code.

References

1. Alon U (2019) An introduction to systems biology: design principles of biological circuits. CRC Press, Boca Raton
2. Blasco X, Herrero JM, Reynoso-Meza G, Martínez Iranzo MA (2017) Interactive tool for analyzing multiobjective optimization results with level diagrams. In: Proceedings of the genetic and evolutionary computation conference companion, p 1689–1696
3. Boada Y, Pitarch J, Vignoni A, Reynoso-Meza G, Picó J (2016) Optimization alternatives for robust model-based design of synthetic biological circuits. IFAC-PapersOnLine 49(7):821–826
4. Boada Y, Reynoso-Meza G, Vignoni A, Picó J (2016) Multi-objective optimization framework to obtain model-based guidelines for tuning biological synthetic devices: an adaptive network case. BMC Syst Biol 10:27

5. Boada Y, Vignoni A, Reynoso-Meza G, Picó J (2016) Parameter identification in synthetic biological circuits using multi-objective optimization. IFAC-PapersOnLine 49(26):77–82
6. Boada Y, Vignoni A, Picó J (2017) Engineered control of genetic variability reveals interplay among quorum sensing, feedback regulation, and biochemical noise. ACS Synth Biol 6 (10):1903–1912
7. Boada Y, Vignoni A, Picó J (2017) Multi-objective identification of synthetic circuits stochastic models using flow cytometry data. In: Proceedings 25th mediterranean conference on control and automation MED, p 1077–1082. https://doi.org/10.1109/MED.2017.7984261
8. Boada Y, Vignoni A, Picó J (2017) Multi-objective optimization for gene expression noise reduction in a synthetic gene circuit. IFAC-PapersOnLine 50(1):4472–4477. 20th IFAC World Congress
9. Boada Y, Vignoni A, Alarcon-Ruiz I, Andreu-Vilarroig C, Monfort-Llorens R, Requena A, Picó J (2019) Characterization of gene circuit parts based on multiobjective optimization by using standard calibrated measurements. ChemBioChem 20(20):2653–2665
10. Boada Y, Vignoni A, Picó J (2020) Multiobjective identification of a feedback synthetic gene circuit. IEEE Trans Control Syst Tech 28 (1):208–223. https://doi.org/10.1109/TCST.2018.2885694
11. Boada Y, Vignoni A, Picó J, Carbonell P (2020) Extended metabolic biosensor design for dynamic pathway regulation of cell factories. Iscience 23(7):101305
12. Chellaboina V, Bhat S, Haddad M, Bernstein D (2009) Modeling and analysis of mass-action kinetics. IEEE Control Syst 29(4):60–78
13. Elowitz MB, Levine AJ, Siggia ED, Swain PS (2002) Stochastic gene expression in a single cell. Science 297(5584):1183–1186
14. Gorochowski TE, di Bernardo M, Grierson CS (2010) Evolving enhanced topologies for the synchronization of dynamical complex networks. Phys Rev E 81(5):056212
15. Higham DJ (2001) An algorithmic introduction to numerical simulation of stochastic differential equations. SIAM Rev 43(3):525–546
16. Higham DJ (2008) Modeling and simulating chemical reactions. SIAM Rev 50(2):347–368
17. Kelley NJ, Whelan DJ, Kerr E, Apel A, Beliveau R, Scanlon R (2014) Engineering biology to address global problems: synthetic biology markets, needs, and applications. Ind Biotechnol 10(3):140–149
18. Kumar P, Sinha R, Shukla P (2020) Artificial intelligence and synthetic biology approaches for human gut microbiome. Crit Rev Food Sci Nutr. https://doi.org/10.1080/10408398.2020.1850415
19. Mallik S, Bhadra T, Seth S, Bandyopadhyay S, Chen J (2018) Multi-objective optimization approaches in biological learning system on microarray data. In: Multi-objective optimization, springer, p 159–180
20. Miettinen K (2012) Nonlinear multiobjective optimization, vol 12. Springer, Berlin
21. Picó J, Vignoni A, Boada Y (2021) Stochastic differential equations for practical simulation of gene circuits. In: Menolascina F (ed) Synthetic gene circuits, methods in molecular biology, vol 2229. Humana, New York, p 41–90
22. Reynoso-Meza G, Sanchis J, Blasco X, Garcia-Nieto S (2014) Physical programming for preference driven evolutionary multi-objective optimization. Appl Soft Comput 24:341–362
23. Ruess J, Lygeros J (2015) Moment-based methods for parameter inference and experiment design for stochastic biochemical reaction networks. ACM Trans Model Comput Simul 25(2):8
24. Sohlberg B, Jacobsen EW (2008) Grey box modelling–branches and experiences. IFAC Proc Volumes 41(2):11415–11420
25. Vignoni A, Bajur A, Knust E, Sbalzarini IF (2018) Multi-objective identification from fluorescence recovery after photobleaching experiments: understanding morphogenetic regulation of epithelial polarity. IFAC-PapersOnLine 51(19):8–11
26. Yaman F, Adler A, Beal J (2018) AI challenges in synthetic biology engineering. In: Proceedings of the AAAI conference on artificial intelligence, vol 32

Chapter 5

Relationship Between Dimensionality and Convergence of Optimization Algorithms: A Comparison Between Data-Driven Normalization and Scaling Factor-Based Methods Using PEPSSBI

Andrea Degasperi, Lan K. Nguyen, Dirk Fey, and Boris N. Kholodenko

Abstract

Ordinary differential equation models are used to represent intracellular signaling pathways in silico, aiding and guiding biological experiments to elucidate intracellular regulation. To construct such quantitative and predictive models of intracellular interactions, unknown parameters need to be estimated. Most of biological data are expressed in relative or arbitrary units, raising the question of how to compare model simulations with data. It has recently been shown that for models with large number of unknown parameters, fitting algorithms using a data-driven normalization of the simulations (DNS) performs best in terms of the convergence time and parameter identifiability. DNS approach compares model simulations and corresponding data both normalized by the same normalization procedure, without requiring additional parameters to be estimated, as necessary for widely used scaling factor-based methods. However, currently there is no parameter estimation software that directly supports DNS. In this chapter, we show how to apply DNS to dynamic models of systems and synthetic biology using PEPSSBI (Parameter Estimation Pipeline for Systems and Synthetic Biology). PEPSSBI is the first software that supports DNS, through algorithmically supported data normalization and objective function construction. PEPSSBI also supports model import using SBML and repeated parameter estimation runs executed in parallel either on a personal computer or a multi-CPU cluster.

Key words Parameter estimation, Relative data, ODE models, Signaling pathways, Data normalization

1 Introduction

Mathematical modeling plays a prominent role in systems and synthetic biology by formalizing hypotheses and predicting systems behavior [1]. Modeling aids the systematic analysis of complex and nonlinear signaling network behaviors and their underlying signaling and regulatory mechanisms [2, 3]. Consequently, many

Quentin Vanhaelen (ed.), *Computational Methods for Estimating the Kinetic Parameters of Biological Systems*, Methods in Molecular Biology, vol. 2385, https://doi.org/10.1007/978-1-0716-1767-0_5,

software tools have been developed to facilitate the development and analysis of mathematical models of biological systems [4, 5].

The predominant modeling method is ordinary differential equations (ODEs), strongly supported by a constantly growing software repertoire, including Copasi [6], CellDesigner [7], PottersWheel [8], the Systems Biology Workbench [9], Data2Dynamics [10], PESTO [11], and AMIGO2 [12]. A widely used pathway model file format, the Systems Biology Mark-up Language (SBML, [13]), allows models to be shared and analyzed across different software.

The development of quantitative and predictive mathematical models requires the estimation of unknown model parameters using experimental data. This task is known as parameter estimation, and it is an optimization problem where an objective function describing the discrepancy between data and model simulations is minimized [10, 14, 15]. Due to nonlinearity of signaling pathways and limited experimental data, local minima are usually present in the objective function. Thus, the parameter estimation problem is usually not convex, and heuristic optimization algorithms are required [16]. For relatively simple parameter estimation problems (i.e., limited number of unknown parameters and equations), often deterministic local search optimization algorithms with multiple random restart are employed successfully, while for more complex problems, hybrid stochastic-deterministic methods, which combine local deterministic search methods with global stochastic search strategies, are more effective [14, 15].

Moreover, the parameter estimation optimization problem may have multiple good solutions, which means that more than one estimated parameter set can reproduce the experimental data equally well (considering the noise in the data). Parameters for which no unique value can be estimated are called *non-identifiable* [17, 18]. A simple method to estimate non-identifiability of parameters is to run the parameter estimation problem multiple times and obtain an ensemble of optimal parameter sets [19]. Subsequently, the ensemble of estimates is used to study the relationship between non-identifiable parameters and suggest new experiments that can improve the identifiability of the parameters [17].

Statistical approaches to parameter estimation based on Bayesian frameworks are also used in systems biology applications [20, 21]. Their main advantage over optimization approaches is that they model the full posterior distribution of the unknown parameters, which reveals their covariation and identifiability, and that they can use Bayesian model selection to discriminate between competing models. Among the disadvantages are the higher computational cost and the need to specify prior distribution of the unknown parameters.

The parameter estimation of signaling pathway models requires high-resolution time-course data and multiple perturbations

[3, 22–24]. Perturbations of the experimental system are, for example, the use of different ligands or varying doses of ligands for receptor and downstream pathway activation. A collection of perturbation experiments is often referred to as a *multi-condition experiment* [8]. When modeling multi-condition experiments, it is often necessary to indicate which parameters in the models assume different values in each condition (local parameters) and which are the same for all conditions (global parameters). For example, the initial concentration of a ligand may assume a different value in each condition, while the binding constant between the ligand and the receptor will be the same for all conditions. Parameter estimation is then performed using data from all conditions, and all unknown local and global parameters are estimated at the same time. Modeling of multi-condition experiments is supported by available software, such as PottersWheel [8], Data2Dynamics [10], and Copasi [6].

Experimental data used for modeling are typically obtained from techniques such as Western blotting [25, 26], multiplexed Elisa [27], or RT-qPCR [28] and are commonly expressed in arbitrary units often referred to as relative data [15, 29]. The use of relative data poses the question of how to compare independent biological replicates of the same experiment [30]. Because each biological replicate has different arbitrary units, a data normalization strategy needs to be applied. Two most commonly used normalization strategies are:

1. Normalization by fixed point or control, where all data points in a replicate are divided by the value at a control data point. Assuming data points $\widehat{y}_i$, normalized data points are $\widetilde{y}_i = \widehat{y}_i / \widehat{y}_{\text{norm}}$, where $\widehat{y}_{\text{norm}} = \widehat{y}_p$ is the control data point.
2. Normalization by sum or average, where all data points in a replicate are divided by the average (or sum) of the data points in that replicate. Assuming data points $\widehat{y}_i$, normalized data points are $\widetilde{y}_i = \widehat{y}_i / \widehat{y}_{\text{norm}}$, where $\widehat{y}_{\text{norm}} = \sum_j \widehat{y}_j / n$ is the average of the data points in a replicate.

These normalization strategies affect the variability of the normalized data and their quantitative use. For example, increased variability in the normalized data raises the rate of false negatives in hypothesis testing. Depending on the data and the selected normalization, one normalization strategy may be more suitable than the other in avoiding excessive increase or decrease in data variability [30].

Simulations from models of signaling pathways typically have well-defined units, such as nano-Molar or other derived units. Thus, when using relative data for parameter estimation, one needs to decide how to scale the units of the model simulations to the units of the corresponding normalized data. Two commonly used approaches are:

1. Scaling factors (SF): Simulations are multiplied by an additional parameter, the scaling factor. Assuming normalized data points $\widetilde{y}_i$ and model simulation outputs $y_i(\theta)$, the deviation between SF $*$ $y_i(\theta)$ and $\widetilde{y}_i$ have to be minimized using a proper objective function to determine or constrain model parameters θ. The SF is unknown and needs to be estimated; typically there is one SF for each observable biochemical species.
2. Data-driven normalization of the simulations (DNS): Simulations are normalized in the same way the data were normalized. For instance, if the data are normalized by dividing on the control, maximal, or the average response point ($\widehat{y}_{\text{norm}}$), i.e., $\widetilde{y}_i = \widehat{y}_i / \widehat{y}_{\text{norm}}$, then $y_i(\theta)$ are also divided by the corresponding simulation output point, i.e., $y_i(\theta)/y_{\text{norm}}(\theta)$. Then, the parameters θ are determined by the minimization of the deviation between normalized data points and normalized simulations (*see* Subheading 3 for more detail). Thus, DNS avoids the estimation of additional parameters.

A major advantage of DNS over SF is that it does not require estimation of additional scaling factors, thereby reducing the dimensionality of the optimization problem. For complex optimization problems, particularly those with a large number of observables, this reduction in dimensionality translates into a significantly faster convergence of both deterministic local optimization algorithms and hybrid deterministic-stochastic global optimization algorithms [15]. In addition, the DNS approach induces a lower degree of practical non-identifiability than SF [15]. These strengths render the DNS approach superior to SF.

A disadvantage of DNS is that without explicit scaling factors, there is less control on the allowed range of actual concentration of the simulated variables. In DNS, such range is determined solely by the model parameters, whose search boundaries need to be regulated so that simulated values are biologically sound. Another disadvantage may be the need to specify an appropriate reference $\widehat{y}_{\text{norm}}$ for the normalization of data and simulations. However, even in the case of SF, such reference is needed to normalize the data, though it is usually not part of the discussion in publications that focus on parameter estimation methods. For example, data normalization was not discussed in recent publications on systems biology benchmarks [31, 32].

Compared to SF, the strengths of DNS outweigh its shortcomings, yet there is currently no parameter estimation software for signaling pathway models that support DNS.

In this chapter, we present PEPSSBI (Parameter Estimation Pipeline for Systems and Synthetic Biology), a novel parameter estimation pipeline that supports DNS.

Importantly, PEPSSBI implements algorithmically supported data normalization and objective function construction (Fig. 1).

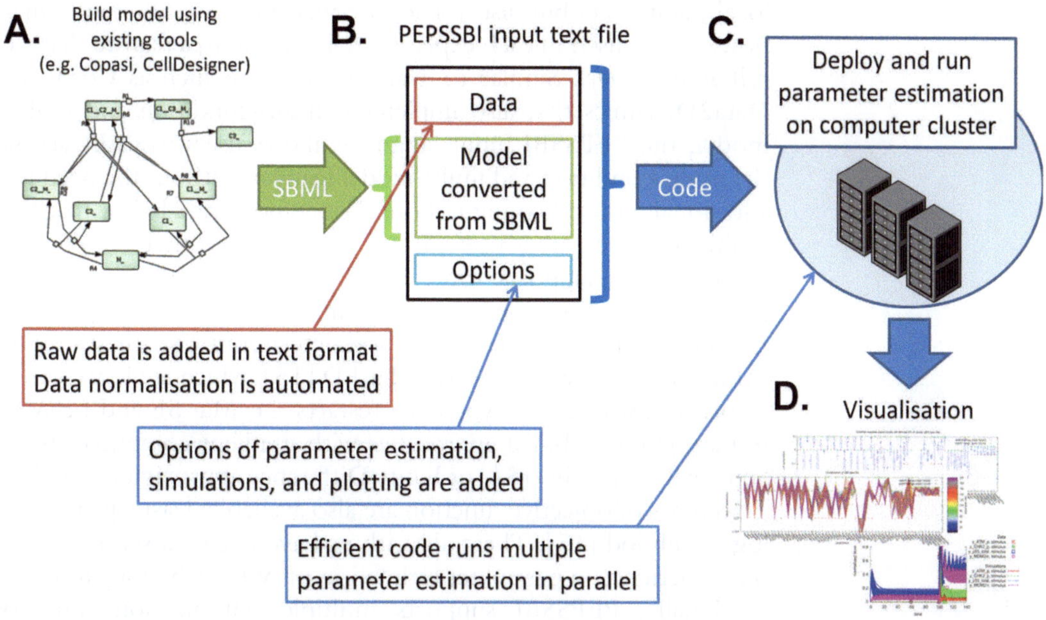

Fig. 1 PEPSSBI: parameter estimation pipeline for systems and synthetic biology. (**a**) SBML models generated with other software can be imported. (**b**) Data are added and automatically managed in the same file. (**c**) PEPSSBI generates code that can be run on a computer cluster for efficient, parallel simulations. (**d**) Visualization of the results is also generated by the software

Without a software that can generate the DNS objective function programmatically, the application of DNS can be time consuming and error prone. For example, addition of new data or new replicates may require changes in the normalization strategy and thus in how $\widehat{y}_{\text{norm}}$, and in turn $y_{\text{norm}}(\theta)$, is calculated. If optimization algorithms that require the gradient of the objective function are used, such gradient is also dependent on the choice of $\widehat{y}_{\text{norm}}$.

Furthermore, PEPSSBI uses an input file written in a simple dedicated language, which incorporates the mathematical model, raw data, and normalization strategies, and is compiled to generate C/C++ code and scripts that can be run on a personal computer or a High-Performance Computing (HPC) infrastructure.

The use of a dedicated input language offers convenience that saves time and reduces possible human errors. For example, changing the data normalization strategy can be done by editing a single line of code. Moreover, by first compiling the input file, and then running the optimization algorithm, this two-step pipeline allows identification of possible mistakes and inconsistencies at compilation time before running the optimization. For example, the compiler will inform users if variable definitions are missing or if the data are incorrectly formatted.

In addition, PEPSSBI includes support for multi-condition experiments and multiple models. PEPSSBI allows not only for local parameters but also for local equations, i.e., different equations can be used in each experimental condition, a feature that we call *multi-model*. Similar to available software such as Copasi or Data2Dynamics that also implement multi-condition and multi-model, the PEPSSBI input language allows users to easily access the multi-condition and multi-model features without the need for additional files and configurations. We demonstrate below that the multi-model feature is useful in the context of synthetic biology.

PEPSSBI supports three optimization algorithms: GLSDC [33] (Genetic Local Search algorithm with distance independent Diversity Control), which is a hybrid deterministic-stochastic algorithm; and LevMar SE and LevMar FD [15], which are local search algorithms with Latin hypercube restarts. LevMar SE and LevMar FD are gradient-based approaches with gradient computed using Sensitivity Eqs. (SE) [34] or Finite Difference approximation (FD). Two types of objective function are also available: least squares and log-likelihood [15]. These algorithms have been chosen based on our experience in parameter estimation of systems biology models.

Finally, PEPSSBI supports multiple optimization runs to obtain an ensemble of optimal parameter sets and allows the user to organize the runs in parallel and sequential jobs.

2 Materials

PEPSSBI (Parameter Estimation Pipeline for Systems and Synthetic Biology) can be downloaded from the bitbucket repository https://bitbucket.org/andreadega/systems-biology-compiler/. A detailed software manual is also available on the bitbucket repository wiki page (https://bitbucket.org/andreadega/systems-biology-compiler/wiki/Home).

2.1 PEPSSBI Architecture

PEPSSBI is composed of two parts (Fig. 2):

1. A compiler, termed the Systems Biology Compiler (SBC). SBC converts a single input file written in the dedicated PEPSSBI input language into C/C++ source code that implements the parameter estimation optimization problem. Other code generated by SBC includes scripts for simulating and plotting the optimal solutions and visualizing the optimal parameter sets. SBC is written in Java and is implemented using the ANTLR (ANother Tool for Language Recognition, [35]) parser generator.
2. The source code and scripts generated by the SBC and the libraries necessary to run it. The source code of the necessary libraries is included in PEPSSBI and is compiled during PEPSSBI first set-up. The libraries include Sundials CVODES [36] for simulating the ODE models, an open source implementation of the Levenberg–Marquardt nonlinear least squares

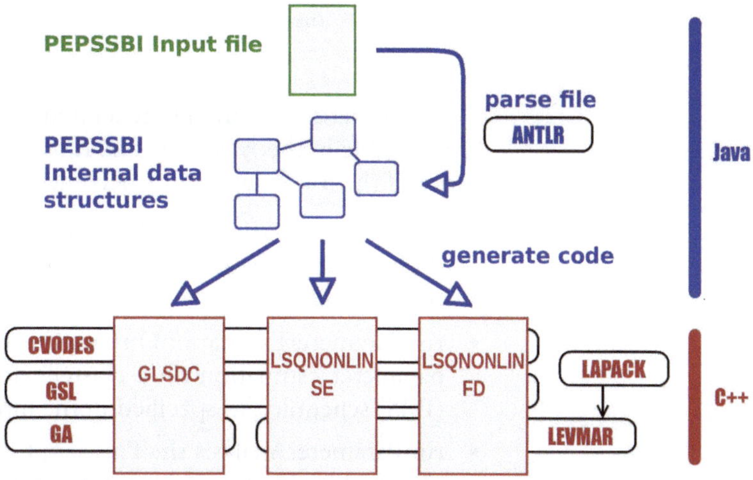

Fig. 2 PEPSSBI software implementation. PEPSSBI consists of a compiler built using the java library ANTLR and the C/C++ code that is generated upon running a PEPSSBI input file through the compiler. The generated code uses the CVODES library to run the ODE models and various other libraries to run the optimization problem

optimization algorithm (levmar, [37]), and an implementation of the GLSDC [33] hybrid stochastic-deterministic algorithm. Additional software that the scripts rely on is the statistical software R [38] and the plotting software gnuplot [39]. The generated code and scripts can be further customized, for example, to tailor the graphics of the plots.

PEPSSBI is a command line software with currently no graphic user interface. To aid the writing and editing of PEPSSBI input files, a syntax highlighter for the PEPSSBI language is available for the Geany text editor [40].

Mathematical models are written in the PEPSSBI input language as differential equations. We also provide software to convert models from the Systems Biology Mark-up Language (SBML) [13] into the PEPSSBI language (SBML2SBC, available online at https://bitbucket.org/andreadega/sbml2sbc).

PEPSSBI is written to run on a Linux operative system or Unix cluster. It also runs on Mac OS, using Homebrew [41], and on Windows, using MinGW-w64 [42] and MSYS2 [43]. We also provide a Docker image [44] that contains PEPSSBI and all necessary software and libraries and that can be run on Linux, Mac OS, or Windows operative systems using Docker virtualization (https://hub.docker.com/r/andreadega/pepssbi/).

2.2 Running PEPSSBI

PEPSSBI is a pipeline organized in two main steps: compile and run. Given an input file *input.sbc*, the compile step is executed by the following command:

```
./runSBC.sh input.sbc
```

Information about the generated files and compilation errors are available in the *results.log* file. If no errors are produced, one or more of the following scripts becomes available:

- runParameterEstimationOnLinuxDesktop.sh: This script runs the parameter estimation on a local computer, as specified in the input file.
- runParameterEstimationOnUnixCluster.sh: This script runs the parameter estimation on a cluster with Portable Batch System (PBS) scheduler, as specified in the input file.
- runParameterAnalysis.sh: This script runs a basic analysis of the obtained optimal parameter sets, such as parameter clustering, parameter correlations and plotting. This script is automatically run after the parameter estimation is complete, and can be rerun manually.
- runSimulations.sh: This script runs model simulations using the obtained optimal parameter sets, or simply runs simulations if no parameter estimation was requested. It also generates plots if requested in the input file. This script is automatically run after the parameter estimation is complete, and can be rerun manually.

3 Methods

We begin by describing relative data normalization in Subheading 3.1, followed by the definition of data-driven normalization of the simulations (DNS) objective functions in Subheading 3.2. We will then introduce PEPSSBI input syntax in Subheading 3.3, followed by case studies that illustrate how to use DNS with PEPSSBI (Subheadings 3.4–3.6), and how the parameter estimation time performance compares when using scaling factors (SF) as opposed to DNS (Subheading 3.7).

3.1 Data Normalization

Assuming relative data, we use $\widehat{y}_{kij}$ to indicate a data point of observable $k \in K$, measured in condition or time point $i \in I$ and in the biological replicate $j \in J$. Values of data points are directly comparable across the index i (e.g., they belong to the same Western blot image obtained using the same antibodies), while they are not comparable across the index j (e.g., different Western blot image, so different proportionality between light intensity and protein amount) and not comparable across the index k (e.g., different Western blot also using a different antibody for a different biochemical species). In order to make the data points comparable across the index j, we apply one of two data normalizations available in PEPSSBI: normalization by fixed point and normalization by average.

Normalization by fixed point is given by:

$$\tilde{y}_{kij} = \frac{\hat{y}_{kij}}{\hat{y}_{kpj}} \qquad \forall k \in K, i \in I, j \in J \tag{1}$$

The index $p \in I$ in Eq. 1 indicates the condition or time point to use as reference. Once this normalization is applied to a given species k, $\tilde{y}_{kpj}$ will have value 1 and standard deviation 0 for all $j \in J$. Notice that an index $p \in I$ cannot be used if there is at least one replicate j where $\tilde{y}_{kpj}$ is missing.

Normalization by average is given by:

$$\tilde{y}_{kij} = \frac{\hat{y}_{kij}}{\sum_{w \in W_k} \hat{y}_{kwj}/|W_k|} \qquad \forall k \in K, i \in I, j \in J \tag{2}$$

The set of indexes $W_k \subseteq I$ in Eq. 2 defines which conditions or time points are included in the average, while $|W_k|$ indicates the size of the set W_k. Note that an index set W_k cannot be used if there is an index $w \in W_k$ such that there is at least one replicate j where $\tilde{y}_{kwj}$ is missing. In PEPSSBI, the set W_k is obtained programmatically: the modeler needs to specify which experimental conditions he/she wants to include, and PEPSSBI will search for all time points in those conditions that do not have missing values across replicates. Note that for each $k \in K$, a different normalization can be used. For more details about data normalization, *see* **Note 1**.

3.2 DNS Objective Functions

Two types of objective functions are available in PEPSSBI: least squares (LS), which is used by default, and log likelihood (LL). Let $y_{ki}(\theta)$ be the value of observable k simulated using parameters θ at condition or time point i. The DNS objective functions for LS (obj_{LS}) and LL (obj_{LL}) are given by:

$$obj_{LS}(\theta) = \sum_{k \in K} \sum_{i \in I} \sum_{j \in J} \left(\frac{y_{ki}(\theta)}{y_{k,\text{ref}}(\theta)} - \tilde{y}_{kij} \right)^2 \tag{3}$$

$$obj_{LL}(\theta) = \sum_{k \in K} \sum_{i \in I} \sum_{j \in J} \left(\frac{y_{ki}(\theta)/y_{k,\text{ref}}(\theta) - \tilde{y}_{kij}}{\sigma_{ki}} \right)^2 + 2 \sum_{k \in K} \sum_{i \in I} \sum_{j \in J} \ln(\sigma_{ki}) \tag{4}$$

In Eqs. 3 and 4, $y_{k,\text{ref}}(\theta)$ is the reference used in the normalization of the simulations and depends on the unknown parameters θ and the data normalizations used for each $k \in K$. In the case of normalization by fixed point, $y_{k,\text{ref}}(\theta) = y_{kp}(\theta)$ for a given condition or time point index $p \in I$. In the case of normalization by average, $y_{k,\text{ref}}(\theta) = \sum_{w \in W_k} y_{kw}(\theta)/|W|$, for a given set of indexes $W_k \subseteq I$.

In Eq. 4, σ_{ki} is called the measurement error, which we model using the linear model $\sigma_{ki} = s_a + s_b \cdot y_{ki}(\theta)/y_{k,\text{ ref}}(\theta)$, in analogy with ref. [34]. The coefficients $s_a > 0$ and $s_b > 0$ are estimated along with θ.

For more details about DNS objective functions and their comparison with the SF approach, *see* the Supplementary Methods 1 of ref. [15].

3.3 The PEPSSBI Input Language

In this section, we illustrate the PEPSSBI input language using an example input file (Fig. 3) which describes a theoretical scenario explained below. For illustration purposes, some lines of code have been omitted from this figure. Full specification of the input file syntax is provided in the software manual available online (https://bitbucket.org/andreadega/systems-biology-compiler/wiki/Home).

In this example, the aim is to estimate parameters of a reaction that activates a kinase *K*, converting it from its inactive form *iK* to its active form *aK*. This activation is mediated by a signaling enzyme *S* (unknown parameters *kcat1* and *Km1*). We also assume that *iK* is spontaneously converted into *aK* at a basal rate *kbasal*. Finally, *aK* can be converted back into *iK* at a known rate *kback*. The reaction mediated by *S* is modeled with Michaelis–Menten kinetics, while the spontaneous conversion and the backward reaction are modeled with mass action kinetics. The scheme with all the reactions considered is shown in Fig. 3.

We consider two experimental conditions. In condition 1, the amount of kinase *K* is known to be 100 nM ($K_{\text{tot}} = 100$), while in condition 2, it is 1000 nM ($K_{\text{tot}} = 1000$). In both the conditions, the signaling molecule *S* is measured at time points 0, 5, 15, and 30 min, assumed to be in nM (nanoMolar). In both the conditions, the level of *aK* is measured at time points 0, 5, 15, and 30 min. These measurements are assumed to be in relative units. The level of *aK* in the two conditions is assumed to be measured together (and thus comparable, such as it would be for data quantified from the same Western blot membrane). We assume four biological replicates for the measurements of *aK* ($n = 4$). Additionally, the level of *aK* at time 0 is unknown and different in the two conditions (unknown parameters *init_aKc1* and *init_aKc2*).

The data for *aK* in Fig. 3 were generated using the above model, by setting specific values for the unknown parameters, simulating the time courses for *aK* and finally adding a 10% coefficient of variation Gaussian noise. The limited number of data points and replicates and the variability are typical features of biological data, e.g., Western blot data. Using these sets of simulated data, we will then attempt to recover the parameters used. The procedure of generating simulated data by fixing the parameter values and then recovering these parameter values through parameter estimation is a common practice for assessing parameter identifiability [17, 29, 32].

```
1.  data aK =
2.  "time" "condition 1" "condition 2"
3.  0 25.5418871571589 58.1780900055649
4.  5 49.1576803942669 107.782379718123         (a)
5.  15 91.7768588016566 167.280391296763
6.  30 103.624540419953 301.601872011214
7.  -
8.  0 29.7413496797732 51.9152866925858
...
23. normaliseByAverage aK "condition 1" "condition 2" (b)
24.
25. data S =
26. "time" "condition 1" "condition 2"
27. 0       10.0 10.0
28. 5       9.5  9.5
29. 15      8.0  8.0
30. 30      7.0  7.0
31.
32. par kcat1 = [0.1,10]
33. par Km1 = [1,100]
34. par kback = 0.0001
35. par kbasal = 0.001          (c)
36. par init_aKc1 = [1,100]
37. par init_aKc2 = [1,100]
38.
39. Ktot =
40. "condition 1" 100
41. "condition 2" 1000    (d)
42.
43. S = dataLinear (e)
44.
45. init aK =
46. "condition 1" init_aKc1
47. "condition 2" init_aKc2  (f)
48.
49. aK' = (S*kcat1*(Ktot-aK))/(Km1+(Ktot-aK))
           - kback*aK + kbasal*(Ktot-aK)          (g)
50.
51. fittingData =
52. aK "condition 1"  "condition 2"  (h)
```

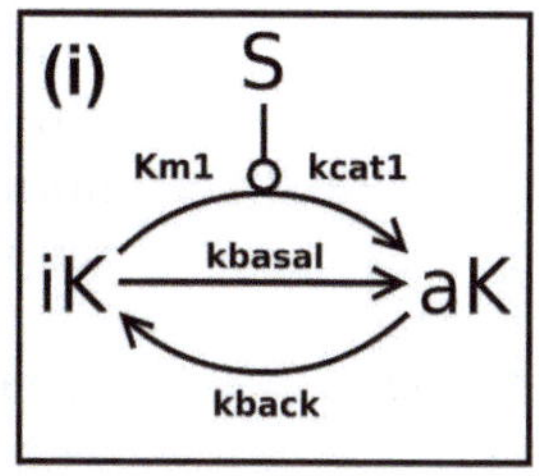

Fig. 3 Example of PEPSSBI input file. (**a**) Data are written as data blocks, where each block is a biological experiment (e.g., quantification of Western blot). (**b**) Data normalization. (**c**) Parameter definitions. (**d**) Definition of model variable K_{tot}. (**e**) Definition of model variable *S*. (**f**) Initial condition for the differential equation for model variable *aK*. (**g**) Definition of the differential equation of model variable *aK*. (**h**) Declaration of which data to use for parameter estimation. (**i**) Reaction scheme of the model. Variable $iK = K_{tot} - aK$

3.3.1 Algorithmically Supported Data Normalization

The keyword *data* followed by a name is used to specify a data entry (Fig. 3, lines 1 and 25). Data can be specified as a tab or space separated values format. The first data column must always be the time of the measurement, while the other columns indicate different experimental conditions (lines 2 and 26). Data points are written in blocks (e.g., one block is from line 3 to 6), with blocks separated by the character "-" (e.g., line 7). Each block in a data entry represents a biological replicate of the same experiment. In the case of relative data, it is important to note that the data values within each block are comparable (e.g., quantified from the same Western blot membrane), while they are not directly comparable across data blocks (e.g., quantified from different membranes of Western blot).

Two types of data normalization strategies are available in PEPSSBI: by *fixed-point* and by *average*. For fixed-point normalization, the data points within a replicate are divided by the value of a chosen reference point, such as a control condition. For normalization by average, the data points are divided by the average of all the data points within the replicate [15, 30]. It is also possible to define an average based on a subset of the data points. At line 23, the keyword *normaliseByAverage* followed by the species name *aK* and the names of the two conditions indicate that the biological replicates (blocks) of data of *aK* should be normalized, i.e., divided by, the average of all data points of both the conditions in the same replicate. When the input file is compiled, the requested normalization will be performed automatically. Moreover, missing data values can be specified and are dealt with automatically. Changing data normalization for *aK* is as simple as changing one line of the input script (e.g., line 23).

3.3.2 Parameter Definitions

Parameters can be either fixed at prescribed values or confined within defined ranges (lines 32–37). All parameters defined as ranges will form the set of unknown parameters that will be estimated by PEPSSBI within the ranges indicated. If only fixed parameters are declared, PEPSSBI will only perform model simulations and plotting.

3.3.3 Multi-condition and Multi-model

In PEPSSBI, we allow local definition of model variables, a feature which we call multi-model. The usefulness of this feature will be illustrated in detail in a later section, when we introduce models of two genetic circuits that share common components. In the current example, at lines 39–41, the definition of the model parameter *Ktot* is split into two definitions, one for each condition. At lines 45–47, the definition of the initial value for *aK*, the initial condition for the differential equation of *aK*, is also split into two, so that local parameters *init_aKc1* and *init_aKc2* can be used. When a model variable has the same definition in all conditions, it is sufficient to define it only once (e.g., definition of the differential equation for *aK* at line 49).

3.3.4 Interpolated Data as Model Input

Many signaling models use measured dynamics of an upstream signaling protein as time-dependent input, which requires interpolation of the input data. For example, in the model of the Epo/-Jak2/STAT5 pathway presented in ref. [45], the time course interpolation of phosphorylated Epo-bound receptor was used as input, while in the model of c-Fos regulation in ref. [3], the time course interpolation of phosphorylated cytoplasmic MEK was used as input. To use data as model input in PEPSSBI, keywords *dataLinear* or *dataSpline* can be used for defining a model variable and perform linear and spline interpolation, respectively. The model variable will then have the value of the interpolated time course of the data entry with the same name and condition. For example, variable *S* (line 43) is defined as a linear interpolation. When simulating the model for condition 1, the data for *S* in condition 1 (lines 25–30) are interpolated.

3.3.5 Parameter Estimation

Using the keyword *fittingData*, one can specify which data to use for parameter estimation. The keyword *fittingData* needs to be followed by a list of species and conditions (e.g., lines 51 and 52). When the input file is compiled, data definitions are matched with corresponding model variables that have the same name and the objective function is constructed. For example, at line 1 we define data for *aK* and at lines 45–49 we define the model variable *aK*. If the data of a certain species are normalized, such as for *aK* in our example, then DNS is applied [15].

When PEPSSBI compiles the input file, it generates C/C++ code that implements the optimization procedure for the specified problem. The code for three optimization algorithms (GLSDC, LevMar SE, and LevMar FD) is generated and either one of them can be used for the optimization. The options necessary for the setup of the parameter estimation procedure, such as how many computer cores to use in parallel, are also indicated in the input file (not shown in Fig. 3). In our example, we used GLSDC and repeated the parameter estimation procedure 200 times, distributed on multiple computer cores. Each run retrieved an estimated optimal parameter set. To reduce the risk of including suboptimal local minima in the results, we considered only the best 100 (lowest objective function value).

3.3.6 Simulations and Optimal Parameters Plots

Using the keyword *simulateOutput* (not shown in Fig. 3), PEPSSBI can be instructed to generate plots of dynamic simulations (Fig. 4a). In addition, dose–response (keyword *doseResponse*) and two-variable plots (keyword *twoVariablesPlot*) can also be generated. When running the parameter estimation, a plot displaying the optimal parameter sets is automatically generated (Fig. 4b). In the current example, estimates are almost identical for all 100 runs (Fig. 4b), indicating that these parameters are identifiable [17].

3.4 Example of Algorithmically Supported Data Normalization and DNS

In this section, we present a case study of the algorithmically supported data normalization and DNS, along with an example of error at the compilation step. To illustrate this case study, we use short extracts from PEPSSBI input files (Fig. 5) to focus only on the relevant lines of code.

A PEPSSBI input file (input1.sbc, Fig. 5a) contains a data entry for phosphorylated extracellular signal regulated kinase (ppERK). This data entry has two data blocks (Fig. 5a, lines 1–11), and normalization by fixed point is requested (Fig. 5a, line 13). The

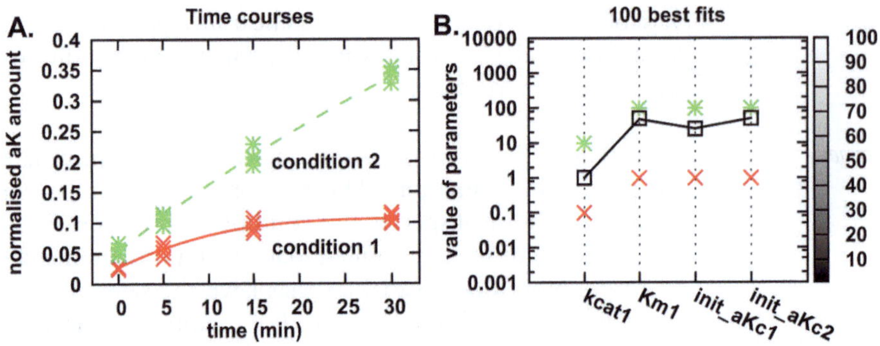

Fig. 4 Parameter estimation for PEPSSBI input file in Fig. 3. (**a**) Visualization of 100 model simulations (lines) obtained with the 100 best parameter sets. Data points are visualized as crosses and asterisks. All the simulations overlap, thus only one line per condition is visible. (**b**) Visualization of the 100 best parameter sets. The parameters of each set are connected with lines with different shades of gray (scale on the right). Only one line is visible because the parameters sets are all very close to each other. The black squares indicate the parameters used to generate the data. Asterisks indicate search upper bounds and crosses indicate search lower bounds

same file also contains a differential equation definition for the ppERK variable (Fig. 5a, line 15). When input1.sbc is compiled, the normalization is applied to the data. In this case, each data block of ppERK is divided by the value of ppERK at time point 5 min. In addition, an objective function is constructed where the corresponding model simulations are also normalized in the same way as the data, i.e., the simulated ppERK is divided by its value at time point 5 min.

If a new biological replicate for ppERK becomes available, the new data block can be added to the input file (input2.sbc, Fig. 5b, lines 13–16). Normally, no further action regarding the normalization is required for completing the data; however, PEPSSBI can also handle missing data values. As an example, assuming that there is a missing value in the new replicate which happens to be the reference time point used for fixed-point normalization (i.e., ppERK at time 5 min, Fig. 5b, line 14), PEPSSBI will generate an error upon compilation of input2.sbc (Fig. 5b), explaining that the normalization cannot be applied due to the missing value. To resolve this error, one can simply change the data normalization strategy (input3.sbc, Fig. 5c) by either changing the reference data point or using normalization by average. Assuming normalization by average is selected, all that is required is to change one line in the PEPSSBI input file (Fig. 5c, line 18). Note that to use an average of data points as the reference value for the normalization, we need to ensure that all the data points used in the average are available in all data blocks. When input3.sbc is compiled, normalization by average will automatically search for a set of data points that is available in all data blocks. In this case, ppERK at times 0, 10, and 15 min. Thus, each data block of ppERK is divided by the average of ppERK

A.

```
data ppERK =                                input1.sbc
"time" "EGF"
0      1
5      15
10     5
15     3
-
0      10
5      120
10     40
15     35

normaliseByFixedPoint ppERK "EGF" 5

ppERK'=...
```

```
./runSBC.sh input1.sbc

No errors reported in log file
```

B.

```
data ppERK =                                input2.sbc
"time" "EGF"
0      1
5      15
10     5
15     3
-
0      10
5      120
10     40
15     35
-
0      5
5      N/A
10     30
15     20

normaliseByFixedPoint ppERK "EGF" 5

ppERK'=...
```

```
./runSBC.sh input2.sbc

[error] could not apply normalisation
by fixed point (ppERK,"EGF",5):
N/A or zero at normalisation point in
at least one of the data blocks.
```

C.

```
data ppERK =                                input3.sbc
"time" "EGF"
0      1
5      15
10     5
15     3
-
0      10
5      120
10     40
15     35
-
0      5
5      N/A
10     30
15     20

normaliseByAverage ppERK "EGF"

ppERK'=...
```

```
./runSBC.sh input3.sbc

No errors reported in log file
```

Fig. 5 Example of use of data normalization in PEPSSBI. (**a**) The input file input1.sbc contains a normalization by fixed point (EGF at 5 min) and compiles correctly. (**b**) The input file input2.sbc presents an additional data block, with missing value at EGF 5 min. An error at compilation time is generated because of the missing value. (**c**) In input3.sbc, the normalization by fixed point is changed to normalization by average, thus changing data normalization and objective function for ppERK, and compiles correctly

at times 0, 10, and 15 min. Finally, the new objective function is constructed, where the simulated ppERK is divided by the average of the simulated values at times 0, 10, and 15 min. Algorithmically supported data normalization and DNS allowed a user to avoid time-consuming and error-prone tasks when working with the input data. In practice, several observables and experimental conditions can be used in biochemical models, which make this level of automation essential for the use of DNS.

3.5 Parameter Estimation of Synthetic Gene Circuits

The multi-model feature of PEPSSBI is particularly useful in the context of synthetic biology. In synthetic biology, gene network components are often re-used (e.g., BioBricks [46]) in different circuit designs. The same biological component can retain the same molecular parameters, such as protein or mRNA degradation rates, in different gene circuits. To mimic this experimental scenario, PEPSSBI can define multiple circuit models at the same time and use local parameters and local equations to describe components that are unique to a specific gene circuit. Unknown parameters can be estimated using data obtained from multiple circuits. In this section, we illustrate the multi-model feature in a synthetic biology application.

In this example, the aim is to identify the parameters of two synthetic gene circuits (Fig. 6). In both circuits, genes X and Y are present, with corresponding mRNA molecules *mRNA_X* and *mRNA_Y* and corresponding proteins *prot_X* and *prot_Y*. Circuit 1 (Fig. 6a) contains a positive feedback between genes X and Y that is activated by molecule S. Circuit 2 (Fig. 6b) contains a negative

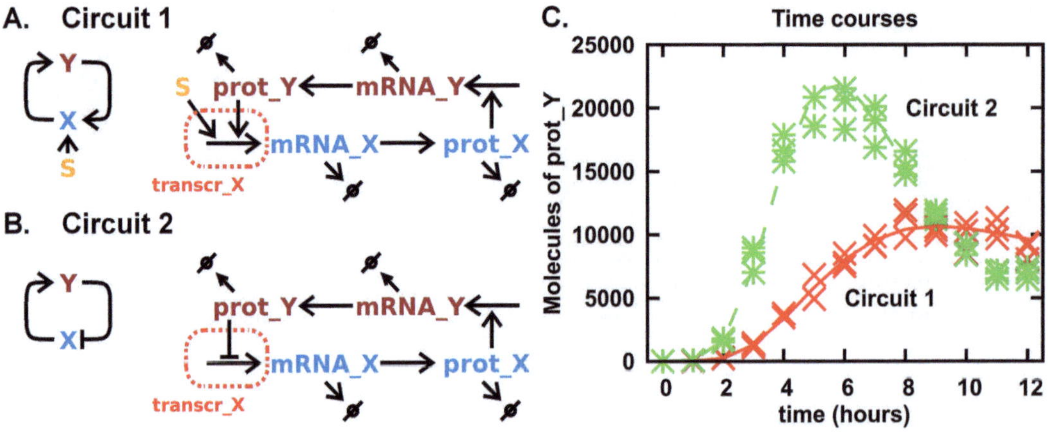

D.

```
transcr_X =
"Circuit 1" r_max_X_s*S/(K_X_s + S) + r_max_X_c1*prot_Y^2/(K_X_c1^2 + prot_Y^2)    (1)
"Circuit 2" r_max_X_c2*(1/(1 + (prot_Y/K_X_c2)^2))

mRNA_X' = transcr_X - k_degm_X*mRNA_X
prot_X' = k_tr_X*mRNA_X - k_degp_X*prot_X                                          (2)
mRNA_Y' = r_max_Y*prot_X^2/(K_Y^2 + prot_X^2) - k_degm_Y*mRNA_Y
prot_Y' = k_tr_Y*mRNA_Y - k_degp_Y*prot_Y

```

Fig. 6 Synthetic gene circuits. (**a**) Synthetic gene circuit 1 is constituted of a positive feedback loop between genes X and Y. The positive feedback is activated by molecule S. (**b**) Synthetic gene circuit 2 is constituted by a negative feedback loop between genes X and Y, in which X activates Y and Y represses X. Red dotted lines indicate the components that are different in the two circuits. (**c**) Simulations (lines) of the models used to generate the data and sample data set for prot_Y in circuits 1 and 2 (crosses and asterisks). (**d**) Lines of the PEPSSBI input file that describe the model equations for circuits 1 and 2. The components in common are specified only once (2), while the transcription of gene X is specified differently for each circuit (1)

feedback, where X activates Y and Y suppresses X. Thus, the two circuits contain shared components they have in common and components that differ. We also assume the same rates of translation and degradation in both circuits. For this example, we generated artificial measurement data for both *prot_X* and *prot_Y* in the form of a time series measured every 1 h from time 0 to 12 h in three biological replicates ($n = 3$). To generate the data, we set specific values for the unknown parameters, simulated the time courses of *prot_X* and *prot_Y* and added a 10% coefficient of variation Gaussian noise. Figure 6c shows the simulations (lines) and the artificial data generated from these simulations (crosses and asterisks). Finally, we assume that all parameters are known except for the parameters regulating the expression of Y mRNA (*mRNA_Y*) induced by *prot_X* (i.e., parameters *r_max_Y* and *K_Y*) and the parameters regulating the inhibition of X by Y in Circuit 2 (i.e., parameters *r_max_X_c2* and *K_X_c2*). The equations for the rates of mRNA transcription are defined according to ref. [47] that use Hill equations.

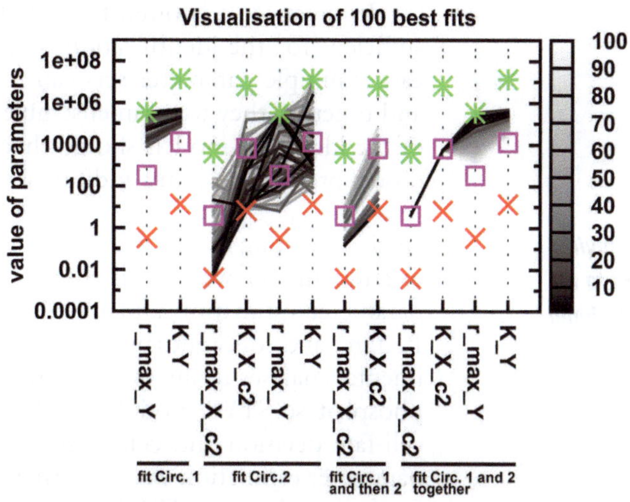

Fig. 7 Parameter estimation of synthetic gene circuits 1 and 2. Results of parameter estimation for circuits 1 and 2. Results are shown as lines connecting parameters in the same estimated set. The set index on the right and shades of gray indicate the rank of the fit, with black as the best fit. Asterisks indicate search upper bounds and crosses indicate search lower bounds. (Fit Circ. 1) Only data from circuit 1 have been used. (Fit Circ. 2) Only data from circuit 2 have been used. (Fit Circ. 1 and then 2) Only data from circuit 2 have been used, fixing r_max_Y and K_Y to the best performing set from (fit Circ. 1). (Fit Circ. 1 and 2 together) Data from both circuits 1 and 2 have been used at the same time

In Fig. 6d, we illustrate the models written in the PEPSSBI input file (the PEPSSBI file is available online, https://bitbucket.org/andreadega/systems-biology-compiler/downloads/examples_from_software_article.zip). Equations that are common to both Circuit 1 and Circuit 2 are specified only once, while the transcription of X (variable *transcr_X*) is specified differently for each circuit. Using data from the two circuits in the same parameter estimation procedure could help to constrain the unknown parameters of the two synthetic circuits. To illustrate this, we estimated the parameters for circuits 1 and 2 separately using only data from one circuit at a time, and then together using data from both circuits at the same time. As in the previous example, we used the GLSDC optimization algorithm, run 200 parameter estimation runs, and selected the best 100.

The results are summarized in Fig. 7. Only the parameters regulating the inhibition of X by Y in Circuit 2 (*r_max_X_c2* and *K_X_c2*) are clearly identifiable. Most importantly, these parameters are identifiable only if both data from circuits 1 and 2 are used together. This indicates that data from Circuit 1 helps to identify parameters that are local to Circuit 2. The estimates of the parameters regulating the activation of Y by X (*r_max_Y* and *K_Y*) were variable, indicating practical non-identifiability.

In practice, it is often the case that the available data are not sufficient for the identification of a unique parameter set. In this case, multiple parameter sets can reproduce the data equally well and together they form an ensemble of models [3]. The suitability of the identified models should then be evaluated by validation of model predictions using additional experimental data.

3.6 Applying PEPSSBI to an Existing Model

In this section, we illustrate how PEPSSBI can be applied to an existing model written in SBML. We reproduce the results of parameter estimation and model predictions in ref. [24], where Reiterer and colleagues used a combination of modeling and experimental analyses to uncover the mechanisms by which the pseudophosphatase STYX modulates ERK dynamics (Fig. 8a), affecting cell-fate decision and cell migration. In the original publication, parameter estimation was performed using time courses of ppERK for a control and a STYX knockdown experiment. Data of ppERK for STYX overexpression were used to validate model predictions.

First, we converted the model from SBML (https://www.ebi.ac.uk/biomodelsmain/BIOMD0000000557) into the PEPSSBI input language automatically using the computer program SBML2SBC. Second, we edited the obtained PEPSSBI input file by adding data, parameter search ranges and the necessary options for parameter estimation and plotting.

In the original work [24], normalization by fixed point at time zero was used for the data, while a scaling factor was also estimated to convert the relative units into ppERK concentration. Here instead we use a normalization by average and use data-driven normalization of the simulations (DNS). Once again, we used the GLSDC optimization algorithm, run 200 parameter estimation runs, and selected the best 100. As can be seen in Fig. 8b, c, we reproduced the results, obtaining good fits and correct predictions (Fig. 3a, c in ref. 24).

3.7 PEPSSBI and Systems Biology Benchmarks

Current parameter estimation benchmarks for systems biology models do not consider the DNS approach. In Villaverde et al. [32], all six examined benchmarks assume that the data is in actual concentration, while in Frohlich et al. [31] only the scaling factor approach is considered for relative data. In both studies data normalization is not discussed and often provided as an average, possibly with standard deviation. Thus, to illustrate the performance of PEPSSBI on existing benchmarks, we adapted benchmarks B2 and B4 from Villaverde et al. [32], defining SF and DNS versions. To do so, once again we used SBML2SBC to convert the benchmark into the PEPSSBI input language and added data and either scaling factors or normalization commands to the input file. Finally, we performed ten optimization runs using GLSDC for each benchmark and for each SF and DNS version.

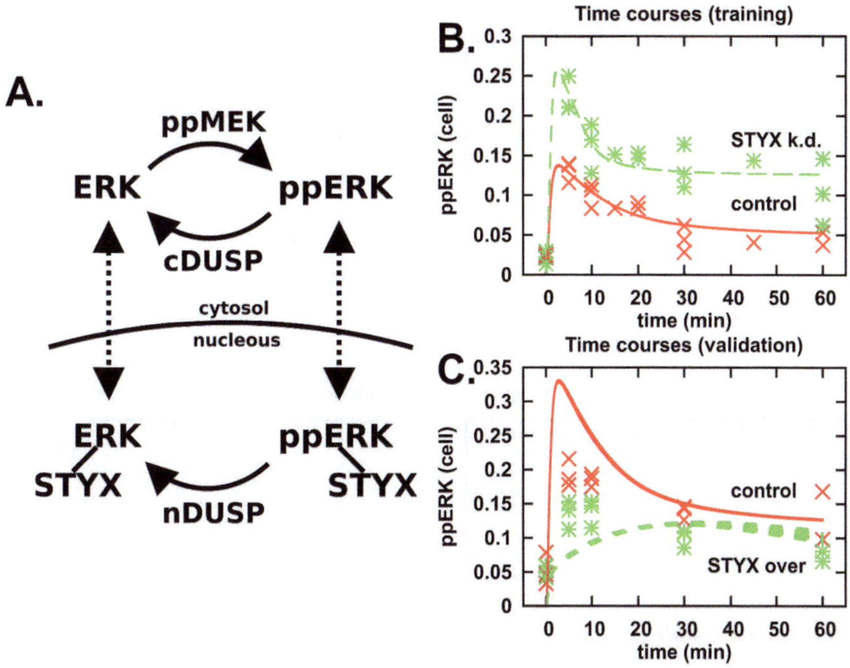

Fig. 8 Parameter estimation of a model of pseudophosphatase STYX and ERK interactions (STYX-1-10). (**a**) Reaction scheme of Model 1 of [24]. (**b**) Model training using control and STYX knockdown data, obtained using PEPSSBI, analogous to Fig. 3a in ref. 24. (**c**) Model validation using control and STYX overexpression data, obtained using PEPSSBI, analogous to Fig. 3c in ref. 24

The time convergence of GLSDC was strongly improved by the use of DNS as opposed to DNS for both B2 and B4 benchmarks (Fig. 9a, b), confirming DNS benefits in combination with hybrid deterministic stochastic and derivative free methods. In addition, we report here the performance of SF and DNS on two models that we have previously published, EGF/HRG-8-74 and STYX-1-10 (Fig. 9c, d). Table 1 summarizes the number of observables and the number of unknown parameters considered in each of the four benchmark models. The PEPSSBI input files for these models are available online (https://bitbucket.org/andreadega/systems-biology-compiler/downloads/examples_from_software_article.zip).

3.8 Conclusions

In this chapter, we have presented PEPSSBI, a new parameter estimation pipeline for systems and synthetic biology, and illustrated how PEPSSBI can be used to perform parameter estimation using data-driven normalization of the simulations (DNS). The advantages of DNS for complex parameter estimation problems have been previously investigated in detail [15], demonstrating both increased convergence speed and better practical parameter identifiability. PEPSSBI is the first software to support DNS natively using a high level of automation.

Using a dedicated input language, PEPSSBI simplifies user interactions when using DNS. The PEPSSBI language reduces

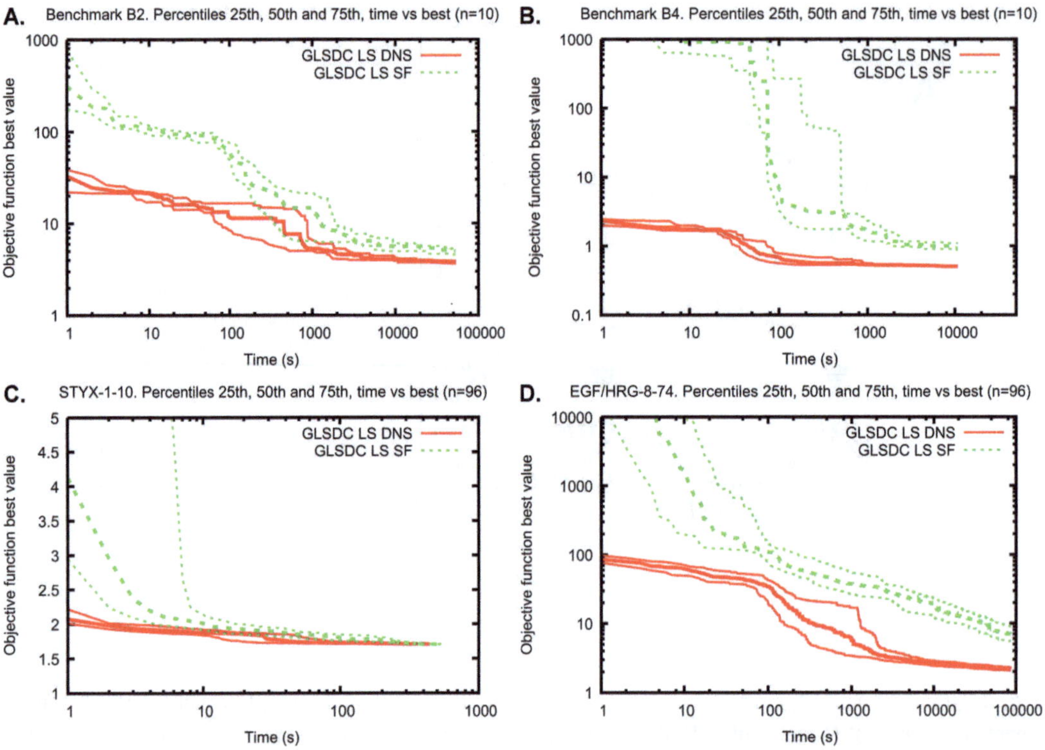

Fig. 9 Time convergence of GLSDC optimization algorithm with least squares (LS) objective function for benchmarks B2 (**a**) and B4 (**b**) from Villaverde et al. [32], and STYX-1-10 (**c**) and EGF/HRG-8-74 (**d**) from Degasperi et al. [15]. Lines are the 25th, 50th, and 75th percentile of $n = 10$ optimization runs. Either data-driven normalization of the simulations (DNS) or scaling factor (SF) objective functions are used

Table 1
Benchmark models used to compare the scaling factor (SF) approach to parameter estimation to the data-driven normalization of the simulations (DNS) approach

Model	Description	Observables	Unknown parameters	References
B2	Dynamic model of the central carbon metabolism of *E. coli*	9	116	[32]
B4	Metabolic model of Chinese hamster ovary (CHO) cells	13	117	[32]
STYX-1-10	Interactions between the kinase ERK and the pseudophosphatase STYX following the activation of ERK kinase MEK in PC12 cells	1	10	[15]
EGF/HRG-8-74	ERK pathway activation and two transcriptional negative feedback loops triggered by the activation of ERK kinase MEK in MCF-7 cells	8	74	[15]

the time and effort to implement different normalizations and minimizes the risks of human errors. Moreover, PEPSSBI provides different data normalization strategies that can be largely automated and defined in a single line of code, while objective function construction is completely automated (Figs. 3 and 5).

Another distinctive feature of PEPSSBI is its support for estimating parameters of multiple models simultaneously, a feature which we termed multi-model. Multi-model expands the well-adopted concept of multi-condition experiments [8], allowing not only single parameters or input functions to change between conditions but also model equations (Fig. 6). Multi-model can be implemented in other software, such as Copasi [48] or Data2Dynamics [10], by using "if then else" statements to control which model equations correspond to which model. The PEPSSBI input language implements multi-model in a simpler and more accessible way as a single input file, without the need to define additional variables or modify additional files.

PEPSSBI's multi-model feature is particularly useful for synthetic biology applications, where recurrent synthetic molecular components are often reused and embedded in multiple networks. PEPSSBI input language was designed so that code for the shared parts needs to be written only once, thus avoiding code repetition. In addition, we have demonstrated that PEPSSBI's multi-model feature results in better parameter constraint. Indeed, combined estimation of two genetic circuits with a shared module yielded tighter and more accurate estimates.

PEPSSBI uniquely complements the current landscape for parameter estimation software by providing user-friendly DNS support and the multi-model feature. Users can easily utilize PEPSSBI by importing existing models constructed with other systems biology tools using SBML. We have provided an example of this by applying PEPSSBI to a published model of STYX–ERK interactions, which was originally constructed using PottersWheel [8] (Fig. 8). Other software programs that can be used to construct models and import them to PEPSSBI include Antimony [49] and BioNetGen [50].

The superior impact that DNS has on the time convergence of optimization algorithms as opposed to the use of scaling factors (Fig. 9) implies that DNS should be considered in future parameter estimation benchmarks. Because biological data are often expressed in relative units, we believe it is also important that data normalization is part of the discussion, to ensure that the gap between methods development and their application is properly bridged.

Finally, PEPSSBI is an open-source software, which is open to improvement by the wider system biology community.

4 Notes

1. When we refer to relative data normalization, we intend the scaling of relative units so that they represent a multiple of a given reference, which could be a control (normalization by fixed point) or the average intensity (normalization by average), so that data points can be compared *across* biological replicates. We assume that there are multiple biological replicates of the same experiment, e.g., Western blot images of replicates, and we assume that the data *within* each replicate are comparable. In the case of Western blot images, it is important to remind that to achieve comparable data *within* the same replicate, values need to be scaled by the corresponding housekeeping protein or loading control from the same lane, in order to adjust for small loading errors. This operation is sometimes also referred to as normalization (loading control normalization), which could lead to confusion.

Acknowledgments

A.D., D.F., and B.N.K. received funding from the European Union Seventh Framework Programme (FP7/2007–2013) SynSignal (grant no. 613879). A.D. received funding from Science Foundation Ireland Industry Fellowship No. 15/IFA/2925. B.N.K. acknowledges support from the EU H2020 SmartNanoTox (grant no. 686098) and NanoCommons (grant no. 731032). L.K.N. is supported by a Victorian Cancer Agency Mid-Career Research Fellowship (MCRF18026); an Investigator Initiated Research Scheme grant from Australia's National Breast Cancer Foundation (IIRS-20-094); and the Metcalf Venture Grants Scheme administered by Cancer Council Victoria, Australia. We would like to thank Dr. Sungyoung Shin and Marco Dalla Torre for the useful comments.

References

1. Kitano H (2002) Computational systems biology. Nature 420(6912):206–210. https://doi.org/10.1038/Nature01254
2. Kholodenko BN, Demin OV, Moehren G, Hoek JB (1999) Quantification of short term signaling by the epidermal growth factor receptor. J Biol Chem 274(42):30169–30181
3. Nakakuki T, Birtwistle MR, Saeki Y, Yumoto N, Ide K, Nagashima T, Brusch L, Ogunnaike BA, Okada-Hatakeyama M, Kholodenko BN (2010) Ligand-specific c-Fos expression emerges from the spatiotemporal control of ErbB network dynamics. Cell 141(5):884–896. https://doi.org/10.1016/j.cell.2010.03.054
4. Ghosh S, Matsuoka Y, Asai Y, Hsin KY, Kitano H (2011) Software for systems biology: from tools to integrated platforms. Nat Rev Genet 12(12):821–832. https://doi.org/10.1038/nrg3096
5. Klipp E (2005) Systems biology in practice : concepts, implementation and application. Wiley-VCH, Weinheim
6. Mendes P, Hoops S, Sahle S, Gauges R, Dada J, Kummer U (2009) Computational modeling of biochemical networks using COPASI.

Methods Mol Biol 500:17–59. https://doi.org/10.1007/978-1-59745-525-1_2

7. Matsuoka Y, Funahashi A, Ghosh S, Kitano H (2014) Modeling and simulation using CellDesigner. Methods Mol Biol 1164:121–145. https://doi.org/10.1007/978-1-4939-0805-9_11
8. Maiwald T, Timmer J (2008) Dynamical modeling and multi-experiment fitting with PottersWheel. Bioinformatics 24(18):2037–2043
9. Sauro HM, Hucka M, Finney A, Wellock C, Bolouri H, Doyle J, Kitano H (2003) Next generation simulation tools: the Systems Biology Workbench and BioSPICE integration. OMICS 7(4):355–372. https://doi.org/10.1089/153623103322637670
10. Raue A, Steiert B, Schelker M, Kreutz C, Maiwald T, Hass H, Vanlier J, Tonsing C, Adlung L, Engesser R, Mader W, Heinemann T, Hasenauer J, Schilling M, Hofer T, Klipp E, Theis F, Klingmuller U, Schoberl B, Timmer J (2015) Data2Dynamics: a modeling environment tailored to parameter estimation in dynamical systems. Bioinformatics 31(21):3558–3560. https://doi.org/10.1093/bioinformatics/btv405
11. Stapor P, Weindl D, Ballnus B, Hug S, Loos C, Fiedler A, Krause S, Hross S, Frohlich F, Hasenauer J, Wren J (2018) PESTO: Parameter EStimation TOolbox. Bioinformatics 34 (4):705–707. https://doi.org/10.1093/bioinformatics/btx676
12. Balsa-Canto E, Henriques D, Gabor A, Banga JR (2016) AMIGO2, a toolbox for dynamic modeling, optimization and control in systems biology. Bioinformatics 32(21):3357–3359. https://doi.org/10.1093/bioinformatics/btw411
13. Hucka M, Finney A, Sauro HM, Bolouri H, Doyle JC, Kitano H, Arkin AP, Bornstein BJ, Bray D, Cornish-Bowden A, Cuellar AA, Dronov S, Gilles ED, Ginkel M, Gor V, Goryanin II, Hedley WJ, Hodgman TC, Hofmeyr JH, Hunter PJ, Juty NS, Kasberger JL, Kremling A, Kummer U, Le Novere N, Loew LM, Lucio D, Mendes P, Minch E, Mjolsness ED, Nakayama Y, Nelson MR, Nielsen PF, Sakurada T, Schaff JC, Shapiro BE, Shimizu TS, Spence HD, Stelling J, Takahashi K, Tomita M, Wagner J, Wang J, Forum S (2003) The systems biology markup language (SBML): a medium for representation and exchange of biochemical network models. Bioinformatics 19(4):524–531
14. Gabor A, Banga JR (2015) Robust and efficient parameter estimation in dynamic models of biological systems. BMC Syst Biol 9:74. https://doi.org/10.1186/s12918-015-0219-2
15. Degasperi A, Fey D, Kholodenko BN (2017) Performance of objective functions and optimisation procedures for parameter estimation in system biology models. NPJ Syst Biol Appl 3:20. https://doi.org/10.1038/s41540-017-0023-2
16. Ashyraliyev M, Fomekong-Nanfack Y, Kaandorp JA, Blom JG (2009) Systems biology: parameter estimation for biochemical models. FEBS J 276(4):886–902. https://doi.org/10.1111/j.1742-4658.2008.06844.x
17. Hengl S, Kreutz C, Timmer J, Maiwald T (2007) Data-based identifiability analysis of non-linear dynamical models. Bioinformatics 23(19):2612–2618. https://doi.org/10.1093/bioinformatics/btm382
18. Raue A, Kreutz C, Maiwald T, Bachmann J, Schilling M, Klingmuller U, Timmer J (2009) Structural and practical identifiability analysis of partially observed dynamical models by exploiting the profile likelihood. Bioinformatics 25(15):1923–1929. https://doi.org/10.1093/bioinformatics/btp358
19. Dalle Pezze P, Le Novere N (2017) SBpipe: a collection of pipelines for automating repetitive simulation and analysis tasks. BMC Syst Biol 11 (1):46. https://doi.org/10.1186/s12918-017-0423-3
20. Eydgahi H, Chen WW, Muhlich JL, Vitkup D, Tsitsiklis JN, Sorger PK (2013) Properties of cell death models calibrated and compared using Bayesian approaches. Mol Syst Biol 9:644. https://doi.org/10.1038/msb.2012.69
21. Liepe J, Kirk P, Filippi S, Toni T, Barnes CP, Stumpf MP (2014) A framework for parameter estimation and model selection from experimental data in systems biology using approximate Bayesian computation. Nat Protoc 9 (2):439–456. https://doi.org/10.1038/nprot.2014.025
22. Bachmann J, Raue A, Schilling M, Bohm ME, Kreutz C, Kaschek D, Busch H, Gretz N, Lehmann WD, Timmer J, Klingmuller U (2011) Division of labor by dual feedback regulators controls JAK2/STAT5 signaling over broad ligand range. Mol Syst Biol 7:516. https://doi.org/10.1038/msb.2011.50
23. Byrne KM, Monsefi N, Dawson JC, Degasperi A, Bukowski-Wills JC, Volinsky N, Dobrzynski M, Birtwistle MR, Tsyganov MA, Kiyatkin A, Kida K, Finch AJ, Carragher NO, Kolch W, Nguyen LK, von Kriegsheim A, Kholodenko BN (2016) Bistability in the Rac1,

PAK, and RhoA signaling network drives actin cytoskeleton dynamics and cell motility switches. Cell Syst 2(1):38–48. https://doi.org/10.1016/j.cels.2016.01.003
24. Reiterer V, Fey D, Kolch W, Kholodenko BN, Farhan H (2013) Pseudophosphatase STYX modulates cell-fate decisions and cell migration by spatiotemporal regulation of ERK1/2. Proc Natl Acad Sci U S A 110(31):E2934–E2943. https://doi.org/10.1073/pnas.1301985110
25. Burnette WN (1981) "Western blotting": electrophoretic transfer of proteins from sodium dodecyl sulfate—polyacrylamide gels to unmodified nitrocellulose and radiographic detection with antibody and radioiodinated protein A. Anal Biochem 112(2):195–203
26. Towbin H, Staehelin T, Gordon J (1979) Electrophoretic transfer of proteins from polyacrylamide gels to nitrocellulose sheets: procedure and some applications. Proc Natl Acad Sci U S A 76(9):4350–4354
27. Tighe PJ, Ryder RR, Todd I, Fairclough LC (2015) ELISA in the multiplex era: potentials and pitfalls. Proteomics Clin Appl 9 (3–4):406–422. https://doi.org/10.1002/prca.201400130
28. Heid CA, Stevens J, Livak KJ, Williams PM (1996) Real time quantitative PCR. Genome Res 6(10):986–994
29. Weber P, Hasenauer J, Allgöwer F, Radde N (2011) Parameter estimation and identifiability of biological networks using relative data. IFAC Proceedings 44(1):11648–11653. https://doi.org/10.3182/20110828-6-IT-1002.01007
30. Degasperi A, Birtwistle MR, Volinsky N, Rauch J, Kolch W, Kholodenko BN (2014) Evaluating strategies to normalise biological replicates of Western blot data. PLoS One 9 (1):e87293. https://doi.org/10.1371/journal.pone.0087293
31. Frohlich F, Loos C, Hasenauer J (2019) Scalable inference of ordinary differential equation models of biochemical processes. Methods Mol Biol 1883:385–422. https://doi.org/10.1007/978-1-4939-8882-2_16
32. Villaverde AF, Henriques D, Smallbone K, Bongard S, Schmid J, Cicin-Sain D, Crombach A, Saez-Rodriguez J, Mauch K, Balsa-Canto E, Mendes P, Jaeger J, Banga JR (2015) BioPreDyn-bench: a suite of benchmark problems for dynamic modelling in systems biology. BMC Syst Biol 9:8. https://doi.org/10.1186/s12918-015-0144-4
33. Kimura S, Konagaya A (2003) High dimensional function optimization using a new genetic local search suitable for parallel computers. IEEE Syst Man Cybern:335–342
34. Raue A, Schilling M, Bachmann J, Matteson A, Schelke M, Kaschek D, Hug S, Kreutz C, Harms BD, Theis FJ, Klingmuller U, Timmer J (2013) Lessons learned from quantitative dynamical modeling in systems biology. PLoS One 8(9):e74335. https://doi.org/10.1371/journal.pone.0074335
35. Parr T (2013) The definitive ANTLR 4 reference. Pragmatic Bookshelf
36. Hindmarsh AC, Brown PN, Grant KE, Lee SL, Serban R, Shumaker DE, Woodward CS (2005) SUNDIALS: suite of nonlinear and differential/algebraic equation solvers. ACM Trans Math Software 31(3):363–396
37. Lourakis MIA (2004) levmar: Levenberg-Marquardt nonlinear least squares algorithms in {C}/{C}++. http://www.ics.forth.gr/~lourakis/levmar/. 2016
38. Team RC (2017) R: a language and environment for statistical computing. R Foundation for Statistical Computing, Vienna. https://www.R-project.org/
39. Williams T, Kelley C (2015) Gnuplot 5.0 reference manual. Samurai Media Limited
40. Wendling C, Treleaven N, Brush M, Tröger E, Lanitz F (2017) Geany text editor version 1.31. https://www.geany.org
41. Howell M (2017) Homebrew: the missing package manager for macOS. https://brew.sh/
42. (2017) mingw-w64: GCC for Windows 64 & 32 bits. https://mingw-w64.org
43. (2017) MSYS2: software distro and building platform for Windows. http://www.msys2.org/
44. (2017) Docker—build, ship, and run any app, anywhere. www.docker.com
45. Swameye I, Muller TG, Timmer J, Sandra O, Klingmuller U (2003) Identification of nucleocytoplasmic cycling as a remote sensor in cellular signaling by databased modeling. Proc Natl Acad Sci U S A 100(3):1028–1033. https://doi.org/10.1073/pnas.0237333100
46. Shetty RP, Endy D, Knight TF Jr (2008) Engineering BioBrick vectors from BioBrick parts. J Biol Eng 2:5. https://doi.org/10.1186/1754-1611-2-5
47. Alon U (2007) An introduction to systems biology: design principles of biological circuits, Chapman & Hall/CRC mathematical and computational biology series, vol 10. Chapman & Hall/CRC, Boca Raton, FL
48. Hoops S, Sahle S, Gauges R, Lee C, Pahle J, Simus N, Singhal M, Xu L, Mendes P, Kummer

U (2006) COPASI—a COmplex PAthway SImulator. Bioinformatics 22 (24):3067–3074. https://doi.org/10.1093/bioinformatics/btl485

49. Smith LP, Bergmann FT, Chandran D, Sauro HM (2009) Antimony: a modular model definition language. Bioinformatics 25 (18):2452–2454. https://doi.org/10.1093/bioinformatics/btp401

50. Faeder JR, Blinov ML, Hlavacek WS (2009) Rule-based modeling of biochemical systems with BioNetGen. Methods Mol Biol 500:113–167. https://doi.org/10.1007/978-1-59745-525-1_5

Chapter 6

Dynamic Optimization Approach to Estimate Kinetic Parameters of Monod-Based Microalgae Growth Models

Siti S. Jamaian, Fathul H. Zulkifli, and Kim S. Ling

Abstract

The biomass concentration of microalgae growth in photobioreactor was predicted using the Monod-based growth models. Kinetic parameters such as maximum specific growth rate and saturation constant of light intensity were evaluated by nonlinear least squares methods that focused on minimizing the sum of squares error (SSE). The importance of good initial guess for the nonlinear least squares method was also discussed. The optimal control problem of the microalgae growth model was determined based on parameter sensitivity. Therefore, a dynamic optimization approach was used where an optimal input design method was formulated to obtain a control function of a problem. The dynamic state equations, additional state equations, cost function, and Hamiltonian function were used to establish a control function of microalgae growth in a photobioreactor. Hence, the biomass production of microalgae can be predicted using numerical methods such as the Taylor series method.

Key words Kinetic parameters estimation, Gauss–Newton method, Levenberg–Marquardt method, Microalgae growth, Dynamic optimization

1 Introduction

Microalgae are aquatic oxygen-evolving photosynthetic autotrophs [1]. The three most important classes of microalgae are the diatoms, green algae, and golden algae [2]. The six different phases of microbial growth in a batch culture include lag, exponential, linear, declining growth, stationary, and death [3]. Microalgae require suitable light conditions, temperature, water, carbon dioxide, and nutrients such as nitrogen and phosphorus to carry out photosynthesis to produce sugar for aerobic chemoheterotrophic biomass growth [4] that involves two general processes which are anabolism and catabolism [5].

Quentin Vanhaelen (ed.), *Computational Methods for Estimating the Kinetic Parameters of Biological Systems*, Methods in Molecular Biology, vol. 2385, https://doi.org/10.1007/978-1-0716-1767-0_6,

The growth kinetic models for microalgae can be affected by a single factor or multiple factors such as physical, chemical, and biological factors that influence microalgae growth [3]. There are many growth kinetic models introduced for the specific growth rate of microalgae such as Teissier, Monod, Moser, Contois, and Andrews models. Moser, Contois, and Andrews models are modified from the Monod model with half-saturation constant [6]. The growth rate and biomass concentration of microalgae can be predicted using the estimated parameters of the kinetic model. Method of moments, maximum likelihood estimation, and least squares method are popular methods used for parameters estimation.

1.1 Least Squares Method

Least squares method can be divided into two categories: linear least squares and nonlinear least squares. The method is widely used for model parameter estimation by fitting the experimental data [7]. The optimal parameters computed minimizes the sum of the squares (SSE) of the residual from the experimental data [8]. In this study, the nonlinear least squares method was applied to estimate the parameters from experimental data due to the nonlinear nature of the growth kinetic model.

The factors affecting microalgae growth are inputs in chemostat which are controlled to maximize the production of biomass. Dynamic optimization approach can be used to solve the optimal control problem of one or more control variables or control parameters of a system [9]. A control variable is a function of time that optimizes the cost function computed in the dynamic optimization. The cost function composed from the terminal condition and the running cost is optimized within the process constraints and the defined lower and upper bounds of input variables [10]. In this study, dynamic optimization approach was used to solve the optimal control problem for photobioreactor.

1.2 Nonlinear Least Squares Method for Parameter Estimation

Nonlinear least squares method is a type of unconstraint optimization method that minimizes SSE as the objective function [11]. It is an iterative procedure that starts with an initial guess for each parameter followed by the convergence of initial values to their optimal values to fit the collected experimental data [7, 12]. Setting the initial guess for each parameter is one of the most difficult and important steps. Good initial guess value can converge with optimal values faster, whereas poor initial guess value may converge to incorrect values [13]. In this study, both Gauss–Newton and Levenberg–Marquardt methods were applied to estimate the parameters of the Monod model, and convergent problem was discussed further. When the arbitrary initial value deviated from the optimal values, comparison with experimental data that correspond to the definition of the parameters was performed [14].

1.3 Optimal Control Problem for Microalgae

In this study, an optimal control problem for microalgae which consisted of state equations, Hamiltonian function, co-state equations, and stationary equations was formulated. The Monod model was substituted into the photobioreactor model to define the microalgae concentration rate of change as the dynamic state equation in an optimal control problem. Furthermore, optimal input design including parametric sensitivities was used to find the control function [15, 16].

1.4 Numerical Methods for Biomass Concentration of Microalgae

The state equations in the formulated optimal control problem are first-order differential equations. Therefore, numerical methods such as first-order and second-order Taylor series method were used to solve the system of state equations to estimate the biomass concentration of microalgae [17].

2 Materials

Table 1 shows the experimental data used in this study which described the change of the specific growth rate, μ, of *Botyrococcus* sp. microalgae in function of different light intensity levels, I, as specified in a previous study [18].

3 Methods

3.1 Monod Model

The Monod model was selected to describe the kinetic growth model of microalgae which depends on the light intensity. The Monod model [19] can be written follows:

Table 1
Specific growth rate versus light intensity

Light intensity (μmol m^{-2} s^{-1})	Specific growth rate (day^{-1})
2.7	0.412
48.6	0.461
94.5	0.959
176.0	1.026
243.0	1.307
324.0	1.160

$$\mu = \mu_{\max} \frac{I}{K_{\mathrm{I}} + I}, \tag{1}$$

where μ is the specific growth rate, $\mu_{\max}$ is the maximum specific growth rate, I is the light intensity, and K_{I} is the saturation constant for light intensity corresponding to the value of I when the value of μ is half of $\mu_{\max}$.

3.2 Gauss–Newton Method

In this study, the sum of squares error (SSE) was the objective function which was defined as:

$$\min \text{SSE} = \sum_{i=1}^{n} \left(\mu_i - \frac{\mu_{\max} I_i}{K_{\mathrm{I}} + I_i} \right)^2 \tag{2}$$

For the Gauss–Newton method, the updated value of parameters, p, was calculated using the following formula [20]:

$$p_{k+1} = p_k - \left[J(p_k)^T J(p_k) \right]^{-1} J(p_k)^T f(p_k) \tag{3}$$

where J is the Jacobian matrix of SSE and can be written as the following formula [21]:

$$J(\mu_{\max}, K_{\mathrm{I}}) = \left(\frac{\partial f_i}{\partial \mu_{\max}}, \frac{\partial f_i}{\partial K_{\mathrm{I}}} \right) = \left(-\frac{I_i}{K_{\mathrm{I}} + I_i}, \frac{\mu_{\max} I_i}{(K_{\mathrm{I}} + I_i)^2} \right) \tag{4}$$

where $f_i(\mu_{\max}, K_{\mathrm{I}}) = \mu_i - \frac{\mu_{\max} I_i}{K_{\mathrm{I}} + I_i}$

The initial guess for parameters $\mu_{\max}$ and K_{I} were specified at 1.307 day^{-1} and 70μmol m^{-2} s^{-1}, respectively (*see* **Note 1**). The iteration was stopped when the desired tolerance value was achieved. In the present study, the desired tolerance value was specified at 0.000001. The convergence problem could arise when using Gauss–Newton method, and in this case, the Levenberg–Marquardt method can be employed (*see* **Note 2**). The results of the Gauss–Newton Method with the initial guess corresponding to the parameter's definition are shown in Table 2. The algorithm used for the parameter estimation of the Monod model by the Gauss–Newton method was implemented with the Matlab software [22], and it is described in the box below. The words in green color are comments to explain the different steps of the algorithm.

3.3 Levenberg–Marquardt Method

The Levenberg–Marquardt method is a nonlinear least squares method which is a combination of Gauss–Newton and gradient-descent methods [23]. The algorithm used for Levenberg–Marquardt and Gauss–Newton method is similar as discussed earlier except the iteration procedure which has been updated. For the Levenberg–Marquardt method, the updated value of p was defined as the following formula [24, 25]:

Table 2
Result of Gauss–Newton method with initial guess corresponding to parameter's definition

k	μ_{max}	K_I	SSE	Error
0	**1.307000**	**70.000000**	**0.281965**	–
1	1.531786	74.193811	0.187543	0.094422
2	1.533219	73.877706	0.187517	0.000026
3	1.533164	73.869295	0.187517	0.000000

Coding of Parameter Estimation of Monod Model by Gauss–Newton Method in Matlab Software

Set stopping criteria and initial guess for iteration:

```
function GaussNewton()
%tol is desired tolerance value
tol = 1e-6;
%set maximum iterations
maxstep = 15;

%S is SSE, set initial SSE=0
oldS=0;

%initial guess for the parameter, p
p = [1.307;70];
n=length(p); %determine numbers of p

%set pold equal to initial guess
```

Input experimental data:

```
%Experimental data
U = [0.412 0.461 0.959 1.026 1.307 1.160];
I = [2.7 48.6 94.5 176 243 324];
m=length(I); %determine numbers of data
```

(continued)

Compute SSE for each iteration and display the table:

```
disp('    i           Umax              KI             SSE             Error')
for k=0:maxstep
 S = 0; %set initial SSE=0
 for i=1:m
 for j=1:n
 %calculate Jacobian,J and its transpose
 J(i,j) = df(I(i),p(1,1),p(2,1),j);
 JT(j,i) = J(i,j);
 end
 end
 %multiply J transpose and J
 Jz = JT*J;
 for i=1:m
 %Let f(p)=r,calculate r and SSE
 r(i,1) = U(i) - p(1,1)*(I(i)/(I(i)+p(2,1)));
 S = S + r(i,1)^2;
 end
 %mulitply Jz inverse by J transpose
 g = Jz\JT;
 %calculate new p
 p = pold-g*r;

 % compute absolute error of SSE after iteration
 % if absolute error of SSE less than tolerance,iteration end
 err(k+1) = S-oldS;
 if (abs(err(k+1)) <= tol);
 break
 end
 %update SSE
 oldS=S;
 fprintf('%5i %14.6f %14.6f %14.6f
%14.6f\n',k,pold(1,1),pold(2,1),S,abs(err(k+1)));
 %set pold to p
 pold = p;
end
fprintf('%5i %14.6f %14.6f %14.6f
%14.6f\n',k,pold(1,1),pold(2,1),S,abs(err(k+1)));
end

%calculate partial derivative
function value = df(I,a1,a2,index)
switch index
 case 1
 %       df/dUmax
 %     = -I/(KI+I)
 value = -I/(a2+I);
 case 2
 %       df/dKI
```

$$p_{k+1} = p_k - \left(J_k{}^T J_k + \beta I\right)^{-1} \nabla f(p_k), \tag{5}$$

where β is damping parameter, I is identity matrix, $\nabla f(p_k)$ is the gradient of f that can be defined as

$$\nabla f(p_k) = J_i{}^T f_i. \tag{6}$$

If the SSE increased as a result of the update, the step should be retracted by resetting the parameter values and SSE to their previous values. Then, β should be increased by a factor of 10, and an update should be repeated using Eq. 5. In contrast, if the SSE decreased as a result of the update, the step should be accepted by updating the values of the parameters and SSE to their new values. Next, β should be decreased by a factor of 10 and an update should be repeated. The parameter values and SSE were updated until the SSE achieved the desired tolerance value [26]. The results obtained when the Levenberg–Marquardt method was used with the initial value corresponding to the parameter's definition are shown in Table 3. The comparison of the experimental data with the Monod model was also presented graphically in Fig. 1. The algorithm to perform the parameter estimation using the Levenberg–Marquardt method is as described in the box below.

3.4 Photobioreactor Model

Following Hermanto [27], the microalgae photobioreactor system is modeled as shown in Fig. 2. The parameters L_{in} and L_{out} are the incoming and outgoing liquid flow in a photobioreactor, respectively. Meanwhile, V is the volume capacity of a photobioreactor, C is the concentration of microalgae in a photobioreactor, C_{in} is the incoming concentration of microalgae, C_{out} is the outgoing

Table 3
Result of Levenberg–Marquardt method with initial guess corresponding to parameter's definition

k	μ_{max}	K_I	SSE	Error
0	**1.307000**	**70.000000**	**0.281965**	–
1	1.508991	70.003735	0.187728	0.094237
2	1.510204	70.058205	0.187720	0.000008
3	1.513063	70.531209	0.187672	0.000048
4	1.524475	72.421402	0.187545	0.000127
5	1.532258	73.717019	0.187517	0.000028
6	1.533127	73.863027	0.187517	0.000000

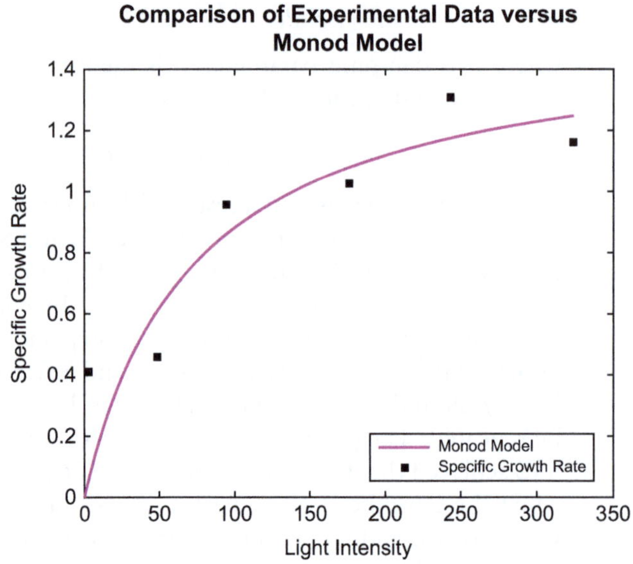

Fig. 1 Comparison of experimental data versus Monod model

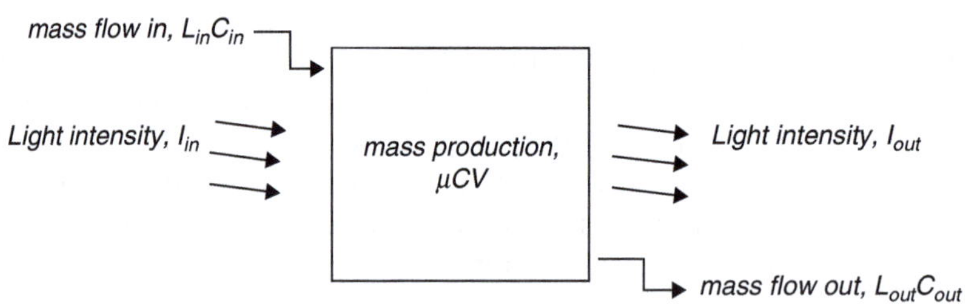

Fig. 2 Microalgae photobioreactor system

concentration of microalgae, and μ is the specific growth rate of the microalgae.

The mass balance in the photobioreactor can be defined as follows:

$$[\text{Mass accumulation}] = [\text{mass flow in}] - [\text{mass flow out}] + [\text{mass production}]$$

$$\frac{d[VC]}{dt} = L_{\text{in}}C_{\text{in}} - L_{\text{out}}C_{\text{out}} + \mu CV \tag{7}$$

By applying the product rule, Eq. 7 can be expressed as follows:

$$V\frac{dC}{dt} + C\frac{dV}{dt} = L_{\text{in}}C_{\text{in}} - L_{\text{out}}C_{\text{out}} + \mu CV \tag{8}$$

Coding of Parameter Estimation of Monod Model by Levenberg–Marquardt Method in Matlab Software

Set stopping criteria and initial guess for iteration:

```
% e is SSE, set initial SSE=0
e=0; old_e=e;
% tol is desired tolerance value
tol = 1e-6;
% Number of maximum iterations
n_iters=20;
% Initial damping parameter,k
k=0.01;

% Number of parameters
Nparams=2;
% initial guess for the parameters
Umaxo=1.307; KIo=70;
NewUmax=Umaxo;
```

Input experimental data:

```
% Experimental data
U = [0.412 0.461 0.959 1.026 1.307 1.160];
I = [2.7 48.6 94.5 176 243 324];
% Number of data
Ndata=length(U);
```

Compute and display SSE for initial guess:

```
% iteration 0: Let f(p)=r,compute r
NewU = NewUmax*(I./(NewKI+I));
r=U-NewU;
% iteration 0: compute SSE
e=dot(r,r);
old_e=e;
disp('    i          Umax              KI              SSE              Error')
% display iteration 0
```

Input partial derivative function for Jacobian matrix:

```
% partial derivative for Jacobian,J
J1=inline('-I/(I+KI)', 'Umax', 'KI','I');
J2=inline('(I*Umax)/(I+KI)^2', 'Umax', 'KI','I');
```

(continued)

Compute SSE for next iterations and display the table:

```
% set updateJ==1,compute J
updateJ=1;
for it=1:n_iters
 if updateJ==1
 J=zeros(Ndata,Nparams);
 for i=1:Ndata
 J(i,:)= [J1(NewUmax,NewKI,I(i)),J2(NewUmax,NewKI,I(i))];
 end
 % iteration 1:compute r
 NewU = NewUmax*(I./(NewKI+I));
 r=U-NewU;
 % iteration 1:compute SSE
 if it==1
 e=dot(r,r);
 end
 end

 % compute the updated parameters
 H_lm=J'*J+(k*eye(Nparams,Nparams));
 dp=-inv(H_lm)*(J'*r(:));
 Umax_lm=NewUmax+dp(1);
 K1_lm=NewKI+dp(2);

 % compute the SSE of the updated parameters
 NewU_lm = Umax_lm*(I./(K1_lm+I));
 r_lm=U-NewU_lm;
 e_lm=dot(r_lm,r_lm);

 % If SSE of the updated parameters is more than the previous one
 % then reset the parameters values and SSE to their previous values
 % and increases k by a factor of 10
 if e_lm>e
 updateJ=0;
 k=k*10;
 % otherwise update the parameters values and SSE to their new values
 % and decreases k by a factor of 10
 else
 k=k/10;
 NewUmax=Umax_lm;
 NewKI=K1_lm;
 e=e_lm;
 updateJ=1;
 end

 % compute absolute error of SSE after iteration
 % if absolute error of SSE less than tolerance,iteration end
 err(it) = e-old_e;
 if (abs(err(it)) <= tol);
 break
 end

 % update SSE
 old_e=e;

fprintf('%5i %14.6f %14.6f %14.6f
```

(continued)

Produce graph of comparison of the experimental data with Monod model:

```
Xe=0:0.1:max(I);
for i=1:length(Xe)
   Ye(i)=NewUmax*(Xe(i)./(NewKI+Xe(i)));
end
figure(1)
plot(Xe,Ye,'m','LineWidth',1.5);
hold on;
scatter(I,U,'k','s','filled');
title('Comparison of Experimental Data versus Monod
Model','Fontweight','bold','Fontsize',10,'Fontname','Time New Roman');
xlabel('Light Intensity','Fontsize',10,'Fontname','Times New Roman');
ylabel('Specific Growth Rate','Fontsize',10,'Fontname','Times New
Roman');
```

The change of volume is given by the difference between the incoming and outgoing liquid flow. The corresponding equation for the temporal evolution of the volume can be written as follows:

$$\frac{dV}{dt} = L_{\text{in}} - L_{\text{out}} \tag{9}$$

Substituting Eq. 9 into Eq. 8, we obtain the following Eq. 10:

$$V\frac{dC}{dt} + C(L_{\text{in}} - L_{\text{out}}) = L_{\text{in}}C_{\text{in}} - L_{\text{out}}C_{\text{out}} + \mu CV \tag{10}$$

Assuming that there is no influx of microalgae into the system, then $C_{\text{in}} = 0$, $C_{\text{out}} = C$. In this case, Eq. 10 can be expressed as:

$$V\frac{dC}{dt} + CL_{\text{in}} = \mu CV \tag{11}$$

From Eq. 11 we have,

$$\frac{dC}{dt} = \mu C - \frac{L_{\text{in}}}{V}C \tag{12}$$

Assuming that the dilution factor, $D = \frac{L_{\text{in}}}{V}$ is equal to 0, the temporal evolution of the microalgae concentration can be expressed as:

$$\frac{dC}{dt} = \mu C \tag{13}$$

Then, the Monod model (Eq. 1) can be substituted into Eq. 13 which becomes:

$$\frac{dC}{dt} = \mu_{\max}\frac{I}{K_{\text{I}} + I}C \tag{14}$$

3.5 Optimal Input Design Method

Input design is the process of designing informative inputs that maximize the information about the system dynamics using minimum cost and resources in a system identification experiment [28]. Therefore, an optimal input design can be formulated as a dynamic optimization problem [29], where in this study, this method was used to solve the optimal control problem of microalgae that maximizes the parameter sensitivity to determine the control function, $u(t)$.

3.5.1 State Equations of Monod Model

Let $C = x_1$ and $I = x_2$, the dynamic state equation as displayed in Eq. 14 become [17]

$$\dot{x}_1 = \mu_{\max} \frac{x_2}{K_I + x_2} x_1 \tag{15}$$

The second state equation is the control function, $u(t)$, where x_2 is the input factor defined as the following:

$$\dot{x}_2 = u(t). \tag{16}$$

Besides, two additional state variables, $x_3 = \frac{\partial x_1}{\partial K_\mathrm{I}}$ and $x_4 = \frac{\partial x_1}{\partial \mu_{\max}}$, were introduced to maximize the sensitivity of the parameters. By assuming that the parameter $\mu_{\max}$ and K_I are time-invariant, the dynamic state equation, Eq. 15 can be differentiated according to the parameters, respectively, generating two additional state equations which can be written as follows [27]:

$$\dot{x}_3 = \frac{\partial f}{\partial x_1} x_3 + \frac{\partial f}{\partial K_\mathrm{I}} = \frac{\mu_{\max}[x_2 x_3 (K_\mathrm{I} + x_2) - x_1 x_2]}{(K_\mathrm{I} + x_2)^2}, \tag{17}$$

$$\dot{x}_4 = \frac{\partial f}{\partial x_1} x_4 + \frac{\partial f}{\partial \mu_{\max}} = \frac{\mu_{\max} x_2 x_4 + x_1 x_2}{K_\mathrm{I} + x_2}, \tag{18}$$

where $f = \dot{x}_1$.

3.5.2 Cost Function

The optimal control problem maximizes the cost function, F, which can be defined as follows [27]:

$$F = \emptyset[x(t_f)] + \int_{t_0}^{t_f} L(x, u, t) dt, \tag{19}$$

where $\emptyset[x(t_f)]$ is the terminal condition and $L(x, u, t)$ is the running cost.

Additional state variables of the parametric sensitivity in the optimal input design problem were computed in time by maximizing the running cost, L [15]. The weighting factors of the parametric sensitivity were also considered for priority during the optimization of L. For the optimization of the trace of the Fisher information matrix (FIM), the quadratic cost function was defined as follows [27]:

$$F = \int_{t_0}^{t_f} \left(q_1 s_1^2 + q_2 s_2^2 + \ldots + q_n s_n^2 \right) dt \tag{20}$$

where n is the number of the additional state variable, s is the parametric sensitivity, and q is the weighting factor for the parametric sensitivity.

Besides, F can be added as a new additional state variable, which is known as the Mayer formulation to optimize itself in time. Therefore,

$$F = x_{n+1} = \int_{t_0}^{t_f} \left(q_1 s_1^2 + q_2 s_2^2 + \ldots + q_n s_n^2\right) dt \tag{21}$$

By comparing Eqs. 19 and 21, we obtained the following Eq. 22:

$$L(x, u, t) = q_1 s_1^2 + q_2 s_2^2 + \ldots + q_n s_n^2 \tag{22}$$

In this study, there are two parameters modeling the sensitivity, x_3 and x_4. Since the good choice for the weighting factors of the parameters sensitivity was $q_1 : q_2 = 1 : 0$ [27], only x_3 was maximized with a maximum running cost using the formula, $L(x, u, t) = x_3^2$.

3.5.3 Co-state Equations for Monod Model

The minimum scalar Hamiltonian function for the optimal control problem, $H(t)$, was defined as follows [27]:

$$\min \; H(t) = L(x(t), u(t), t) + \lambda_i^{\mathrm{T}}(t)\dot{x}_i(t) \tag{23}$$

where λ is the time-varying Lagrange multiplier vector.

Since maximizing L is equivalent to minimizing $-L$ [15],

$$\begin{aligned} H(t) &= -x_3^2(t) + \lambda_1(t)\dot{x}_1(t) + \lambda_2(t)\dot{x}_2(t) + \lambda_3(t)\dot{x}_3(t) \\ &= -x_3^2(t) + \lambda_1(t)\mu_{\max}\frac{x_2}{K_I + x_2}x_1 + \lambda_2(t)u(t) + \lambda_3(t) \\ &\times \frac{\mu_{\max}[x_2 x_3(K_{\mathrm{I}} + x_2) - x_1 x_2]}{(K_{\mathrm{I}} + x_2)^2} \end{aligned} \tag{24}$$

the Hamiltonian function was used to compute the co-state equations as follows:

$$\dot{\lambda}_1 = -H_{x_1}^{\mathrm{T}} = -\frac{\mu_{\max}x_2}{K_{\mathrm{I}} + x_2}\lambda_1 + \frac{\mu_{\max}x_2}{(K_{\mathrm{I}} + x_2)^2}\lambda_3 \tag{25}$$

$$\begin{aligned} \dot{\lambda}_2 &= -H_{x_2}^{\mathrm{T}} \\ &= \frac{\mu_{\max}x_1 x_2}{(K_{\mathrm{I}} + x_2)^2}\lambda_1 - \frac{\mu_{\max}x_1}{K_{\mathrm{I}} + x_2}\lambda_1 \\ &\quad - \frac{\mu_{\max}[x_3(K_{\mathrm{I}} + 2x_2) - x_1]}{(K_{\mathrm{I}} + x_2)^2}\lambda_3 \\ &\quad + 2\frac{\mu_{\max}[-x_1 x_2 + x_2 x_3(K_{\mathrm{I}} + x_2)]}{(K_{\mathrm{I}} + x_2)^3}\lambda_3 \end{aligned} \tag{26}$$

$$\dot{\lambda}_3 = -H^{\mathrm{T}}_{x_3 = 2x_3 - \frac{\mu_{max}x_2}{K_{\mathrm{I}} + x_2}\lambda_3} \tag{27}$$

3.5.4 Stationary Equations for Monod Model

The stationary equation used for optimal control problem in this study can be written as follows:

$$\frac{\partial H}{\partial u} = \lambda_2 \tag{28}$$

Since $u(t)$ does not appear in Eq. 28, it is a singular optimal control problem. Therefore, Eq. 28 was differentiated repeatedly with respect to t until $u(t)$ appears explicitly [30].

$$\begin{aligned}\frac{\partial}{\partial t}\left(\frac{\partial H}{\partial u}\right) &= \dot{\lambda}_2 \\ &= \frac{\mu_{\max} x_1 x_2}{(K_\mathrm{I} + x_2)^2}\lambda_1 - \frac{\mu_{\max} x_1}{K_\mathrm{I} + x_2}\lambda_1 \\ &\quad - \frac{\mu_{\max}[x_3(K_\mathrm{I} + 2x_2) - x_1]}{(K_\mathrm{I} + x_2)^2}\lambda_3 \\ &\quad + 2\frac{\mu_{\max}[-x_1 x_2 + x_2 x_3(K_\mathrm{I} + x_2)]}{(K_\mathrm{I} + x_2)^3}\lambda_3\end{aligned} \tag{29}$$

$$\frac{\partial^2}{\partial t^2}\left(\frac{\partial H}{\partial u}\right) = \frac{2}{(K_\mathrm{I} + x_2)^4}\mu_{\max}$$

$$[-K_\mathrm{I} x_3(K_\mathrm{I} + x_2)(x_3(K_\mathrm{I} + x_2) - u\lambda_3) + x_1(x_3$$

$$(K_\mathrm{I}{}^2 - x_2{}^2) + u(K_\mathrm{I}(K_\mathrm{I} + x_2)\lambda_1 + (-2K_\mathrm{I} + x_2)\lambda_3))] \tag{30}$$

Equations 28–30 form a system of stationary equations which are equal to zero. This system can be rewritten in the matrix form: $A\lambda^\mathrm{T} = 0$. Since λ cannot be zero, matrix A must be singular to solve the system [31]. As a result, $u(t)$ can be obtained by allowing the determinant of the matrix A to be equal to zero [32].

$$\mathrm{Det}(A) = -\frac{2K_\mathrm{I}{}^2 \mu_{\max}{}^2 u x_1^2}{(K_\mathrm{I} + x_2)^6} = 0 \tag{31}$$

Therefore, the control function, $u(t) = 0$ and thus $\dot{x}_2 = 0$.

All the computations for the optimal input design method were performed using the Mathematica (Wolfram Research, Inc., Version 11.3, Champaign, IL) software. The algorithm for performing these steps is summarized in the box below:

3.6 Taylor Series Method

The first-order and second-order Taylor series methods were applied to solve the state equations which are given by Eqs. 15–17 to compute the biomass concentration of microalgae. The first-order Taylor series method is also called the Euler method, and it can be written as follows [33]:

Coding of Optimal Input Design Method in Mathematica Software

Input state equations:

```
In[ ]:= x1'[t] = Umax * x1[t] * x2[t] / (KI + x2[t])
        x2'[t] = u[t]
        x3'[t] = Umax * (-x1[t] * x2[t] + x2[t] * x3[t] * (KI + x2[t])) / (KI + x2[t]) ^2
```

Input Hamiltonian function:

```
In[ ]:= H[t] = -x3[t] ^2 + λ1[t] * x1'[t] + λ2[t] * x2'[t] + λ3[t] * x3'[t]
```

Compute the co-state equations:

```
In[ ]:= λ1'[t] = -D[H[t], x1[t]]
        λ2'[t] = -D[H[t], x2[t]]
        λ3'[t] = -D[H[t], x3[t]]
```

Compute the system of stationary equations:

```
In[ ]:= eqn = Simplify[Table[D[D[H[t], u[t]], {t, i}], {i, 0, 2}]]
```

Find matrix A:

```
In[ ]:= A = Table[Coefficient[eqn[[i]], {λ1[t], λ2[t], λ3[t]}], {i, 3}]
```

Display matrix A:

```
In[ ]:= A // MatrixForm
```

Compute determinant of matrix A:

```
In[ ]:= Simplify[Det[A]]
```

Compute control function:

```
In[ ]:= OPTu = Simplify[Solve[Det[A] == 0, u[t]]]
```

$$x_m(t_{i+1}) = x_m(t_i) + h\, x_m{}'(t_i, x_{1,i}, x_{2,i}), \quad 1 \le m \le n \tag{32}$$

whereas the formula for the second-order Taylor series method is given by Eq. 33,

$$x_m(t_{i+1}) = x_m(t_i) + h\, x_m{}'(t_i, x_{1,i}, x_{2,i}) + \frac{h^2}{2!} x_m{}''(t_i, x_{1,i}, x_{2,i}), \quad 1 \le m \le n \tag{33}$$

where i is the number of iteration, h is the step size, and n is the number of the first-order differential equation.

In this study, the parameters of the Monod model were specified as $\mu_{\max} = 1.5331$ and $K_I = 73.8630$ which were estimated using the Levenberg–Marquardt method. Besides, h was specified as 0.5 and the time required for microalgae growth was specified at 14 days. The initial value of biomass concentration, x_1, light intensity, x_2, and parameter sensitivity, x_3, were specified as 1000, 324, and 0, respectively. Eq. 32 with $x_m{}' = \dot{x}_m$ and Eq. 33 with $x_m{}'' = \frac{\partial \dot{x}_m}{\partial x_m}$ were applied to compute the biomass concentration of microalgae. Tables 4 and 5 show the predicted biomass concentrations of *Botyrococcus* sp. microalgae for 14 days which were generated by the Euler method and the second-order Taylor series method, respectively. Figures 3 and 4 show the effect of time on the biomass concentration of microalgae computed using first-order and second-order Taylor series method, respectively. The algorithms for the Euler method and the second-order Taylor Series method implemented using the Matlab software are described in the following box.

4 Notes

1. In this chapter, arbitrary values are sometimes used as default initial values for the parameters. We refer to two arbitrary values. The first arbitrary value is equal to 1 and the second arbitrary value is equal to 0. A good initial guess corresponding to the definition of the parameter can be selected from the experimental data [5]. Therefore, the initial guess for $\mu_{\max}$ was specified as the maximum value of μ in the experimental data, whereas initial guess for K_I was the value of I when the value of μ was half of the $\mu_{\max}$ in the dataset [4]. The results generated using the Gauss–Newton method and the Leven-

Coding of Euler Method in Matlab Software

Iteration:

```
% compute number of subintervals
N=14;t0=0;h=0.5;
n=(N-t0)/h;

% set values of parameters and initial value of x
Umax=1.5331;KI=73.8630;
x1=1000;x2=324;x3=0;

% set initial t
t=t0;
% let X be the vector of [x1,x2,x3]
X=[x1;x2;x3];

% f1 is first derivative
f1=@(X,t) [(Umax*X(2)*X(1))/(KI+X(2));0;(Umax*(-
X(1)*X(2)+X(2)*(KI+X(2))*X(3)))/(KI+X(2))^2];
% iteration
for i=1:n
 X(:,i+1)=X(:,i)+h*f1(X(:,i),t);
```

Produce table:

```
t=t0:h:N;
fprintf('\tt \t\t\t\tx1 \t\t\t\tx2  \t\t\t\tx3\n')
fprintf('%5.1f %20.6f %20.6f %20.6f\n',[t;X])
```

Produce graph of biomass concentration for microalgae:

```
plot(t,X(1,:),'LineWidth',1.5);
xlabel('Time','Fontname','Time New Roman');
ylabel('Biomass concentration for microalgae','Fontname','Time New
Roman');
title({'Biomass Concentration for Microalgae by ';'Euler
Method'},'Fontweight','bold','Fontname','Time New Roman');
```

Coding of Second-Order Taylor Series Method in Matlab Software

Iteration:

```
% compute number of subintervals
N=14;t0=0;h=0.5;
n=(N-t0)/h;

% set values of parameters and initial value of x
Umax=1.5331;KI=73.8630;
x1=1000;x2=324;x3=0;

% set initial t
t=t0;
% let X be the vector of [x1,x2,x3]
X=[x1;x2;x3];

% f1 is first derivative, f2 is second derivative
f1=@(X,t) [(Umax*X(2)*X(1))/(KI+X(2));0;(Umax*(-
X(1)*X(2)+X(2)*(KI+X(2))*X(3)))/(KI+X(2))^2];
f2=@(X,t)[Umax*X(2)/(KI+X(2));0;(Umax*X(2)*(KI+X(2)))/((KI+X(2))^2)];
% iteration
for i=1:n
 X(:,i+1)=X(:,i)+h*f1(X(:,i),t)+(h^2*f2(X(:,i),t))/2;
```

Produce table:

```
t=t0:h:N;
fprintf('\tt \t\t\t\tx1 \t\t\t\tx2  \t\t\t\tx3\n')
fprintf('%5.1f %20.6f %20.6f %20.6f\n',[t;X])
```

Produce graph of biomass concentration for microalgae:

```
plot(t,X(1,:),'LineWidth',1.5);
xlabel('Time','Fontname','Time New Roman');
ylabel('Biomass concentration for microalgae','Fontname','Time New
Roman');
title({'Biomass Concentration for Microalgae by ';'Second Order Taylor
Series Method'},'Fontweight','bold','Fontname','Time New Roman');
```

berg–Marquardt method with a first arbitrary initial guess, μ_{max} and K_I that were specified at 1.0 are shown in Tables 6 and 7, respectively. From the results displayed in Tables 2, 3, 6, and 7, it can be concluded that the initial guess corresponding to the parameter definition was good with a smaller SSE of 0.281965 when compared to the first arbitrary initial guess. Hence, the iteration for both methods with the initial guess corresponding to the parameter definition converged faster compared to the first arbitrary initial guess.

Table 4
Biomass concentration of microalgae versus time by Euler method

Time (day)	Biomass concentration (g ℓ^{-1})
0	1000.000000
1	2638.157218
2	6959.873506
3	18,361.240523
4	48,439.839213
5	127,791.911448
6	337,135.153555
7	889,415.538709
8	2,346,418.023024
9	6,190,219.643298
10	16,330,772.631422
11	43,083,145.689010
12	113,660,111.762810
13	299,853,244.217261
14	791,060,000.497344

2. If the second arbitrary initial guess for both μ_{max} and K_I were specified at zero, the Gauss–Newton method showed a convergence problem where the Jacobian matrix was singular or close to singular. However, the Levenberg–Marquardt method performed well as shown in Table 8.

Table 5
Biomass concentration of microalgae versus time by second-order Taylor series method

Time (day)	Biomass concentration (g ℓ^{-1})
0	1000.000000
1	2638.566757
2	6961.363474
3	18,365.580833
4	48,451.699173
5	127,823.609425
6	337,219.187344
7	889,637.642594
8	2,347,004.377530
9	6,191,766.948209
10	16,334,855.074580
11	43,093,916.225432
12	113,688,526.540751
13	299,928,207.278316
14	791,257,765.247468

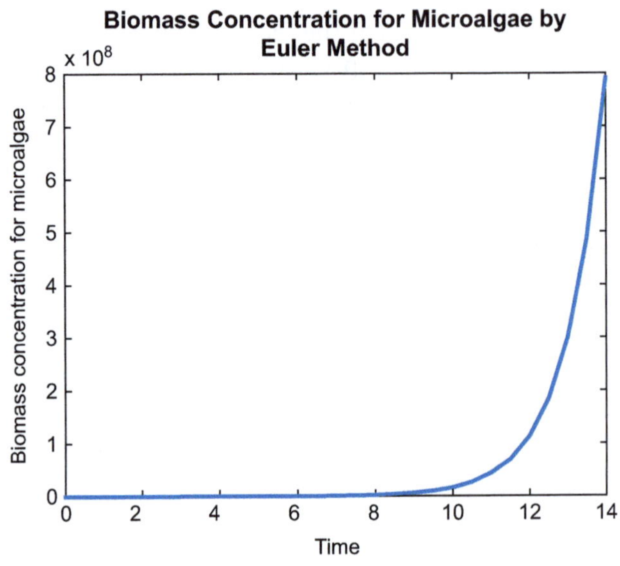

Fig. 3 Biomass concentration for microalgae by Euler method

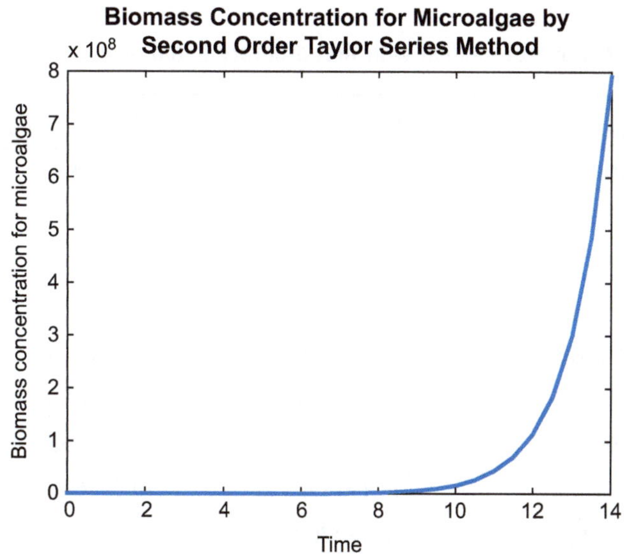

Fig. 4 Biomass concentration for microalgae by second-order Taylor series method

Table 6
Result of Gauss–Newton method with first arbitrary initial guess

k	μ_{max}	K_I	SSE	Error
0	**1.000000**	**1.000000**	**0.495456**	–
1	1.015309	2.879044	0.371017	0.124439
2	1.037327	4.872861	0.338130	0.032886
3	1.060439	7.073609	0.322348	0.015782
4	1.090499	10.236845	0.306534	0.015814
5	1.144439	16.534261	0.278704	0.027830
6	1.258471	31.314595	0.229109	0.049595
7	1.417944	54.611697	0.193805	0.035304
8	1.512041	70.129956	0.187717	0.006088
9	1.531854	73.638981	0.187518	0.000199
10	1.533117	73.861392	0.187517	0.000001

Reproduced from [21]

Table 7
Result of Levenberg–Marquardt method with first arbitrary initial guess

k	μ_{max}	K_I	SSE	Error
0	**1.000000**	**1.000000**	**0.495456**	–
1	1.000394	2.442905	0.384340	0.111115
2	1.028775	4.248653	0.345005	0.039336
3	1.052092	6.274844	0.327161	0.017844
4	1.078613	8.950390	0.312638	0.014522
5	1.121154	13.741384	0.290688	0.021950
6	1.209978	24.834796	0.247765	0.042923
7	1.364871	46.491597	0.201578	0.046187
8	1.490463	66.422708	0.188340	0.013238
9	1.529316	73.184344	0.187523	0.000817
10	1.533014	73.843809	0.187517	0.000006
11	1.533158	73.868228	0.187517	0.000000

Table 8
Result of Levenberg–Marquardt method with second arbitrary initial guess

k	μ_{max}	K_I	SSE	Error
0	**0.000000**	**0.000000**	**5.408471**	–
1	0.886023	0.000000	0.682547	4.725924
2	1.000554	1.847700	0.413696	0.268851
3	1.025266	3.801943	0.351403	0.062293
4	1.047694	5.838767	0.330126	0.021278
5	1.072833	8.338367	0.315656	0.014470
6	1.110733	12.524588	0.296093	0.019563
7	1.187418	21.913695	0.257762	0.038331
8	1.333473	41.868936	0.207920	0.049842
9	1.473684	63.607444	0.189128	0.018793
10	1.526520	72.684056	0.187536	0.001591
11	1.532880	73.820614	0.187517	0.000020
12	1.533153	73.867488	0.187517	0.000000

References

1. Guiry MD (2012) How many species of algae are there? J Phycol 48(5):1057–1063
2. Carlsson AS, Van Beilen JB, Möller R, Clayton D (2007) Micro-and macro-algae: utility for industrial applications. University of York, York
3. Lee E, Jalalizadeh M, Zhang Q (2015) Growth kinetic models for microalgae cultivation: a review. Algal Res 12:497–512
4. Markou G, Angelidaki I, Georgakakis D (2012) Microalgal carbohydrates: an overview of the factors influencing carbohydrates production, and of main bioconversion technologies for production of biofuels. Appl Microbiol Biotechnol 96(3):631–645
5. Blanken W, Postma PR, de Winter L, Wijffels RH, Janssen M (2016) Predicting microalgae growth. Algal Res 14:28–38
6. Mulder C, Hendriks AJ (2014) Half-saturation constants in functional responses. Glob Ecol Conserv 2:161–169
7. Johnson ML, Faunt LM (1992) [1] Parameter estimation by least-squares methods. In: Methods in enzymology, vol 210. Academic Press, London, pp 1–37
8. Björck Å (1996) Numerical methods for least squares problems. Society for Industrial and Applied Mathematics, Philadelphia, PA
9. Poe WA, Mokhatab S (2016) Modeling, control, and optimization of natural gas processing plants. Gulf Professional Publishing, Houston, TX
10. Hadiyanto H, Esveld DC, Boom RM, Van Straten G, Van Boxtel AJB (2008) Control vector parameterization with sensitivity based refinement applied to baking optimization. Food Bioprod Process 86(2):130–141
11. Lai WH, Kek SL, Tay KG (2017) Solving nonlinear least squares problem using Gauss-Newton method. Int J Innov Sci Eng Technol 4:258–262
12. Motulsky H, Christopoulos A (2004) Fitting models to biological data using linear and nonlinear regression: a practical guide to curve fitting. Oxford University Press, Oxford
13. Mahanta DJ, Borah M, Saikia P (2014) Study on kinetic models for analysing the bacterial growth rate. AIJRSTEM 8(1):2328–3491
14. Ruckstuhl A (2010) Introduction to nonlinear regression. IDP Institut für Datenanalyse und Prozessdesign. ZHAW Zürcher Hochschule für Angewandte Wissenschaften. https://stat.ethz.ch/~stahel/courses/cheming/nlreg10E.pdf
15. Keesman KJ, Stigter JD (2002) Optimal parametric sensitivity control for the estimation of kinetic parameters in bioreactors. Math Biosci 179(1):95–111
16. Keesman KJ, Stigter JD (2003, September) Optimal input design for low-dimensional systems: an haldane kinetics example. In: 2003 European control conference (ECC). IEEE, Washington, DC, pp 1583–1588
17. See JJ, Jamaian SS (2019) Kinetic parameter estimation for microalgae growth model using dynamic optimization approach. Commun Comput Appl Math 1(1)
18. Gani P, Sunar NM, Matias-Peralta H, Jamaian SS (2016) Effects of different culture conditions on the phycoremediation efficiency of domestic wastewater. J Environ Chem Eng 4 (4):4744–4753
19. Monod J (1949) The growth of bacterial cultures. Annu Rev Microbiol 3(1):371–394
20. Eriksson J (1996) Optimization and regularization of nonlinear least squares problems. Verlag nicht ermittelbar, Jerusalem
21. See JJ, Jamaian SS, Salleh RM, Nor ME, Aman F (2018, April) Parameter estimation of Monod model by the least-squares method for microalgae Botryococcus braunii sp. J Phys Conf Ser 995:012026
22. Herman W. Applications of the Gauss-Newton Method. https://ccrma.stanford.edu/~wherman/tulane/gauss_newton.pdf. Accessed 15 Aug 2020
23. Madsen K, Nielsen HB, Tingleff O (2004) Methods for non-linear least squares problems
24. Ranganathan A (2004) The Levenberg-Marquardt algorithm. Tutorial on LM Algorithm 11(1):101–110
25. Mittrapiyanuruk P (2006) A memo on how to use the Levenberg-Marquardt algorithm for refining camera calibration parameters. Website: http://cobweb.ecn.purdue.edu/~kak/courses-i-teach/ECE661/HW5 LM handout.pdf.
26. Yu H, Wilamowski BM (2011) Levenberg-Marquardt training. Ind Electron Handb 5 (12):1
27. Hermanto MB (2009) Identification of algae growth kinetics. Thesis Syst Control
28. Kumar A, Jan NM, Narasimhan S (2018) Economically optimal input design approach for identification of constrained processes. Ind Eng Chem Res 57(20):6956–6967
29. Balsa-Canto E, Alonso AA, Banga JR (2010) An iterative identification procedure for

dynamic modeling of biochemical networks. BMC Syst Biol 4(1):11

30. Powers WF (1980) On the order of singular optimal control problems. J Optim Theory Appl 32(4):479–489
31. Kirk DE (2004) Optimal control theory: an introduction. Courier Corporation, USA
32. Stengel RF (1994) Optimal control and estimation. Courier Corporation, USA
33. Atkinson K, Han W, Stewart DE (2011) Numerical solution of ordinary differential equations, vol 108. John Wiley & Sons, Hoboken, NJ

Chapter 7

Automatic Assembly and Calibration of Models of Enzymatic Reactions Based on Ordinary Differential Equations

Jure Stojan, Milan Hodošček, and Dušanka Janežič

Abstract

Enzymatic reactions have been studied for more than a 100 years. Indeed, isolation of enzymes from biological materials is no longer the main source of enzymes today, as they are now largely produced using recombinant technology, or can even be synthesized from scratch. Studies of the three-dimensional structures of enzymes can provide answers to many questions, but the kinetics of enzymatic reactions is the only method that can lead to better understanding of their function. The complexity of high-throughput analysis of progress curves of data obtained can only be achieved through numerical solutions of a suitable system of ordinary differential equations. We have developed the freely available server ENZO: a web tool for derivation and evaluation of kinetic models of enzyme-catalyzed reactions (http://enzo.cmm.ki.si/). ENZO can be used for simultaneous analysis of a series of progress curves obtained under many different conditions. In this chapter, we exemplify the principles and possibilities of this type of high-throughput analysis.

Key words Kinetic models, Enzyme-catalyzed reactions, Ordinary differential equations, ENZO

1 Introduction

After more than 100 years of studies of enzymes, their kinetics characterization still provides the best description of their actual function [1–4]. During this time, scientists have put enormous efforts into the development of techniques for isolation and purification of proteins, which have enabled extensive and specific investigations to be carried out on particular enzymes. Indeed, great progress has been made through the development of methods in molecular biology [5–7], which have allowed for the production of wild-type and mutated enzymes. Additionally, the frequent availability of large amounts of pure protein have promoted new investigational methods, such as structure determination in particular [8–10]. As it turns out, it is the combination of structural and kinetic studies that provides the most effective way to uncover the mechanisms of enzymatic reactions [11–13].

Quentin Vanhaelen (ed.), *Computational Methods for Estimating the Kinetic Parameters of Biological Systems*, Methods in Molecular Biology, vol. 2385, https://doi.org/10.1007/978-1-0716-1767-0_7,

Classical enzyme kinetics started with the monumental paper of Michaelis and Menten [14]. Although, they measured the time-course of the reaction, their analysis was based on information from the initial portions of the reaction curve. They had to sacrifice the majority of the information carried by each measured curve, as 100 years ago evaluation of mechanism-specific kinetic parameters from progress curves was not possible, due to the lack of modern computing resources. Moreover, in most cases, the required kinetic integrated rate equations were impossible to derive at all, let alone in their explicit forms. This is also true for the best-known equation of kinetics, the Michaelis–Menten equation. Many attempts at progress curve analysis were initiated over the past century [15], and in 2013, a very good estimate based on the Lambert omega function was presented by Goličnik [16]. Unfortunately, this equation can only be used for reaction mechanisms where only the substrate or product concentrations change, while all of the other participants need to follow zero-order reaction kinetics [16, 17].

As many enzymatic reactions are inhibited by the product that results, and more and more relevant inhibitors function through slow or tight binding phenomena, numerical methods have become the analytical methods of choice. This all started in 1983, with the excellent contribution by Yamaoka and Nakagawa [18]. They were the first to combine the numerical integration method for solving the differential equation system that underlies the hypothetical reaction mechanism with a nonlinear regression algorithm, with the aim to prove the postulated reaction mechanism and to obtain the best estimates for the characteristic parameters. In 1996, Kuzmic introduced DynaFit [19], a versatile suite for analysis of kinetic data, in terms of initial rates, equilibrium studies, and analysis of progress curve data.

In 1997, we presented a computer program for the evaluation of time-resolved kinetic data that used a semi-implicit numerical integration algorithm to calculate theoretical curves, the coefficients of which were subsequently fitted to the corresponding experimental progress curves [20]. This used a Gauss–Newton regression algorithm of a modified computer program originally developed by Duggleby [21]. Our program was later extended by a comprehensive graphical interface, which constructs differential equations from reaction schemes drawn up, which we have named ENZO [22] (http://enzo.cmm.ki.si/).

In this present contribution, we explain the objectives of progress curve analysis and discuss its advantages and disadvantages compared to initial rate data analysis. To illustrate this, we present some typical examples where progress curve experiments and analyses can distinguish reaction mechanisms, while the corresponding initial rate analyses cannot.

2 Comparisons Between Experimental Initial Rate Data and Complete Progress Curves

As was originally written by Markus and Plesser [23], six progress curves carry the information equivalent to 128 initial rate measurements (*see* **Note 1**). However, what is actually meant by the term "progress curves"? It is clear that the progress of a reaction is tracked by the transformation of the substrate(s) into the product (s), such that each reaction participant can be identified. The same is true for the determination of initial rates, although these are followed within the time it takes to use a negligible amount of substrate and produce an insignificant amount of product. During this period, the measured curve is usually linear, and so the rate is simply its slope.

However, if the reaction is allowed to continue until the entire substrate is used up or until equilibrium is reached, any analysis must take these changes into account. Here it is possible to understand the main advantages of the information contained in the so-called complete progress curves: during the course of the reaction, the reactant concentrations change continuously for each individual measurement, while one initial rate measurement is performed on a single selected set of reactant concentrations.

Usually, the strongest disadvantage of progress curve data is considered to be systematic error due to errors in adding the reactants. This is true to a certain extent when applied to the enzyme involved. A small difference in the amount of the enzyme added can have significant effects on the overall rate, and thus on the time until the end of the reaction. However, this can be largely avoided by repeating the time sweeps at different multiples of the added enzyme. In this way, these commonly used repetitions increase the reliability of the data, although the linear dependence of the reaction rate also needs to be checked (*see* **Note 2**). If a nonlinear dependency is found, it is often the result of an incorrectly used reaction mechanism, rather than improper technical handling. On the other hand, with the measurements carried out to the end of the reaction, this yields an exact concentration of the added reactants in thermodynamic irreversible reactions [24], or the ratios when the equilibrium reaction is measured [12].

When inhibited reactions are investigated, the initial rate data can easily give the false impression that the inhibitory compound has no effects. This is particularly true when irreversible or slow tight binding inhibitors are studied, which can show identical initial slopes, sometimes even at high concentrations. Again, only a single measurement of the progress curve can show the opposite, and a few additional experiments are sufficient to provide sound quantitative characterization [24]. On the other hand, if quasi-classical fast-acting inhibitors are considered, an initial rate analysis appears to be sufficient. Sometimes, however, an inhibitor can act quickly, and its effects can become enhanced with time. Indeed, measurement of the progress curve would reveal such unusual behavior [17].

Nevertheless, when conducting experiments with initial rates, it would be very convenient to follow at least a few measurements to their conclusion, as the information contained is of great importance.

2.1 The Essence of Progress Curve Analysis Using Numerical Approaches

With their origins being in physical chemistry, enzyme kinetics tries to describe the reaction mechanisms using mathematical equations in their explicit forms. For this reason, researchers have searched for analytical tricks to reach solutions and have postulated explanations of enzyme actions. However, it was soon realized that even the simplest equation in enzyme kinetics, the Michaelis–Menten equation, cannot be evaluated unless it is transformed into its double reciprocal form [25]. The estimation of the two parameters in the Michaelis–Menten equation was reached by drawing a line through the points using a ruler. Unfortunately, the fit of the parameters determined with the original Michaelis–Menten hyperbola were seldom checked according to the measured experimental data.

However, at the beginning of the computer era, a long-known nonlinear regression algorithm was implemented in programs, and much more reliable evaluations followed. During this period, analytical solutions of putative reaction mechanisms (*see* **Note 3**) in the form of explicit equations were successfully tested and evaluated according to the experimental data collected. With increasing knowledge of the structure and actions of enzymes, however, it became clear that for many of these, the experimental information could not be explained by classical kinetic equations. Furthermore, the number of initial rate measurements required to reliably characterize even a simple reversible reaction with two substrates rapidly increased to hundreds [12].

On the other hand, much more valuable information can be obtained from only a few progress curves. Although progress curve measurement can take longer than initial rate measurements, this can be compensated for by increasing the enzyme concentration, for instance, on a stopped-flow apparatus. However, the benefit of the considerable amount of information that comprises progress curves can only be accessed and correctly understood if the appropriate analysis is applied. Although it is sometimes possible to solve the system of differential equations analytically, it becomes clear that numerical solutions using a computer would enormously simplify the task, especially in cases where an appropriate reaction mechanism is being searched for from among many alternative reaction mechanisms (*see* **Note 4**).

Solutions for such analyses have been put forward by several authors [18–20], and in all cases, the general principles have been similar (*see* **Note 5**):

1. After working out a putative reaction scheme, the corresponding differential equations must be derived (*see* **Note 6**).
2. A numerical integration algorithm is used to calculate the time-course of each participating species, using the first estimates of all of the kinetic parameters under consideration (*see* **Note 7**).
3. Finally, an iterative nonlinear least squares regression protocol is used to find the parameters of the differential equations that can best reproduce the actual experimental data, using theoretically calculated time-courses of the corresponding species.

Although in most cases the experimental data represent the time-course of only one species in a reaction, usually as the change in the substrate or product, such a program can simultaneously include the curves for all of the participating species for which data are available. For instance, if the substrate, the intermediate, and the product absorb light at different wavelengths, this does not include only the qualitative mechanistic information, whereby all three of the curves can be used in simultaneous evaluation of the reaction scheme being modeled (*see* **Note 8**). If such an analysis is successful, the truth is much closer than the myth, as expressed poetically by Gibson [26], the inventor of the stopped-flow apparatus (*see* **Note 9**).

3 Exemplifying the Power of Numerical Analysis of Progress Curve Data

A thorough kinetic analysis of an enzyme requires a very elaborate step-by-step procedure that takes into account the possible reaction mechanisms with regard to the enzyme function and structure. On the other hand, the development of powerful experimental techniques for the production of discovered enzymes, and especially for synthesis of inhibitors, requires high-throughput characterization. Nowadays, most of this work is computerized, and even kinetic data acquisition can be robotized. However, evaluation of all this information cannot be fully automated.

In light of these new requirements, we recently proposed a method for rapid simultaneous mechanistic and quantitative descriptions of newly developed enzyme inhibitors, using the web application ENZO [24]: http://enzo.cmm.ki.si/kinetic.php?uwd=170623269&load=true. The web tool ENZO [22] can be run on a freely accessible public web server, and it automatically generates differential equations from the reaction scheme drawn up, and subsequently fits the coefficients of the differential equations, to thus reproduce the experimental progress curves theoretically. In this way, several different or alternative reaction schemes can be evaluated rapidly, and the optimal one for the respective enzymatic reaction can be found.

The above-mentioned presentation [24] covers single-step evaluation of the actions of various inhibitors, as classical reversible, irreversible, slow binding, and slow tight binding. The graphical interface that generates the differential equations from the reaction schemes drawn up allow for rapid assessment of any rational reaction mechanism. The use is straightforward and self-explanatory, so that it can also be performed by non-experts.

We conclude this chapter by suggesting an experiment and the analysis of progress curves for testing reaction mechanisms, by checking the dependence of the activity on the enzyme, substrate, and product concentrations. For this purpose, hydrolysis of the artificial substrate butyrylthiocholine by purified human butyrylcholinesterase is detected in the presence of a surplus of the detecting reagent, 5,5′-dithiobis-2-nitrobenzoic acid (DTNB), by following the yellow color produced through the fast reaction between the product thiocholine and the DTNB [27]. We then perform two subsequent experiments:

(a) The hydrolysis is measured at four different enzyme concentrations [28].

(b) After complete hydrolysis of the initially present substrate by the added enzyme at a given concentration, negligible amounts of highly concentrated substrate were added seven times, with each addition after the complete hydrolysis of the previous addition (*see* Fig. 1; for clarity, only four additions are depicted; *see* also Fig. 3).

It has been shown many times in the literature that at low substrate concentrations, the kinetics of human butyrylcholinesterase is hyperbolic [29]. Therefore, applying the reaction mechanism shown on the scheme depicted in Fig. 2 gives the following set of five differential equations:

$$dE/dt = -(k_{cat}/K_M)*E*S - k_2*E*P + k_3*EP + k_{cat}*EA; \quad (1)$$

$$dEA/dt = (k_{cat}/K_M)*E*S - k_{cat}*EA; \quad (2)$$

$$dEP/dt = k_2*E*P - k_3*EP; \quad (3)$$

$$dS/dt = -(k_{cat}/K_M)*E*S; \quad (4)$$

$$dP/dt = -k_2*E*P + (k_{cat}/K_M)*E*S + k_3*EP. \quad (5)$$

These should reproduce theoretically all of the experimental curves in Fig. 1, simultaneously. The analysis is carried out by drawing up reaction scheme depicted in Fig. 2 with the ENZO server (Fig. 3), which automatically derives these equations (Eqs. 1–5; Fig. 4).

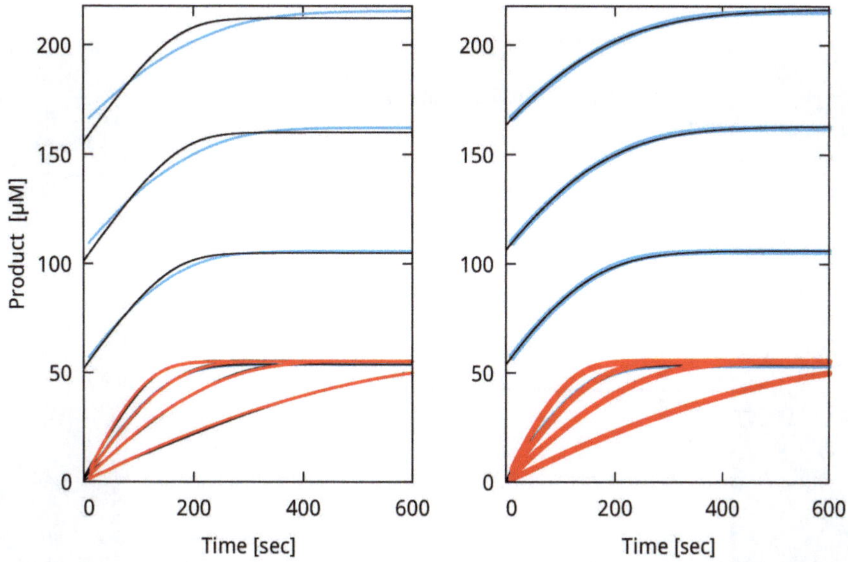

Fig. 1 Progress curves for hydrolysis of butyrylthiocholine by purified human butyrylcholinesterase. Red curves, measurements at one substrate concentration and four enzyme concentrations; blue curves, measurements at a single enzyme concentration, with the substrate added in aliquots, several times subsequently after the complete hydrolysis of the previous addition; black curves, theoretical curves obtained through the ENZO analysis using differential equations derived from reaction scheme depicted in Fig. 2 without the product inhibition step (left side), and with the complete scheme depicted in Fig. 2 (right panel), respectively. For clarity, only four blue curves are shown; the complete experimental data and the real-time analysis can be run through the following link: http:/enzo.cmm.ki.si/kinetic.php?uwd=200417663&load=true (click Set Parameters, Start)

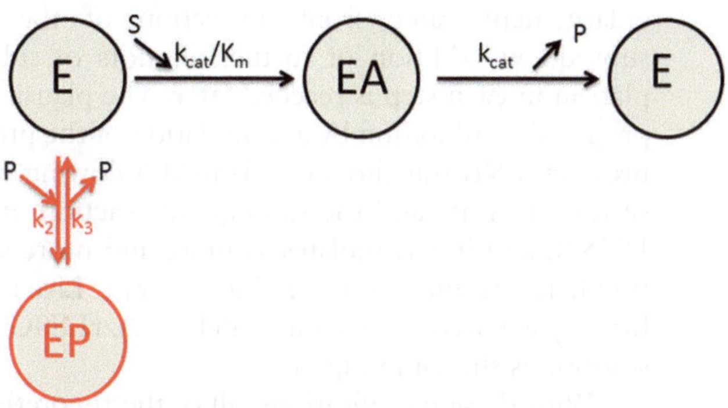

Fig. 2 E, enzyme; S, substrate; EA, enzyme–substrate intermediate; P, product; K_P is shown as k_3/k_2. Black, the Van Slyke–Cullen reaction Scheme [30]; red, the product inhibition step

In the first step, however, we used just the Van Slyke–Cullen mechanism [30], which does not include product inhibition, and thus no formation of an enzyme–product complex (EP). Indeed, the fits were not satisfactory (Fig. 1, left panel). The complete reaction summarized in Fig. 2, however, results in perfect

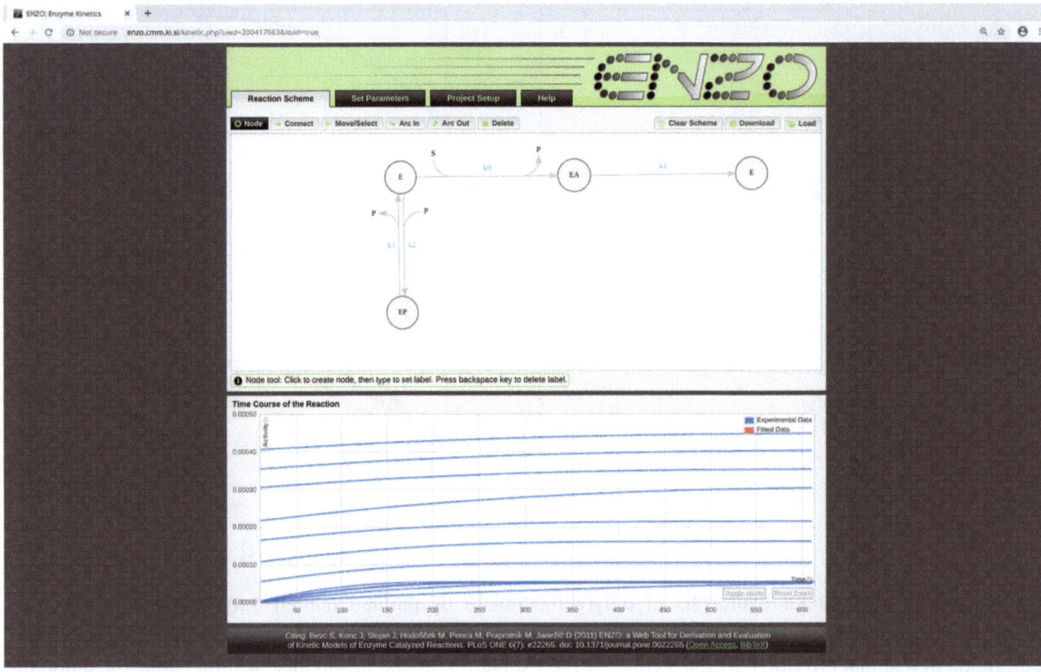

Fig. 3 Screenshot that depicts the reaction scheme where E is the enzyme, S the substrate, EA is an enzyme–substrate intermediate, and P represents products. In the analysis, we try to find two or three characteristic parameters: K_m (k_1/k_0), k_{cat} (k_1) and K_P (k_3/k_2). Therefore, this simple Van Slyke–Cullen reaction mechanism [30] with or without the product inhibition step can be implemented and the data submitted to the ENZO web server

agreement between experiment and theory. The idea for this enlargement came from inspection of the curves with the subsequent addition of further aliquots of substrate, where the plateau in each step is reached later. The plausible cause for this is progressive inhibition by accumulation of the product. Indeed, the product TNB-thiocholine is formed following enzymatic cleavage of the substrate and the subsequent reaction of the product with DTNB, and it accumulates as more and more substrate is hydrolyzed. The result is shown in Fig. 1, right. Live fitting can be run at https://www.youtube.com/watch?v=AzrLF0GrM2M and the solution is shown in Fig. 4.

With these modifications, all of the theoretical curves can correctly reproduce the course of the experimental data, which confirms the hypothesis that the product TNB-thiocholine inhibits butyrylthiocholine hydrolysis by the purified human butyrylcholinesterase. From inspection of the experimental data, and according to the K_P determined, we can conclude that there is significant product inhibition in this system. As cholinesterases are in general also inhibited by their substrate, the complexity of the events in an enzyme reaction can be understood [13], along with the need for numerical analytical tools, as only these can handle the task.

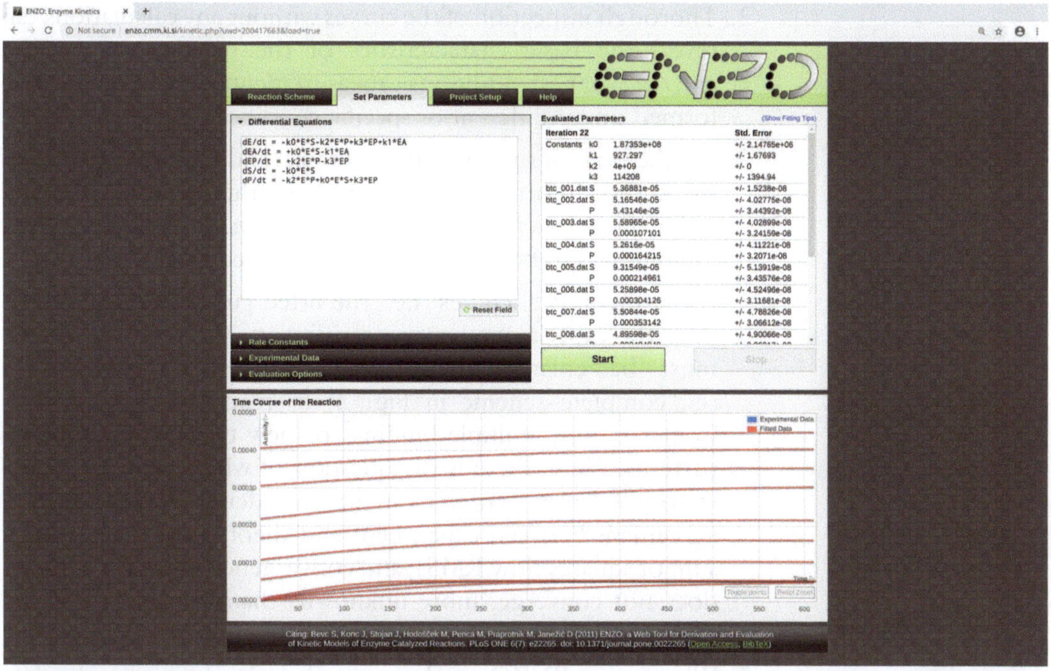

Fig. 4 Screenshot of the final result obtained by ENZO, with the improved fit that includes progressive product inhibition

4 Notes

Here we propose an unconventional kinetic analysis in basic biochemical research. Instead of the classical initial rate approach, we describe high-throughput data acquisition in the form of progress curves, which are followed by straightforward quantitative evaluation. For this evaluation, we have implemented a numerical integration algorithm that can solve systems of differential equations associated with nonlinear regression, as a web application called ENZO.

To expand the field of potential users, an interface has been added that generates differential equations from reaction schemes that are drawn up, thus avoiding cumbersome mathematical derivations, and typing and debugging. By using this tool, it is possible to concentrate on specific scientific questions, instead of spending time on technical problems. To increase the effectiveness when entering the field, we suggest similar reasoning for the kinetic data acquisition and the analysis as in the example given above. For clarity, this can be summarized in the following nine procedural notes:

1. Progress curves should always be collected until the complete conversion of the reactants or equilibrium is reached.

2. Thorough inspection of the curves without any calculations can reveal the basic characteristics of the reaction mechanism and/or suggest additional measurements that can be performed under different or specific conditions.
3. On the basis of the inspection analysis, and together with all of the known information obtained from the literature, the first rational reaction scheme (cf. scheme in Fig. 2, without product inhibition) can be constructed and checked by ENZO.
4. If the theoretical curves (i.e., the fitted curves) do not reproduce the experimental curves satisfactory, the reaction scheme must be enlarged by either adding new intermediates (cf. complete scheme in Fig. 2) or reaction steps, such as isomerization, multiple binding, or new types of modulation patterns. Here, there is the question of what might be "satisfactory"? Several criteria can be considered, such as optical agreement between experiment and theory, low final residual sum of squares (low RMS), and convergence of the fit. If the fit does not converge, more unknowns than information is the plausible cause, which is known as overfitting.
5. When using ENZO, a scheme needs to be drawn up, the data needs to be submitted, and all border conditions need to be filled (i.e., the concentrations of all of the participants), as well as initial estimates of all of the rate constants. These initial estimates can indeed be problematic, especially when very complex reaction mechanisms are tested [31]. At this point, it can be helpful to correlate the time scale of the inspected events with the parameter estimates, while taking into account, of course, the order of the reaction steps and the concentrations used.
6. The ENZO application allows simultaneous managing of experimental data that represent different reaction intermediates (e.g., in compartmental analysis; [18]), and therefore each progress curve should be indicated accordingly.
7. The evaluation should START with the simulation (with all border conditions and rate constants FIXed), and later, the constraints can be released one at a time.
8. The final choice is the simplest reaction scheme that provides good agreement between the calculated and experimental curves, but more importantly, the results must reflect a rational interpretation of the events during the course of the reaction.
9. There is always the need to remain alert to check and repeat curves that appear to be off line.

Acknowledgments

This work is supported by the Slovenian Research Agency (ARRS) grants P1-0170 (J.S.); L7-8269 (M.H.); J1-1715 (D.J.), and J1-9186 (D.J). We gratefully acknowledge the support of NVIDIA Corporation with the donation of the GPU hardware (two Tesla K80, one Tesla K40) that was used in this study.

References

1. Laidler KJ, Bunting PS (1973) The chemical kinetics of enzyme reactions. Clarendon Press: Oxford University Press, Oxford
2. Fersht A (1985) Enzyme structure and mechanism. W.H. Freeman, San Francisco
3. Segel IH (1993) Enzyme kinetics: behavior and analysis of rapid equilibrium and steady-state enzyme systems. John Wiley & Sons, New York
4. Cornish-Bowden A (2012) Fundamentals of enzyme kinetics. John Wiley & Sons, New York
5. Gräslund S, Nordlund P, Weigelt J, Hallberg BM, Bray J, Gileadi O et al (2008) Protein production and purification. Nat Methods 5 (2):135–146
6. Grange JM, Fox KR, Morgan NL (eds) (1990) Genetic manipulation: techniques and applications. John Wiley & Sons, New York
7. Bio-Rad Explorer: protein expression and purification series. explorer.bio-rad.com
8. Wüthrich K (1990) Protein structure determination in solution by NMR spectroscopy. J Biol Chem 265(36):22059–22062
9. Kendrew JC, Bodo G, Dintzis HM, Parrish RG, Wyckoff H, Phillips DC (1958) A three-dimensional model of the myoglobin molecule obtained by X-ray analysis. Nature 181 (4610):662–666
10. Cressey D, Callaway E (2017) Cryo-electron microscopy wins chemistry Nobel. Nature 550 (7675):167
11. Dvir H, Silman I, Harel H, Rosenberry TL, Sussman JL (2010) Acetylcholinesterase: from 3D structure to function. Chem Biol Interact 187(1–3):10–22
12. Çassetta A, Krastanova I, Kristan K, Brunskole Švegelj M, Lamba D, Lanišnik Rižner T, Stojan J (2012) Insights into subtle conformational differences in the substrate-binding loop of fungal 17β-hydroxysteroid dehydrogenase: a combined structural and kinetic approach. Biochem J 441(1):151–160
13. Colletier JP, Fournier D, Greenblatt HM, Stojan J, Sussman JL, Zaccai G, Silman I, Weik M (2006) Structural insights into substrate traffic and inhibition in acetylcholinesterase. EMBO J 25(12):2746–2756
14. Michaelis L, Menten ML (1913) Die Kinetik der Invertinwirkung. Biochem Z 49:335–369. (see translation by Goody RS, Johnson KA. 2013)
15. Duggleby RG, Morrison JF (1978) Progress curve analysis in enzyme kinetics. Model discrimination and parameter estimation. Biochim Biophys Acta 526(2):398–409
16. Goličnik M (2013) The integrated Michaelis-Menten rate equation: déjà vu or vu jàdé? J Enzyme Inhib Med Chem 28(4):879–893
17. Stojan J, Pavlic MR (1991) On the inhibition of cholinesterase by D-tubocurarine. Biochim Biophys Acta 1079(1):96–102
18. Yamaoka K, Nakagawa T (1983) A nonlinear least squares program based on differential equations, MULTI (RUNGE), for microcomputers. J Pharmacobiodyn 6(8):595–606
19. Kuzmic P (1996) Program DYNAFIT for the analysis of enzyme kinetic data: application to HIV proteinase. Anal Biochem 237 (2):260–273
20. Stojan J (1997) Analysis of progress curves in an acetylcholinesterase reaction: a numerical integration treatment. J Chem Inf Comput Sci 37:1025–1027
21. Duggleby RG (1984) Regression analysis of nonlinear Arrhenius plots: an empirical model and a computer program. Comput Biol Med 14(4):447–455
22. Bevc S, Konc J, Stojan J, Hodoščck M, Penca M, Praprotnik M, Janežič D (2011) ENZO: a web tool for derivation and evaluation of kinetic models of enzyme catalyzed reactions. PLoS One 6(7):e22265
23. Markus M, Plesser T (1980) In: Endrenyi L (ed) Kinetic data analysis: design and analysis of enzyme and pharmacokinetic experiments. Plenum Press, New York, pp 317–339
24. Stojan J (2017) Rapid mechanistic evaluation and parameter estimation of putative inhibitors

in a single-step progress-curve analysis: the case of horse butyrylcholinesterase. Molecules 22 (8):1248

25. Lineweaver H, Burk D (1934) The determination of enzyme dissociation constants. J Am Chem Soc 56(3):658–666
26. Gibson QH (1988) Chapter 3. Rapid reaction methods in biochemistry. In: van Deenen LLM, Neuberger A (eds) Modern Physical Methods in Biochemistry, Part B, Volume 11B, 1st edn. Elsevier Science
27. Ellman GL, Courtney KD, Andres V Jr, Featherstone RM (1961) A new and rapid colorimetric determination of acetylcholinesterase activity. Biochem Pharmacol 7:88–95
28. Stojan J (2019) The mechanism and benefit of human butyrylcholinesterase activation by what would otherwise be inhibitors. Chem Biol Interact 308:350–356
29. Rosenberry TL (1975) Acetylcholinesterase. Adv Enzymol 43:103–218
30. Kuzmic P (2009) Application of the Van Slyke-Cullen irreversible mechanism in the analysis of enzymatic progress curves. Anal Biochem 394 (2):287–289
31. Škedelj V, Fonović UP, Molek P, Magnet S, Mainardi JL, Blanot D, Gobec S, Stojan J, Zega A (2019) Kinetic mechanism of *Enterococcus faecium* d-aspartate ligase. Biochimie 158:217–223

Chapter 8

Data Processing to Probe the Cellular Hydrogen Peroxide Landscape

Fernando Antunes and Paula Brito

Abstract

Hydrogen peroxide (H_2O_2) regulates signaling pathways by modulating the activity of redox-sensitive proteins denominated redox switches. The magnitude of the transient variations in localized H_2O_2 pools during signaling events and how these variations impact redox switches present in the cell remain elusive. A canonical model with two chemical reactions comprising the oxidation/reduction cycle of a redox switch is described. The model is dimensionless with respect to the redox switch concentration. Thus, the time-series data required to apply the equations deduced is the percentage of oxidation of a redox switch, avoiding the application of absolute concentrations that are often difficult to measure experimentally. Here, we describe detailed protocols for the processing of experimental data with the canonical model to probe the absolute concentrations of H_2O_2 found in the vicinity of redox switches and probes, as well as the kinetic parameters that describe the reduction and oxidation of redox switches. The protocols are an analytical tool that helps to depict the cellular hydrogen peroxide signaling landscape, giving new insights on H_2O_2 signaling mechanisms, and hold the potential to be a framework for a future redox kinetomics analytical platform.

Key words Hydrogen peroxide, Redox biology, Kinetic parameters, Oxidation–reduction cycle, Rate constants

1 Introduction

Redox biology has witnessed major advances in the last decades, with the identification of virtually hundreds of redox targets that play key roles in the cell and whose biological activity is controlled by the redox state of cysteine residues [1, 2]. On the other hand, important knowledge gaps remain because many of the identified redox targets have unknown oxidation–reduction kinetic parameters and because the absolute levels of oxidants achieved under basal conditions, in experiments, or in clinically relevant conditions, are also mostly unknown. This lack of quantitative data impacts the advance of the field as expressed, for example, in a remarkable lack of successful cases of translational research, not least because the dose–response curves observed in this field are often complex with

Quentin Vanhaelen (ed.), *Computational Methods for Estimating the Kinetic Parameters of Biological Systems*, Methods in Molecular Biology, vol. 2385, https://doi.org/10.1007/978-1-0716-1767-0_8,

oxidants showing biphasic effects. In other words, the biological effect triggered by the oxidant is strongly dependent on its concentration, and opposite effects—i.e., activating instead on inhibiting, showing a beneficial action instead of a harmful effect—are observed depending on the doses applied [3–6]. Obtaining quantitative data is difficult, and specific experiments tailored to those determinations must be designed. Thus, analytical tools that process data obtained in typical redox biology experiments to determine kinetic parameters of redox targets, or the absolute intracellular concentration of oxidants, would be highly desired. Here, we present such tools.

2 Data Processing

2.1 Theoretical Model

A minimal model constituted by the following two equations was set up to describe the oxidation–reduction cycle of a redox-sensitive target [7, 8].

$$\text{Target}_{\text{rd}} + \text{H}_2\text{O}_2 \rightarrow \text{Target}_{\text{ox}} + \text{H}_2\text{O} \quad (1)$$

$$\text{Target}_{\text{ox}} \rightarrow \text{Target}_{\text{rd}} \quad (2)$$

In reaction 1, an oxidant (e.g., hydrogen peroxide, H_2O_2) oxidizes a target, and in reaction 2, the oxidized form of the target is reduced. Hydrogen peroxide concentration changes according to the conditions of each experiment, but this concentration is assumed to be constant during the experiment, i.e., independent of time, and thus the following differential equation is obtained:

$$\frac{d\text{Target}_{\text{rd}}}{dt} = k_{\text{rd}}(1 - \text{Target}_{\text{rd}}) - k_{\text{ox}}\text{Target}_{\text{rd}} \quad (3)$$

Here, $\text{Target}_{\text{rd}}$ is the fraction of the target protein in the reduced state, t is time, k_{ox} and k_{rd} are the pseudo first-order rate constants for reactions 1 and 2, which are assumed to follow mass-action kinetics, and d/dt stands for the differential operator. The master equation describing the time course of $\text{Target}_{\text{rd}}$ is given by the analytical solution of Eq. 3:

$$\text{Target}_{\text{rd}}|_t = \frac{k_{\text{rd}}}{k_{\text{ox}} + k_{\text{rd}}} + e^{-(k_{\text{rd}}+k_{\text{ox}})\times t}\left(\text{Target}_{\text{rd}}|_0 - \frac{k_{\text{rd}}}{k_{\text{ox}} + k_{\text{rd}}}\right) \quad (4)$$

where $\text{Target}_{\text{rd}}|_t$ and $\text{Target}_{\text{rd}}|_0$ are the fraction of reduced target at time t and at $t = 0$ (i.e., at the beginning of the experiment), respectively. By letting t to tend to infinite, the steady-state value of the reduced fraction of the target is reached, $\text{Target}_{\text{rd}}|_{\text{ss}}$, and Eq. 5 is obtained:

$$\text{Target}_{\text{rd}}|_{\text{ss}} = \frac{k_{\text{rd}}}{k_{\text{ox}} + k_{\text{rd}}} \quad (5)$$

By fitting Eq. 4 to experimental time courses in which the reduction state of a redox target is followed, rate constants k_{ox} and k_{rd} are determined.

2.2 Data Handling

2.2.1 Determination of k_{ox} and k_{rd}

To illustrate the application of Eq. 4, the rate constants k_{ox} and k_{rd} for HyPer, a redox protein used to detect H_2O_2 [9], are determined from an experiment in which HyPer oxidation is followed subjecting K562 cells to 2.5 μM of H_2O_2 (Fig. 1).

The rate constants k_{ox} and k_{rd} contain valuable kinetic information concerning the redox dynamics of the redox target. The kinetic parameter k_{ox} is a pseudo first-order apparent rate constant that depends on the reactivity of the redox target with H_2O_2, $k_{target+H_2O_2}$, and the concentration of H_2O_2:

$$k_{ox} = k_{target+H_2O_2} \times [H_2O_2] \quad (6)$$

When $k_{target+H_2O_2}$ is known, the value of the concentration of H_2O_2 can be obtained directly from Eq. 6, from the determination of k_{ox}. This concentration refers to the H_2O_2 concentration present in the vicinity of the redox target. Conversely, if H_2O_2 concentration is known, one can estimate $k_{target+H_2O_2}$ for a redox target for which this rate constant is unknown, which applies to most redox targets. Next, detail protocols on how to treat the data in these two scenarios are described.

2.2.2 Analysis of Hydrogen Peroxide Landscape Based on Redox Genetic Probes

Whose molecular design is based on the chemistry of redox targets identified in cells [9, 12, 13]. These probes have known $k_{target+H_2O_2}$ and so they can be used to probe the H_2O_2 concentration landscape. If the two sources of H_2O_2, basal and the H_2O_2 added during the experiment are made explicit in Eq. 6, Eq. 7 is obtained.

$$k_{ox} = k_{target+H_2O_2} \times \left([H_2O_2]_{basal} + [H_2O_2]_{addin}\right) \quad (7)$$

The value of $[H_2O_2]_{addin}$ is unknown in most cases, and it will a be a function of some experimental variable, *ExpVar*:

$$k_{ox} = k_{target+H_2O_2} \times \left([H_2O_2]_{basal} + f(ExpVar)\right) \quad (8)$$

By varying *ExpVar*, it will be possible to estimate $[H_2O_2]_{basal}$ and in some cases undercover the function *f*(*ExpVar*) so that the H_2O_2 intracellular values are fully characterized in the experimental setting. Exposure to H_2O_2 can either be in the form of an extracellular application or a stimulus of the intracellular H_2O_2 production, for example, by adding D-amino acid to cells in which the H_2O_2 producer yeast D-amino acid oxidase was genetically encoded [14]. In the former case, *ExpVar* is the extracellular concentration of H_2O_2 added ($[H_2O_2]_{ex}$) and, $f=[H_2O_2]_{ex}/\text{gradient}$ where gradient is the ratio between extracellular and intracellular concentrations, Eq. 8 becomes:

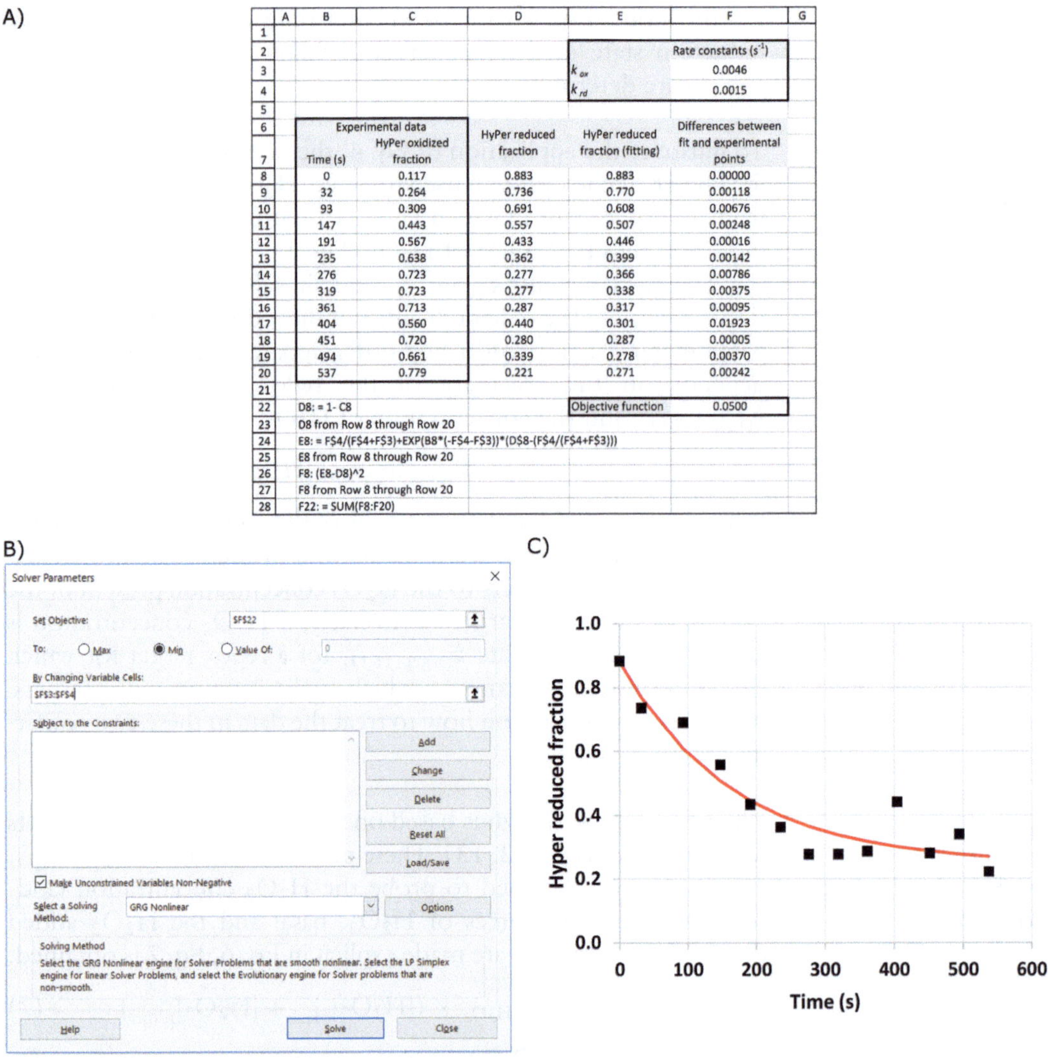

	A	B	C	D	E	F	G
1							
2						Rate constants (s⁻¹)	
3					k_{ox}	0.0046	
4					k_{rd}	0.0015	
5							
6		Experimental data					
7		Time (s)	HyPer oxidized fraction	HyPer reduced fraction	HyPer reduced fraction (fitting)	Differences between fit and experimental points	
8		0	0.117	0.883	0.883	0.00000	
9		32	0.264	0.736	0.770	0.00118	
10		93	0.309	0.691	0.608	0.00676	
11		147	0.443	0.557	0.507	0.00248	
12		191	0.567	0.433	0.446	0.00016	
13		235	0.638	0.362	0.399	0.00142	
14		276	0.723	0.277	0.366	0.00786	
15		319	0.723	0.277	0.338	0.00375	
16		361	0.713	0.287	0.317	0.00095	
17		404	0.560	0.440	0.301	0.01923	
18		451	0.720	0.280	0.287	0.00005	
19		494	0.661	0.339	0.278	0.00370	
20		537	0.779	0.221	0.271	0.00242	
21							
22		D8: = 1- C8			Objective function	0.0500	
23		D8 from Row 8 through Row 20					
24		E8: = F$4/(F$4+F$3)+EXP(B8*(-F$4-F$3))*(D$8-(F$4/(F$4+F$3)))					
25		E8 from Row 8 through Row 20					
26		F8: (E8-D8)^2					
27		F8 from Row 8 through Row 20					
28		F22: = SUM(F8:F20)					

Fig. 1 Calculation of the pseudo first-order rate constants k_{ox} and k_{rd} from the oxidation profile of HyPer in K562 cells caused by an extracellular 2.5 μM H_2O_2 addition. (**a**) From the experimental fraction of oxidized HyPer (column C) taken from [10], the reduced fraction is calculated (column D) and fitted to Eq. 4. The resulting fitted values are shown in column E, and the square difference between fitted values and experimental data is calculated in column F, with the sum of the squared differences calculated in cell F22. (**b**) The Solver Tool from Excel [11] is used to perform the nonlinear fitting using as objective the minimization of F22 cell ("Min" option should be ticked), having as variables the two cells with the values of the rate constants k_{ox} and k_{rd} (F3 and F4). "Make Unconstrained Variables Non-Negative" should be ticked as the rate constants cannot be negative. (**c**) Nonlinear fit of Eq. 4 (line, column E in panel **a**) to HyPer reduction data points (■, column D in panel **a**)

$$k_{ox} = k_{target+H_2O_2} \times \left([H_2O_2]_{basal} + [H_2O_2]_{ex}/\text{gradient}\right) \quad (9)$$

As shown in Fig. 2, k_{ox} is plotted against $[H_2O_2]_{ex}$ to estimate both the $[H_2O_2]_{basal}$ and the gradient.

	A	B	C	D	E	F	G	H	I	J
1										
2		H_2O_2 (μM)	k_{ox} (s^{-1})							
3		0.0								
4		0.5	0.0018							
5		1.0	0.0025							
6		1.5	0.0025							
7		2.0	0.0037							
8		2.5	0.0046							
9										
10		$k_{target+H2O2}$ ($M^{-1}s^{-1}$)	5.0E+05							
11		slope	0.00137							
12		intercept	0.00096							
13										
14		Gradient	366							
15		$[H_2O_2]_{basal}$ (M)	1.9E-09							
16										
17		C14:= C10/C11/1000000								
18		C15:= C12/C10								

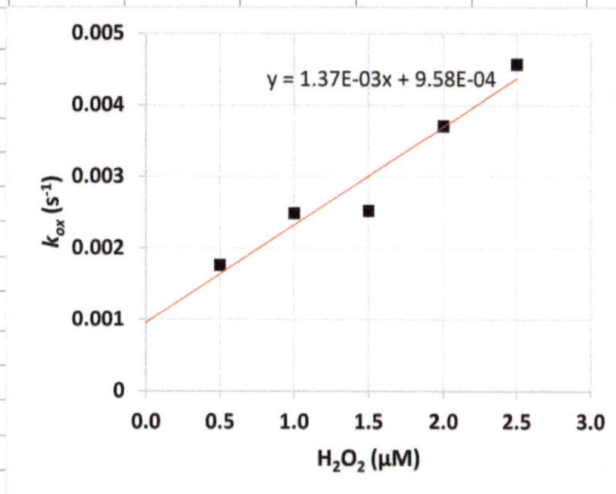

Fig. 2 Characterization of the H_2O_2 landscape based on the pseudo first-order rate constant, k_{ox}, for the oxidation of HyPer in K562 cells. The rate constant k_{ox} (column C, cells 4–8) obtained from [10] was determined for various external H_2O_2 concentrations (column B, cells 4–8) as described in Fig. 1, and the respective plot was done to determine the slope (cell C11) and the intercept (cell C12). By applying Eq. 9, and imputing the value of the rate constant for the reaction between HyPer and H_2O_2—$5 \times 10^5\ M^{-1}\ s^{-1}$ [15] (cell C10)—the gradient (cell C14) and the $[H_2O_2]_{basal}$ (cell C15) are determined

The plot of Fig. 2 is linear, but in general this may not happen. For example, a non-constant gradient between extracellular and intracellular concentrations, for different H_2O_2 concentrations, is reflected in a nonlinear plot. In such case, gradient has to be calculated for each experimental point using Eq. 9, provided $[H_2O_2]_{basal}$ is known.

Estimation of $[H_2O_2]_{basal}$ when $f(ExpVar)$ is unknown can be done by applying Eq. 10, which is obtained from Eqs. 5 and 6 by making $[H_2O_2] = [H_2O_2]_{basal}$, i.e., no H_2O_2 sources are present except for the basal H_2O_2 production.

$$[H_2O_2]_{basal} = (1/\text{Target}_{rd|basal} - 1) \times k_{rd\,basal}/k_{target+H_2O_2} \quad (10)$$

In this equation, $\text{Target}_{rd}|_{basal}$ is the fraction of the target reduced under basal conditions and $k_{rd\ basal}$ is the k_{rd} value under basal conditions. $k_{rd\ basal}$ is obtained from the plot of k_{rd} as a function of *ExpVar*, by extrapolating k_{rd} to $ExpVar = 0$ (*see* Fig. 3). Once $k_{rd\ basal}$ is known, by imputing $\text{Target}_{rd}|_{basal}$ into Eq. 10, $[H_2O_2]_{basal}$ is determined as shown in Fig. 3.

In general, the application of the protocol described here to experiments that measure the oxidation state of a redox target with known $k_{target+H_2O_2}$ provides an analytical tool to characterize the H_2O_2 landscape, by determining $[H_2O_2]_{basal}$ and, in some cases, $f(ExpVar)$. Thus, the level of H_2O_2 achieved when exposing cells to *ExpVar* becomes known, advancing the current state of the art in

	A	B	C	D	E	F	G	H	I
1									
2		H_2O_2 (μM)	k_{rd} (s^{-1})						
3		0.0							
4		0.5	0.0068						
5		1.0	0.0050						
6		1.5	0.0038						
7		2.0	0.0026						
8		2.5	0.0015						
9									
10		$k_{target+H2O2}$ ($M^{-1}s^{-1}$)	5.00E+05						
11		$Target_{rd}\|_{basal}$	0.88						
12		intercept	0.0079						
13									
14		$k_{rd\ basal}$ (s^{-1})	0.0079						
15		$[H_2O_2]_{basal}$ (M)	2.1E-09						
16									
17		C14:= C12							
18		C15:=(1/C11-1)*C14/C10							

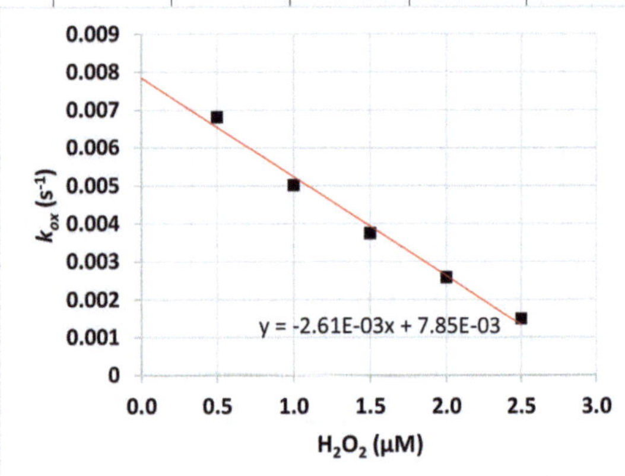

Fig. 3 Determination of the H_2O_2 basal concentration based on the pseudo first-order rate constant, k_{rd}, for the reduction of HyPer in K562 cells. The rate constant k_{rd} (column C, cells 4–8) obtained from [10] was determined for various external H_2O_2 concentrations (column B, cells 4–8) as described in Fig. 1, and the respective plot was done to determine the intercept (cell C12), i.e., k_{rd} at basal conditions (cell C14). By applying Eq. 10, and imputing the value of the rate constant for the reaction between HyPer and H_2O_2—$5 \times 10^5\ M^{-1}\ s^{-1}$ [15] (cell C10)—and the basal level of HyPer in the reduced form (cell C11), the $[H_2O_2]_{basal}$ (cell C15) is determined

which investigators trigger H_2O_2 alterations without knowing what are the absolute H_2O_2 intracellular changes. Thus, quantitative relationships between the biological effects elicited and the absolute H_2O_2 concentrations triggered experimentally are within reach.

2.2.3 Determining Rate Constants for the Oxidation of Redox Target with Unknown Reactivity Toward Hydrogen Peroxide

Once the H_2O_2 variations are characterized by the approach described in the previous section, kinetic parameters like $k_{target+H_2O_2}$ can be estimated from k_{ox} for redox targets for which this information is missing, which is the case for most redox targets. As an initial step, the protocol shown in Fig. 1 is applied to quantify k_{ox} and k_{rd}. Then, $k_{target+H_2O_2}$ is estimated directly from Eq. 6 by imputing k_{ox} and H_2O_2 basal and added H_2O_2 during the experiment, obtained in the previous section.

This procedure is easily scaled up to experiments with large sets of data in which experimental oxidation profiles are acquired simultaneously for different redox targets. k_{ox} and k_{rd} are estimated for each of the redox targets whose experimental data is available, by applying the protocol shown in Fig. 1 to each redox target individually. Nevertheless, redox targets respond differently to H_2O_2 [16], and so estimation of k_{ox} and k_{rd} for all redox targets in a single experiment may not be possible. For example, a redox target with a very high reactivity may already be fully oxidized after the first time point is acquired, or conversely, a redox target with low reactivity may not undergo a significant oxidation during the experiment. To overcome this issue, experiments should be repeated with different H_2O_2 concentrations, so that oxidation profiles that capture the redox dynamics of all redox targets are acquired. Therefore, the determination of many rate constants based on redox proteomics data is possible.

3 Discussion

The equations and protocols described here are tools that may be applied to analyze experiments in which the oxidation levels of redox targets are measured. This analysis is performed to determine the absolute levels of oxidants that are present in the vicinity of the redox targets and also the rate constants operating the oxidation–reduction cycle of these targets. Several success cases of application of this analytical tool have been accomplished. The basal concentration of H_2O_2 in K562 as well as the intracellular concentration achieved when exposing these cells to external H_2O_2 was characterized using HyPer as probe [10]. The application of these equations gave also the first hints on how much H_2O_2 needs to increase in the muscle cell to transduce signaling [17]. The H_2O_2 levels linked to signaling and toxicity were determined using in vivo oxidation status of redox targets [18]. Finally, it was predicted that bicarbonate is essential for protein-tyrosine phosphatase 1B (PTP1B) oxidation under in vivo conditions [8], which was later confirmed experimentally in EGF-stimulated cells [19]. So far, these tools were never applied to large-scale experiments, but their application to redox proteomics data will open the possibility to determine large sets of rate constants simultaneously and to establish a redox kinetomics analytical platform.

Acknowledgments

This research was financed by Fundação para a Ciência e a Tecnologia (FCT) through projects UIDB/00100/2020 (CQE) and UIDB/04138/2020 (IMED).

References

1. Held JM (2020) Redox systems biology: harnessing the sentinels of the cysteine redoxome. Antioxid Redox Signal 32:659–676. https://doi.org/10.1089/ars.2019.7725
2. Jones DP, Sies H (2015) The redox code. Antioxid Redox Signal 23:734–746. https://doi.org/10.1089/ars.2015.6247
3. Gao Y-J, Hirota S, Zhang D-W et al (2003) Mechanisms of hydrogen-peroxide-induced biphasic response in rat mesenteric artery. Br J Pharmacol 138:1085–1092. https://doi.org/10.1038/sj.bjp.0705147
4. Loo AEK, Wong YT, Ho R et al (2012) Effects of hydrogen peroxide on wound healing in

mice in relation to oxidative damage. PLoS One 7. https://doi.org/10.1371/journal.pone.0049215

5. Matias AC, Marinho HS, Cyrne L et al (2011) Biphasic modulation of fatty acid synthase by hydrogen peroxide in Saccharomyce cerevisiae. Arch Biochem Biophys 515:107–111. https://doi.org/10.1016/j.abb.2011.08.009
6. Iwakami S, Misu H, Takeda T et al (2011) Concentration-dependent dual effects of hydrogen peroxide on insulin signal transduction in H4IIEC hepatocytes. PLoS One 6: e27401. https://doi.org/10.1371/journal.pone.0027401
7. Brito PM, Antunes F (2014) Estimation of kinetic parameters related to biochemical interactions between hydrogen peroxide and signal transduction proteins. Cell Biochem 2:82. https://doi.org/10.3389/fchem.2014.00082
8. Antunes F, Brito PM (2017) Quantitative biology of hydrogen peroxide signaling. Redox Biol 13:1–7. https://doi.org/10.1016/j.redox.2017.04.039
9. Belousov VV, Fradkov AF, Lukyanov KA et al (2006) Genetically encoded fluorescent indicator for intracellular hydrogen peroxide. Nat Methods 3:281–286. https://doi.org/10.1038/nmeth866
10. Lyublinskaya O, Antunes F (2019) Measuring intracellular concentration of hydrogen peroxide with the use of genetically encoded H_2O_2 biosensor HyPer. Redox Biol 24:101200. https://doi.org/10.1016/j.redox.2019.101200
11. Harris DC (1998) Nonlinear least-squares curve fitting with Microsoft Excel solver. J Chem Educ 75:119. https://doi.org/10.1021/ed075p119
12. Morgan B, Van Laer K, Owusu TNE et al (2016) Real-time monitoring of basal H_2O_2 levels with peroxiredoxin-based probes. Nat Chem Biol 12:437–443. https://doi.org/10.1038/nchembio.2067
13. Pak VV, Ezeriņa D, Lyublinskaya OG et al (2020) Ultrasensitive genetically encoded indicator for hydrogen peroxide identifies roles for the oxidant in cell migration and mitochondrial function. Cell Metab 31:642–653.e6. https://doi.org/10.1016/j.cmet.2020.02.003
14. Haskew-Layton RE, Payappilly JB, Smirnova NA et al (2010) Controlled enzymatic production of astrocytic hydrogen peroxide protects neurons from oxidative stress via an Nrf2-independent pathway. Proc Natl Acad Sci U S A 107:17385–17390. https://doi.org/10.1073/pnas.1003996107
15. Bilan DS, Pase L, Joosen L et al (2013) HyPer-3: a genetically encoded H(2)O(2) probe with improved performance for ratiometric and fluorescence lifetime imaging. ACS Chem Biol 8:535–542. https://doi.org/10.1021/cb300625g
16. Fu L, Liu K, Sun M et al (2017) Systematic and quantitative assessment of hydrogen peroxide reactivity with cysteines across human proteomes. Mol Cell Proteomics 16:1815–1828. https://doi.org/10.1074/mcp.RA117.000108
17. Nikolaidis MG, Margaritelis NV, Matsakas A (2020) Quantitative redox biology of exercise. Int J Sports Med 41(10):633–645. https://doi.org/10.1055/a-1157-9043
18. Domènech A, Ayté J, Antunes F, Hidalgo E (2018) Using in vivo oxidation status of one- and two-component redox relays to determine H_2O_2 levels linked to signaling and toxicity. BMC Biol 16:61. https://doi.org/10.1186/s12915-018-0523-6
19. Dagnell M, Cheng Q, Rizvi SHM et al (2019) Bicarbonate is essential for protein-tyrosine phosphatase 1B (PTP1B) oxidation and cellular signaling through EGF-triggered phosphorylation cascades. J Biol Chem 294:12330–12338. https://doi.org/10.1074/jbc.RA119.009001

Chapter 9

Computational Methods for Structure-Based Drug Design Through System Biology

Aman Chandra Kaushik, Shakti Sahi, and Dong-Qing Wei

Abstract

The advances in computational chemistry and biology, computer science, structural biology, and molecular biology go in parallel with the rapid progress in target-based systems. This technique has become a powerful tool in medicinal chemistry for the identification of hit molecules. The recent developments in target-based systems have played a major role in the creation of libraries of compounds, and it has also been widely applied for the design of molecular docking methods. The main advantage of this method is that it hits the fragment that has the strongest binding, has relatively small size, and leads to better compounds in terms of pharmacokinetic properties when compared with virtual screening (VS) and high-throughput screening (HTS) hits. De novo design is an essential aspect of target-based systems and requires the synthesis of chemical to allow the design of promising compound.

Key words Target-based systems, Computational chemistry, Structural biology, Molecular biology

1 Introduction

The advances in computational chemistry and biology, computer science, structural biology, and molecular biology go in parallel with the rapid progress in target-based system methods [1]. These techniques have become powerful tools in medicinal chemistry for identifying hit molecules. The development of target-based systems has played a major role in the generation of compound libraries and molecular docking methods using three-dimensional macromolecule biological structures [2, 3]. Many examples of small molecules obtained using the method of target-based systems can be found in the literature [4–6]. Target-based systems reduce the time and cost spent for the discovery and development of new drugs. Nevertheless, this method still faces major challenges including high false-positive rate, the lack of flexibility of the target during docking, and inaccurate scoring functions used to compute target–ligand binding energy

Quentin Vanhaelen (ed.), *Computational Methods for Estimating the Kinetic Parameters of Biological Systems*, Methods in Molecular Biology, vol. 2385, https://doi.org/10.1007/978-1-0716-1767-0_9,

[7, 8]. The drugs available in the market or currently undergoing clinical trials were the least partially influenced by the method of target-based systems [9, 10]. The development of "omics" techniques has become the benchmark of the twenty-first century and is the main driver toward the discovery of new potential drug and target [11]. One cannot deny that the cost associated with the development of a drug has increased dramatically while the returns on investments have undergone a continuous decline [9]. As a result, the number of new molecular entities (NMEs) approved by the FDA is lower than ever, and delays in the manufacturing of NME appears in the pharmaceutical industry. Among the reasons that explain these worrying trends is the role played by the existence of adverse side effects and the reduced efficiency of many potential compounds [12]. Another aspect that should be taken into account to explain the difficulties to find new promising compounds is that most of the easy compounds have been found and that nowadays one is looking to cure much more complex diseases and that the interplay between potential drugs and the targets associated to these diseases in terms of efficacy, selectivity, and potential side effects is much more complicated to handle [13]. The discovery of system-based drugs is a promising direction of research for such complex systemic diseases. The study of system biology should take into account the properties of the biological molecules [14, 15]. Tools like biological modeling and data analysis can help to build a better understanding of the biological functions of a given molecule. Consequently, system biology is a conceptual approach of great interest for the prediction and rationalization of drug effects and side effects [16–20].

2 System Biology and Drug Target Identification

Cardiovascular disease, cancer, and diabetes are among the most complex diseases [21], and finding efficient drugs and compounds to treat those conditions pose a major challenge to the drug discovery industry [22]. Diseases having complex etiology cannot be treated with a single site. This is due to the repetitive structure of the disease network. A drug designed for a specified target area induces side effects. This includes modulations that have repercussions on the functional networks of the underlying biological systems. The binding of drugs known as network drugs causes structural and functional changes that have an immediate effect on the binding partners of the target. These complex protein–protein interactions cause side effects and long-range effects in the cellular environment. On the other hand, a highly selective drug suffers from low usefulness due to its inherent biological properties [23]. There are various limitations of host and pathogens in biochemistry. The tendency to resist mutations is one of them [13]. The only solution to such restrictions is the development and discovery of a drug through system biology [24, 25]. Protein–protein interaction (PPI) network, signaling modeling, and metabolic pathways are the mapping tools commonly used to understand the interactions of biological macromolecules [26, 27]. These tools form the cornerstones of the system-based drug design. This type of drug could be used in different situations. For instance, it is known that inflammatory diseases are associated with topological changes of the arachidonic acid network which is left in an undesirable state. The goal of using drug network theory is to restore the arachidonic acid network to its normal state [28]. Drug network theory can also be used to take into account the interplay between the dynamics of signaling pathways and the diseases. Under certain circumstances, the signal transduction like pS3 and NFκB induces dynamic patterns that can lead to the onset of diseases. These are the two areas where drug network theory and system based approaches are applied for target identification.

The new directions and developments in target-based systems are also being discussed in this chapter. Upregulating and downregulating activities are equally important from a systemic point of view. These can be realized with allosteric regulators [29, 30]. Allosteric drugs have many advantages compared to orthosteric drugs. They can help to deal with the dynamics of the alteration of the network for the related targets. The drug-binding affinity and binding kinetics also significantly affect drug efficacy.

2.1 System-Based Drug Discovery

The selection of drugs is the central issue for the discovery of network-based drugs [31]. This requires not only a comprehensive characterization of the disease but also a good understanding of the difference between diseased and normal states [32]. These dynamic models respond to the perturbation with regulation series and feedbacks. They form the foundation of the dynamics of target-based systems [33].

2.2 Disease as a State of the Network

The search algorithm is needed for the identification of optimal intervention of drug targets in drug-related networks [34, 35]. This can be well illustrated with the fact that our laboratory has worked to develop an algorithm for the investigation of anti-inflammatory drugs targeting the pathways regulating the metabolism of the arachidonic acid. We introduced a Monte Carlo model that simulated the ordinary differential equations for annealing-based search algorithm. This algorithm is defined as a multiple target optimal intervention (MTOI) used for the identification of targets [36]. The algorithm provides the potential of target combinations with their high efficiency and side effects. Multiple interventions are necessary for the redundancy of any sites in a biological network [25, 37]. The prediction and rationalization of drugs and their combinations are the advantages of a target-based systems approach. The goal of the modeling of a three-node enzymatic network is to analyze the synergistic or antagonistic combinations of drugs. These combinations of drugs depend on the topology of the network and provide useful insights for the design of new combinations of drugs. This can become a powerful method for identifying perturbations of interest within the interacting nodes of a given biological network.

2.3 Modulating Network Dynamics

Bi-stability and oscillatory behaviors are common dynamics often exhibited by biological networks. These dynamic behaviors appear in systems eliciting specific regulatory structures and have been shown to be of paramount importance for many biological functions. For instance, the pathway to DNA damage-induced apoptotic with bifurcation points is helpful for the arrest and apoptosis of cell growth. The proteins in the pathway correspond to high-frequency oncogenic mutations. The circadian disorders are caused by the dynamic perturbations or disruption of normal oscillatory dynamics. This change can be associated with tumorigenesis in network models eliciting bi-stable dynamics. Through the unraveling of the network dynamics, the rational design of modulating drugs should be feasible. For complex systemic diseases like cancer or circadian disorders, such drugs could be an attractive solution to modulate the characteristics of the dynamics of the regulating networks. An interesting characteristic of the signaling networks is that the networks typically respond to different signals with different dynamics that will affect the signaling molecule. The transcription factors like p53 and NF-κB in mammalian systems are interesting examples of different behavioral responses in mammalian systems. They are attractive targets for drugs as well as important signal transducers that are involved in the onset and progression of many diseases including cancer. Besides the signaling molecules, the signaling dynamics are also potential targets that could be used within the design of pharmaceutical interventions. This concept is further discussed by researchers who postulated the signaling dynamics, with a selective unwanted response to specific signals. This also maintains a leaving response for other intact signals. This was tested with the TNF signaling and the associated response relationship to the selective target. This also provides a possible therapeutic paradigm. Next-generation discovery will arise with the network-based discovery of drugs with many challenges and opportunities. By exploiting the basis of this opportunity, many pharmaceutical companies produced novel alternatives for complex diseases like Charcot-Marie-Tooth disease. For the successful design of large-scale studies required for the network-based discovery of a drug, the measurement and prediction of responses associated with the drug is essential [18]. The phenotyping approach and computational modeling have opened new ways for the use of network theory.

3 Allosteric Binding Sites as Drug Targets

For specific targets of protein, conventional drugs have less advantage than allosteric drugs [38]. This includes the existence of side effects and more complex regulatory mechanisms. The allosteric drugs increase the target activity, a process more difficult with the orthosteric drugs. Enzyme catalysis rate and substrate binding strength measure the target activity involved. The allosteric drugs is a better solution for the sites where small molecules are not suitable for a design like interfaces of PPI and orthosteric sites [39–41]. This can be explained with the example of lymphocyte function antigen 1 (LFA-1) associated with allosteric inhibitory. TGF-β receptor is inhibited by the compound naringenin. The disruption of PPIs with naringenin has been investigated with the help of molecular dynamics simulations of the allosteric mechanisms. We cannot detect the potential sites of allosteric and activities of the regulated target in most cases [39]. The experimental screening with high throughput is responsible for the allosteric regulation of molecules. This involves the design of allosteric drugs with rationality for binding sites being the first step. Only 907 modulator complexes of the allosteric site have been collected in the allosteric database with non-fibrous protein sites [42, 43]. To identify the site of the allosteric protein, support vector machine (SVM) and MutInf methods have been developed [44]. MutInf is a software package that can be used to analyze data from molecular dynamics simulations to identify statistically significant correlated motions and calculate the residue-by-residue conformational entropies. The prediction of allosteric sites is needed for the identification of pockets for the binding of small molecules. This can be done with the CAVITY program (a web server that offers protein cavity detection and various functional analyses), dynamic simulation on the long time-scale human M2 receptor that is a class-A GPCR, and based on the FTMAP analysis strategy for allosteric sites suggested by Miao et al. [45]. These molecular dynamics simulations are expensive and deal with changes that occur on conformational short scale. Hence, there is a need for an efficient coarse-grained design. This chapter demonstrates the two-state model for the prediction of sites. The allosteric deals with the conformational shift process; hence, such two states for a single protein are constructed with perturbations for binding sites that are helpful for the CAVITY ensemble. These allosteric inhibitors and activators are discovered for *Escherichia coli D*-phosphoglycerates dehydrogenase of predicted sites. Besides, one can also use the normal mode analysis method for this. The non-redundant test on the protein was done by Panjkovich et al. using normal coarse-grained method analysis [46]. For about 70% of allosteric sites, noteworthy variations are obtained that explain the flexible nature of the protein.

Another study involved the use of LIGSITEcsc for the detection of sites that are a potential binder. LIGSITEcsc is an extension of the LIGSITE algorithm for the identification of pockets on protein surfaces. This information is used for the prediction of ligand-binding sites. LIGSITEcsc is based on the notion of surface-solvent-surface events and on the degree of conservation of the involved surface residues. This further added to the octahedron for the representation of ligand to sites. The significant change in a protein deciphered the potential sites. This approach is used by the PARS web server. The PARS web server provides a method that queries protein dynamics and structural conservation to identify pockets on a protein structure that may exert a regulatory effect on the binding of a small-molecule ligand. Another web server, the Allo web server, developed by Huang et al., uses SVM and FPocket algorithms for the forecast of allosteric locations [44]. There is a need for this type of method that has long-distance effects as well as detailed interactions of local sites as binding of the drug can change the target with its structural and dynamic property [47]. This can be helpful for the building of primary partners of targets and interactions of the complex while exhibiting the effects that are of long-range and minimal side effects. These techniques could be further used for directing the new solutions for designing suitable orthosteric drug sites that target the protein network with structural information [48]. Besides that, intrinsically disordered proteins (IDPs) are another drug-binding protein that has conformational dynamic ensembles. This class of protein is found in all organs [49]. The signaling proteins and transcription factors possess IDP domains. They are considered as potential drug targets for the design of drugs for different human diseases. They are also very important from a network point of view in PPIs. Furthermore, they have multiple binding sites for drug-treated diseases like cancer and used for allosteric proteins on signaling. The reported examples of IDP targeting designs are difficult to find [50]. The oncoprotein c-Myc IDP domain that targets the inhibitors was tested by experimental screening. But due to its compositional complex structure, targeting IDPs is not easy. Jin et al. suggested that this C-Myc targeting has insights design targeting IDPs in the future [51].

4 Modulating Drug Target Gene Expression

Current models for the design of drug targeting a specific protein and the study of the associated protein expression are based on the comparison between healthy and diseased tissues [52]. The approaches that are target based develop the treatment with the interaction of activities and functions and directly modulating the targets. Besides that, the drug also targets the expression and modulates the activity by influencing the cellular molecular concentrations. These are termed target expression modulators (TEMs) [53]. This is a tool for the effective and direct treatment of disease. The method is working by modulating the upregulated activity of the target protein. But the challenge is posed by those protein targets that are lacking binding sites and the process of discovery of a drug that computationally responds to disruption of PPIs. With the posttranscriptional or transcription process, targeting proteins can take advantage of present TEMs. Compounds that affect such processes belong to the following classes: regulatory elements with a small molecule on nearby sequences of gene promoter; transcription apparatus that interferes with small molecule and affects the stability of mRNA product; gene expression of a small molecule that has epigenetic expression and influences histones modification, DNA methylation, and micro-RNA transcription [54]; proteostasis system of a small molecule that targets the increase and decrease of concentration of protein [55]. Polo-like kinase 1 (Plk 1) expression has been shown with genistein and vanillin like small molecule. This expression plays a role within the DNA repair system that is supported by the downregulated Plk1 tool. This supports future drug research. This is an essential tool in the biological network for the degradation of protein and expression of the gene.

5 The Influence of the Kinetics of the Drug Binding Process

Kinetics plays an important role in drug efficiency [56] and is influenced by the energy of protein binding. It can be related to the residence time of drugs that is correlated with the efficiency of drug binding. This can be covalently done with slow-designing dissociation. The outcomes are primarily responsible for binding kinetics for the designing of drugs. Previous studies were based on computational analysis dealing with pharmacological conditions like inhibition of enzymes, binding receptor, and metabolic network. The drugs have better effects on the inhibitory enzyme, peak receptor with higher occupancy, and multi-target with the better performance [57]. The relationship with the efficacy should be considered separately with the different kinetics [58]. The binding energy also depends on the topology of the network and how it might impact the kinetics of a target located within the network. There is a need to consider the binding energy of the molecule and the system-level requirement for the successful discovery of the drug. The relationship between the structure of the target and the kinetics of the binding drug can be studied using experimental and computational approaches. This works well for different systems. Such work was carried out for binding free energy of Huperzine by Bai et al. where kinetic constants are predicted [59]. Another study by Schneider et al. focused on the relationship between the compound structure and the kinetic properties in the context of the CDK8/CycC system. The network dynamics as well as the kinetics of PPIs were studied due to their importance for the design of drug with good binding affinities.

6 New Opportunities for SBDD Approaches in the System Biology Era

A huge potential is offered by the system-based discovery of a drug for dealing with diseases and evolving drugs [60]. The following points are some of the progressions in the discovery of drugs in the future.

6.1 De Novo Drug Design

Many docking protein-ligand methods with VS have been proposed during the last decade [61]. It deals with the limitation of available compounds that have identified hits but covers only several chemicals and drugs. Computational VS is an interesting technique for the novel design of drug for molecular entities. This approach does not suffer from structural limitations, and scaffolds with more efficient pharmacological profiles can be obtained. This is illustrated with the chemical space named de novo which is a design of an ideal drug candidate. This has the potential to fulfill the system-based requirement of drug design. There are several de novo design programs for drugs involving BOMB, Ligbuilder, MCSS, Ludi, BOMB, and SPROUT [62]. The de novo is not yet widely used within the chemistry community. This is due to the low success rate of this design method compared to the conventional VS program [63]. This lower performance can be related to the high-throughput approach and HTS or VS. The experimentation process with de novo is time-consuming and core labors are required with technical supports. This is also difficult for the analysis until the synthesis of chemicals is involved in the process of design. To fulfill this, a successful practical approach needs to be designed for a drug model with flexible accessibility [64]. The process requires a purchase of larger set of compounds than the ones typically required for VS. For example, the multipurpose tool named LigBuilder that is built on the pre-developed program RASSE is for the de novo that is based on the structure. This chapter deals with the embedded chemical reaction table with a component design that is being synthesized with a retrosynthesis analyzer. It involves the detection of cavity procedure with the shaping potential of binding sites for the ligand and drug. It also deals with the evaluations that are in large numbers just to improve the success rates of the design. Ligbuilder 2.0 tool has illustrated it with the optimization of an inhibitor named Cyclophilin A. Such an inhibitor can be designed with a single round with great influence experimentally [65]. This proves the authentication of the Ligbuilder tool. The screening based on fragments optimizes the hits for design methods for de novo. This screening focus on the moiety compounds of small nature that binds various sites, then connecting the multiple of these the scaffold is developed [66].

6.2 Multi-Target Drug Design

Multi-pathogenic system deals with complex diseases more effectively than the single drug targets [18, 67]. These complex factors are involved in cancer, diabetes, central nervous system, and inflammations for instance. The analysis related to biology provides multi-targeted solutions for controlling the network. These combinational and multi-target solutions are necessary for the treatment of drugs. These are also defined with the intervention of compounds that deal with pharmacological multiple targets in cancer, inflammations, AIDS, and type 2 diabetes. The mixtures of multi-component are extracted from traditional medicines called ethnopharmacology [68]. Traditional Chinese methods of medicine are used for understanding the network, pharmacology, and simulations. The demerits of combinational drugs are patient compliance especially for treatment of a disease like hypertension and drug–drug interactions. This also includes the very high number of combinations of drugs and their pharmacokinetics and pharmacodynamics. Despite weak effects on their targets, multi-target drugs work well for complex diseases. The process for the design of target-based systems is complicated. Indeed, it includes multiple targets and increased risks. Moreover, it is not very much cost-effective for clinical systems like any other traditional method, but when it is used as a single drug, these consequences can be avoided. For the past few years, ligand discovery like linker strategy and combinational framework has been put into practice. This is accompanied by cross virtual screening. This has been reported in the 5-LOX and mPGES-1 with the combinational framework of the pharmacophore method. Such being a successful dual-target with LTA_4H and PLA2 inhibitors [69]. This is based on the integration approach of the fusion of similar structures or sub-structure of multiple compounds. These results can be explained by the presence of the smaller molecules that are organized in combined states and directly related to the compositional flexible chain. Such chains have fewer efficacies than preclinical complexes with meager pharmacokinetics. The cross-screening reduces the chance of success for distinct unrelated binding sites and targets. It is highly integrated for multi-targets to make complexes for the specified targets. Hence, there is a need for such drugs that work well for unrelated goals. Previous studies on the Ligbuilder for de novo drug model optimized the molecular algorithm and multi-target design of inhibitors for specific compound collection. Besides that, an ensemble and fragment linking strategy and algorithm respectively have also been discussed for the growth strategy of multi-targets. This is useful for the amalgamation of framework and inhibitor combinations to obtain high efficiency. The Ligbuilder is further validated by the COX-2 and LTA4H experimentally for target design and inhibitors optimizations [70]. It is used for proteins and binding pockets. The chemical space available for the single ligand is much larger than the multi-ligand targets. Hence, de novo is the better choice for multi-targets. Their optimizations are more complicated because of their group activity profile. This is in part due to the stepwise optimization used to refine the multi-target to local or minima. The optimization landscape makes such techniques more expensive and strategy less effective. Such work cannot be done manually other than the Ligbuilder 3. This makes the procedure more challenging for single targets and reduces the rate of NME production.

7 Repositioning of Drug Analogs

Repositioning, also known as repurposing, is now a widely used approach for the development of drugs [71]. The repurposed drugs have already passed the toxicity tests and other evaluations. Consequently, the risks of failure due to safety concerns are reduced [72]. The saved R&D costs can be repurposed for bringing the drug to the market. The exploration of new uses of drugs between clinic and laboratory is fairly common. This idea has been used since the 1990s using serendipitous methods. During the last decade, researches have been carried out for the development of new computational tools to optimize the success of drug repositioning procedures [73, 74]. The process of drug repurposing is complex. Only 40% of the drug is available among 9000 off-patent repositionings. This comprehensive data yields the drug availability for a comprehensive library screen. Hence, one has introduced the repositioning analogs of the drug for the search of potentials such as isomers of optical, the precursor of a drug, and drug targeting the disease [74, 75]. These have a structure similar to marketed drugs with the potential of becoming a future drug. It becomes an abundant source for developing new leads with the variations of existing drugs in the markets. Such analogs can be ignored because of their lower potential over the targeted disease [76].

8 Design Drugs to Target Specific Protein–Protein Interactions (PPIs)

This makes the node of signaling and transmits cues which is the pathophysiological molecular network for the biological integration to achieve outputs [77]. This also promotes the progression of pathogenesis and disease. The disruption of PPIs is the pathway of transmission of a signal in a novel way. This can be modified with the small molecule strategy with the inhibitions of such for the modulations of protein activity. This also becomes a tool for the generation of new drugs. Examples include cardiovascular Titrobifan, glycoprotein IIb/IIIa, and gp120, an inhibitor of CCR5 with PPI approaches [78]. The difference between the PPI and the traditional format is challenging like the lack of technologies for controlling the growth of shallow protein and the discovery of such PPIs in the area of target-based systems discovery of drugs. These strategies include (a) PPI inhibitors for the predictions of structures [79], (b) the target-based systems for identification of format, especially the peptide for rational optimization [80], and (c) hot spot identifications that will bind the fragments on PPIs surface [80]. Such fragments or de novo are helpful for this. This results in the production of lipophilicity and molecular weight of high ligands. Non-Ro5-compliant is the existing inhibitors of PPIs that have a chemical space with lower ligands efficiency. Poor binding efficiency index (BEI) and surface efficiency index (SEI) are presented by various potential drugs [81]. This can be further extended with a large rate of attributes for compounds having PPIs. This is a lack of safety profiles and appropriate development in preclinical research. Despite such unsatisfactory values, Navitoclax in phase II presents an effective profile for oral dosing. This is the demonstration of the fact that new delivery properties are required for the PPI standards. This is not suitable for the design of small molecule and allosteric modulations for PPI targets demonstrated by LFA-1, enzymes of nitric oxide synthase, and factor for nerve growth. PPIs may not be done only with static structure binding for the allocations computational binding sites. This further needs the dynamics for druggable pockets that reveal the potential for pocket software sites.

9 Conclusions

The method described herein constitutes an approach for the discovery of next-generation drugs. The predictions of the control strategy, simulations of the biological network require a good knowledge of the regulatory mechanisms which can be achieved through the interference of system regulations. This proves to be an effective new strategy for target-based systems. This chapter provides a new direction to the system-centric idea for the R&D of future drugs. The section starts with the identification of the scope of a new drug. This will further go with the inclusion of a system-centric idea. The second section involves the new perspective of system biology for target-based systems.

References

1. Geete A (2002) Computational systems biology. Nature 420(6912):1–83
2. Jansen JM, Martin EJ (2004) Target-biased scoring approaches and expert systems in structure-based virtual screening. Curr Opin Chem Biol 8(4):359–364
3. Friedrich R (2008) Computational analysis of ligand relationships within target families. Curr Opin Chem Biol 12(3):352–358
4. Oxford scholarship online three-dimensional electron microscopy of macromolecular assemblies : visualization of biological molecules in their native state, no. September 2014. 2010
5. Frank J, Radermacher M (1986) Three-dimensional reconstruction of nonperiodic macromolecular assemblies from Electron micrographs. In: Advanced techniques in biological Electron Microscopy III, Berlin.

Springer Berlin Heidelberg, Heidelberg, pp 1–72

6. Nussinov R, Wolfson HJ (2006) Efficient detection of three-dimensional structural motifs in biological macromolecules by computer vision techniques. Proc Natl Acad Sci U S A 88(23):10495–10499
7. Belenko S (1998) Research on drug courts: A critical review. Nat Drug Court Ins Rev 1 (1):1–42.
8. Payne DJ, Gwynn MN, Holmes DJ, Pompliano DL (2007) Drugs for bad bugs: confronting the. Nat Rev Drug Discov 6:29–40
9. Dimasi JA, Feldman L, Seckler A, Wilson A (2010) Trends in risks associated with new drug development: success rates for investigational drugs. Clin Pharmacol Ther 87 (3):272–277
10. Lalonde RL et al (2007) Model-based drug development. Clin Pharmacol Ther 82 (1):21–32
11. Singh SP, Urooj S (2015) Combined rotation- and scale-invariant texture analysis using radon-based polar complex exponential transform. Arab J Sci Eng 40(8):2309–2322
12. Jakobsson P, Thoren S, Morgenstern R, Samuelsson B (2002) Identification of human prostaglandin E synthase: a microsomal, glutathione-dependent, inducible enzyme, constituting a potential novel drug target. Proc Natl Acad Sci U S A 96(13):7220–7225
13. Lomenick B et al (2009) Target identification using drug affinity responsive target stability (DARTS). Proc Natl Acad Sci U S A 106 (51):21984–21989
14. Arrell DK, Terzic A (2010) Network systems biology for drug discovery. Clin Pharmacol Ther 88(1):120–125
15. Yan Q (2010) Systems biology in drug discovery and development. Drug Discov Today 662:351
16. Singh SP, Urooj S, Lay-Ekuakille A (2016) Breast cancer detection using PCPCET and ADEWNN: a geometric invariant approach to medical X-ray image sensors. IEEE Sensors J 16(12):4847–4855
17. Singh SP, Urooj S (2016) An improved CAD system for breast cancer diagnosis based on generalized pseudo-Zernike moment and Ada-DEWNN Classifier. J Med Syst 40(4):105
18. Ramsay RR, Popovic-Nikolic MR, Nikolic K, Uliassi E, Bolognesi ML (2018) A perspective on multi-target drug discovery and design for complex diseases. Clin Transl Med 7(1):3
19. Schenone M, Wagner BK, Clemons PA, Program B (2017) Target identification and mechanism of action in chemical biology and drug discovery. Nat Chem Biol 9(4):232–240
20. Butcher EC (2005) Can cell systems biology rescue drug discovery? Nat Rev Drug Discov 4 (6):461–467
21. Couvreur P, Vauthier C (2006) Nanotechnology: intelligent design to treat complex disease. Pharm Res 23(7):1417–1450
22. Morens DM, Folkers GK, Fauci AS (2004) The challenge of emerging and re-emerging infectious diseases. Nature 430(6996):242–249
23. Hood L, Perlmutter RM (2004) The impact of systems approaches on biological problems in drug discovery. Nat Biotechnol 22 (10):1215–1217
24. Singh SP, Urooj S (2016) Geometric invariant feature extraction of medical images using Hu's invariants. In: 3rd international conference on computing for sustainable global development (INDIACom), pp 1560–1562
25. Singh SP, Urooj S (2018) A new computational framework for fast computation of a class of polar harmonic transforms. J Signal Process Syst 91:915–922
26. Shen J et al (2007) Predicting protein-protein interactions based only on sequences information. Proc Natl Acad Sci U S A 104 (11):4337–4341
27. De Las RJ, Fontanillo C (2010) Protein–protein interactions essentials: key concepts to building and analyzing interactome networks. PLoS Comput Biol 6(6):e1000807
28. Camacho DM, Collins KM, Powers RK, Costello JC, Collins JJ (2018) Next-generation machine learning for biological networks. Cell 173(7):1581–1592
29. Urooj S, Singh SP (2015) Rotation invariant detection of benign and malignant masses using PHT. In: 2nd international conference on computing for sustainable global development (INDIACom), pp 1627–1632
30. Singhb SP, Urooj S (2015) Wavelets: biomedical applications. Int J Biomed Eng Technol 19 (1)
31. Cho CR, Labow M, Reinhardt M, van Oostrum J, Peitsch MC (2006) The application of systems biology to drug discovery. Curr Opin Chem Biol 10(4):294–302
32. Csermely P, Korcsmáros T, Kiss HJM, London G, Nussinov R (2013) Pharmacology & therapeutics structure and dynamics of molecular networks: a novel paradigm of drug discovery. Pharmacol Ther 138:333–408
33. Davidov EJ, Holland JM, Marple EW, Naylor S (2003) Advancing drug discovery through systems biology. Drug Discov Today 8 (4):175–183

34. Tim R (2002) The 'risk environment': a framework for understanding and reducing drug-related harm. Int J Drug Policy 13:85–94
35. Leaman R, Wojtulewicz L, Sullivan R, Skariah A, Yang J, Gonzalez G (2007) Towards internet-age pharmacovigilance: extracting adverse drug reactions from user posts to health-related websites. dl.acm.org, no. Cdc, pp 117–125
36. Yang K, Bai H, Ouyang Q, Lai L, Tang C (2008) Finding multiple target optimal intervention in disease-related molecular network. Mol Syst Biol 4:228
37. Urooj S, Singh SP, Ansari AQ (2018) Computer-aided detection of breast cancer using pseudo Zernike moment as texture descriptors. In: Sensors and image processing. Advances in intelligent systems and computing. Springer, Singapore, pp 85–92
38. Christopoulos A (2002) Allosteric binding sites on cell-surface receptors: novel targets for drug discovery. Nat Rev Drug Discov 1:198–210
39. Rees S, Morrow D, Kenakin T (2002) GPCR drug discovery through the exploitation of allosteric drug binding sites. Recept Channels 8(5–6):261–268
40. Shan Y, Kim ET, Eastwood MP, Dror RO, Seeliger MA, Shaw DE (2011) How does a drug molecule find its target binding site? J Am Chem Soc 133(24):9181–9183
41. Shaw DE, Shan YB, Kim ET, Eastwood MP, Dror RO, Seeliger MA (2011) How does a drug molecule find its target binding site? J Am Chem Soc 133(24):9181–9183
42. Oikonomakos NG, Skamnaki VT, Tsitsanou KE, Gavalas NG, Johnson LN (2000) A new allosteric site in glycogen phosphorylase b as a target for drug interactions. Structure 8 (6):575–584
43. Suto MJ, Zacharias N (2004) Neuronal nicotinic acetylcholine receptors as drug targets. Expert Opin Ther Targets 8(2):61–64
44. Huang W et al (2013) Structural bioinformatics Allosite: a method for predicting allosteric sites. Bioinformatics 29(18):2357–2359
45. Miao Y, Nichols SE, McCammon JA (2014) Mapping of allosteric druggable sites in activation-associated conformers of the M2 muscarinic receptor. Chem Biol Drug Des 83 (2):237–246
46. Panjkovich A (2004) Comparison of different melting temperature calculation methods for short DNA sequences. Bioinformatics 21 (6):711–722
47. Dror RO et al (2011) Pathway and mechanism of drug binding to G-protein-coupled receptors. Proc Natl Acad Sci U S A 108 (32):13118–13123
48. Geddes-Mcalister J, Shapiro RS (2018) New pathogens, new tricks: emerging, drug-resistant fungal pathogens and future prospects for antifungal therapeutics. Ann N Y Acad Sci 1435(1):57–78
49. Luque I, Freire E (2000) Structural stability of binding sites: consequences for binding affinity and allosteric effects. Proteins Struct Funct Bioinf 41(S4):63–71
50. Bowman GR, Geissler PL (2012) Equilibrium fluctuations of a single folded protein reveal a multitude of potential cryptic allosteric sites. Proc Natl Acad Sci U S A 109 (29):11681–11686
51. Jin F, Yu C, Lai L, Liu Z (2013) Ligand clouds around protein clouds: a scenario of ligand binding with intrinsically disordered proteins. PLoS Comput Biol 9(10):e1003249
52. Hopkins AL (2008) Network pharmacology: the next paradigm in drug discovery. Nat Chem Biol 4(11):682–690
53. Atlung T, Ingmer H (1997) H-NS: a modulator of environmentally regulated gene expression. Mol Microbiol 24(1):7–17
54. Mathieu O, Bender J (2004) RNA-directed DNA methylation. J Cell Sci 117:4881–4888. jcs.biologists.org
55. Díaz-Villanueva JF, Díaz-Molina R, García-González V (2015) Protein folding and mechanisms of proteostasis. Int J Mol Sci 16 (8):17193–17230
56. Blaschke TF, Meffin PJ, Melmon KL, Rowland M (1975) Influence of acute viral hepatitis on phenytoin kinetics and protein binding. Clin Pharmacol Ther 17(6):685–691
57. Copeland R (2016) The drug–target residence time model: a 10-year retrospective. Nat Rev Drug Discov 15:87–95. nature.com
58. Vauquelin G (2016) Impact of target binding kinetics on in vivo drug efficacy: k off, k on and rebinding. Br J Pharmacol 173 (15):2319–2334
59. Bai F et al (2013) Free energy landscape for the binding process of Huperzine A to acetylcholinesterase. Proc Natl Acad Sci U S A 110 (11):4273–4278
60. Brown JB, Okuno Y (2012) Systems biology and systems chemistry: new directions for drug discovery. Chem Biol 19(1):23–28
61. Merk D, Friedrich L, Grisoni F, Schneider G (2018) De novo design of bioactive small

molecules by artificial intelligence. Mol Inform 37(1):1700153

62. Blaschke T, Olivecrona M, Engkvist O, Bajorath J, Chen H (2018) Application of generative autoencoder in de novo molecular design. Mol Inform 37(1):1700123
63. Gupta A, Müller AT, Huisman BJH, Fuchs JA, Schneider P, Schneider G (2018) Generative recurrent networks for de novo drug design. Mol Inform 37(1):1700111
64. Schneider G, Clark DE (2019) Automated de novo drug design: are we nearly there yet? Angew Chem Int Ed 58:10792–10803
65. Yuan Y, Pei P, Lai L (2011) LigBuilder 2: a practical de novo drug design approach. J Chem Inf Model 51(5):1083–1091
66. Bloomingdale P, Nguyen VA, Niu J, Mager DE (2018) Boolean network modeling in systems pharmacology. J Pharmacokinet Pharmacodyn 45(1):159–180
67. Bolognesi ML et al (2007) Multi-target-directed drug design strategy: from a dual binding site acetylcholinesterase inhibitor to a trifunctional compound against Alzheimer's disease. J Med Chem 50(26):6446–6449
68. Heinrich M (2010) Ethnopharmacology in the 21st century – grand challenges. Front Pharmacol 1:8
69. Narayanankutty A, Reséndiz-Hernández JM, Falfán-Valencia R, Teran LM (2013) Biochemical pathogenesis of aspirin exacerbated respiratory disease (AERD). Clin Biochem 46 (7–8):566–578
70. Shang E, Yuan Y, Chen X, Liu Y, Pei J, Lai L (2014) De novo design of multitarget ligands with an iterative fragment-growing strategy. J Chem Inf Model 54(4):1235–1241
71. Marzo T et al (2018) Auranofin and its analogues show potent antimicrobial activity covering multiresistant pathogens: structure-activity relationships. ChemMedChem 13:2448–2454
72. Palos I et al (2017) Repositioning FDA drugs as potential cruzain inhibitors from Trypanosoma cruzi: virtual screening, in vitro and in vivo studies. Molecules 22(6):1015
73. Sayé M, Gauna L, Reigada C, Valera-Vera E, Miranda MR, Pereira CA (2018) Crystal violet chemical analogues identified by drug repositioning present trypanocidal activity in Trypanosoma cruzi through inhibition of the proline transporter TcAAAP069. Manuscrito enviado
74. Santos MA, Chand K, Chaves S (2016) Recent progress in repositioning Alzheimer's disease drugs based on a multitarget strategy. Future Med Chem 8(17):2113–2142
75. Wu Z et al (2017) SDTNBI: an integrated network and chemoinformatics tool for systematic prediction of drug-target interactions and drug repositioning. Brief Bioinform 18 (2):333–347
76. Marzo T et al (2018) Auranofin and its analogues show potent antimicrobial activity against multidrug-resistant pathogens: structure–activity relationships. ChemMedChem 13 (22):2448–2454
77. Jubb H, Higueruelo AP, Winter A, Blundell BL (2012) Structural biology and drug discovery for protein-protein interactions. Trends Pharmacol Sci 33(5):241–248
78. Ivanov A, Khuri FR, Fu H (2013) Targeting protein-protein interactions as an anticancer strategy. Trends Pharmacol Sci 34:393–400. Available: https://www.sciencedirect.com/science/article/pii/S0165614713000783. Accessed 13 June 2019
79. Fuller JC, Burgoyne NJ, Jackson RM (2009) Predicting druggable binding sites at the protein-protein interface. Drug Discov Today 14(3–4):155–161
80. Marcos JF, Muñoz A, Pérez-Payá P, Misra S, López-García B (2008) Identification and rational design of novel antimicrobial peptides for plant protection. Annu Rev Phytopathol 46 (1):273–301
81. Abad-Zapatero C (2007) Ligand efficiency indices for effective drug discovery. Expert Opin Drug Discov 2(4):469–488

Chapter 10

Model Setup and Procedures for Prediction of Enzyme Reaction Kinetics with QM-Only and QM:MM Approaches

Michał Glanowski, Sangita Kachhap, Tomasz Borowski, and Maciej Szaleniec

Abstract

The enzyme-catalyzed reactions are traditionally studied with experimental kinetic assays. The modern theoretical modeling techniques provide a complementary way to investigate these catalytic reactions. Experimental assay frequently does not allow an unequivocal answer to the factors controlling the reaction mechanism. On the other hand, the theoretical experiments provide a precise understanding of the molecular-level steps involved in catalytic reactions. However, modeling requires at least structural data on the enzyme and reactant, and the complexity of the enzyme systems can still be a challenge.

In this chapter, we are going to describe how to apply theoretical modeling methods, such as MD simulation, QM-only cluster models of enzyme active site, or QM:MM multiscale modeling to study enzyme kinetics and even to predict kinetic isotope effect (KIE). We present a full protocol that starts from the PDB structure of the enzyme, through MD simulation of enzyme: substrate complex and statistical analysis of MD trajectory, selection of a model of the active site, and study of reaction pathways. We show how theoretical predictions basing on QM-only cluster models, QM:MM model, or multiple QM:MM models derived from QM:MM:MD simulations can be correlated with experimental kinetic results. Finally, we show how one can calculate intrinsic KIE associated with an individual molecular step.

Key words Enzyme-catalyzed reactions, Quantum mechanics, Molecular mechanics, Molecular dynamics, Gibbs free energy

1 Introduction

In this chapter, we provide detailed procedures with examples enabling modeling of the enzymatic reaction. The workflow (Fig. 1) starts with the structure of the enzyme in PDB format which is then prepared for MD simulation (methods 2 and 3.1) and subjected to molecular dynamic (MD) simulation (Subheading 3). The analysis of trajectory from MD simulation (Subheading 3.4)

Supplementary Information The online version of this chapter (https://doi.org/10.1007/978-1-0716-1767-0_10) contains supplementary material, which is available to authorized users.

Quentin Vanhaelen (ed.), *Computational Methods for Estimating the Kinetic Parameters of Biological Systems*, Methods in Molecular Biology, vol. 2385, https://doi.org/10.1007/978-1-0716-1767-0_10,

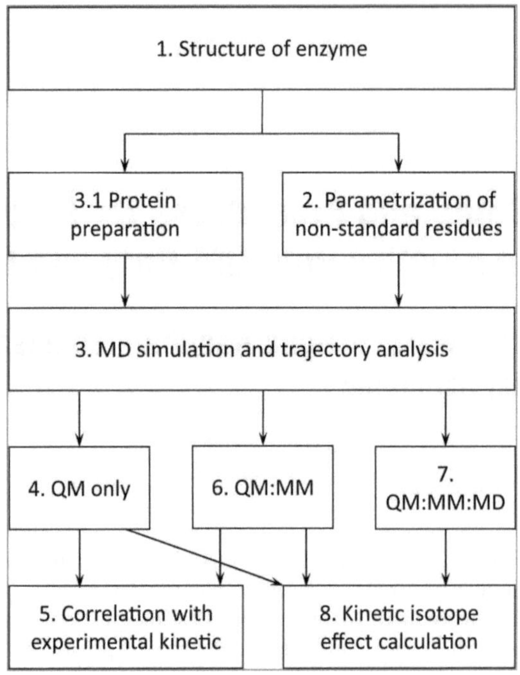

Fig. 1 Schematic workflow of modeling and organization of the chapter

allows the construction of different models like quantum mechanics-only (QM-only) cluster models of the active site (Subheading 4), quantum mechanics:molecular mechanics (QM:MM) models (Subheading 6), or QM:MM models that will be used for QM:MM:MD calculations (Subheading 7). The obtained results from such calculations can be used for correlation with the results of the kinetic experiment (Subheading 5) and prediction of the kinetic isotope effect (Subheading 8).

2 Parametrization of Nonstandard Residues

2.1 Parametrization of an Organic Molecule

In order to establish molecular mechanics (MM) charges for non-standard molecule, we have two options:

- Find these parameters in an external database (like RESP ESP charge DdataBase R.E.DD.B, https://upjv.q4md-forcefieldtools.org/REDDB/index.php).
- Use *Restrained Electrostatic Potential* (RESP) method to calculate point charges on our own.

In this section, we explain how to apply the RESP method using a simple example: dihydrotestosterone (also known as androstanolone, DHT), which we would like to model into the active site of 3-ketosteroid dehydrogenase from *Rhodococcus*

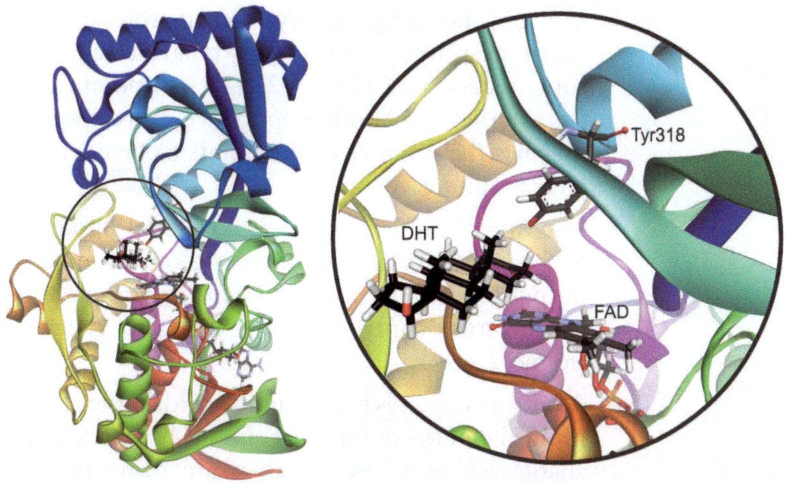

Fig. 2 Structure of 3-ketosteroid dehydrogenase from *R. erythropolis* [1] and magnification of its active site with docked dihydrotestosterone (DHT)

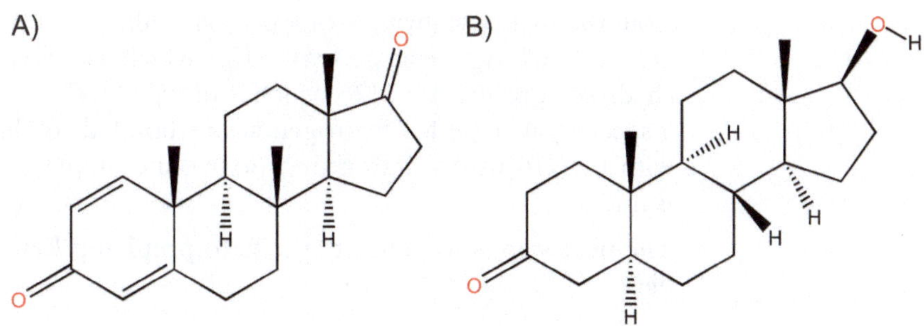

Fig. 3 (**a**) Androsta-1,4-diene-3,17-dione, (**b**) dihydrotestosterone (DHT)

erythropolis (PDB code 4c3y, Fig. 2) [1]. This molecule is not chemically bonded to the rest of the model, it contains only light atoms, and its conformational flexibility is limited. Another factor that simplifies the procedure is the fact that for DHT, we do not expect any pair of atoms to have exactly the same charge due to molecular symmetries. On the other hand, DHT differs a little from the ligand present in the active site in the crystal structure of 4c3y (androsta-1,4-diene-3,17-dione)—*see* Fig. 3. We have to bear this in mind when preparing the model.

The RESP procedure is as follows:

1. Perform geometry optimization of DHT at B3LYP/6-31G(d, p) level of theory. Make sure that the stationary point is found. It is also worth checking whether the conformation adopted by the molecule is obtainable in the active site of the enzyme. Examples of input and output files are located in 2.1. DHT_parametrization/optimization directory in the Electronic Supplementary Material (SI_Revised) on link.springer.com.

2. Using optimized geometry, calculate the electrostatic potential (**auto_gesp.inp**). To do that we have to add to the Gaussian route section iop(6/33 = 2)—to save points and potentials to the external file. Another option we use is iop(6/42 = 6) which specifies the number of the points to be calculated in a single layer around each atom.
3. Use Antechamber from Assisted Model Building with Energy Refinement (AMBER) package to obtain mol2 (all input and output files are in SI in 2.1.DHT_parametrization/antechamber directory): antechamber -i keto.gesp -fi gesp -o keto.mol2 -fo mol2 -c resp.
4. After this step, it is important to validate the results in the **keto.mol2** file. Two points that should be checked are the atomic charges (if nonphysical values are being predicted, we may suspect some error in gesp calculation) and atom types. In this example, we expect that all carbon atoms are of "c3" type except for the ketone group atom, which is of type "c." The oxygen atom from the ketone group is of "o" type, and the one from the hydroxyl group—of type "oh." Almost all hydrogens are of "hc" type except two: H27 which is of type "ho" (hydroxyl group) and H26 which is of type "h1"—this type is a special gaff type for hydrogen atoms bonded to aliphatic C with 1 electron-withdraw group. So in our example, everything is fine.
5. The next step is to generate the **keto.prepi** and **keto.frcmod** files:

```
antechamber -i keto.mol2 -fi mol2 -o keto.prepi -fo prepi
parmchk2 -f prepi -i keto.prepi -o keto.frcmod
```

6. Now we should open and check the keto.frcmod file. If this molecule requires any parameter that is not defined in the GAFF [2] force field, it should be listed there. In keto.frcmod file, there is only one nonempty section:

```
IMPROPER
c3-c3-c -o 1.1 180.0 2.0 Using the default value
```

   As we see, Antechamber suggested the default value for the missing improper.
7. Use the Kabsch method to find the best coordinates of the molecule. To use this method, we have to select dihydrosterone atoms that correspond to atoms from crystal structure—in our example, these are all carbon atoms and the oxygen atom of one of the ketone groups. Kabsch algorithm can be used to transform coordinates of dihydrotestosterone in such a way as to minimize root mean square deviation (RMSD) between the coordinates of selected atoms. To do this, we execute the fit.tcl

script in Visual Molecular Dynamics (VMD) (all input and output files are located in SI 2.1.DHT_parametrization/fit).

```
set mol2id [ mol new "keto.mol2" type "mol2" first 0 last -1
step 1 waitfor 1 ]
set templateId [ mol new "anb.xyz" type "xyz" first 0 last -1
step 1 waitfor 1 ]

set templateSel [ atomselect $templateId "all" ]
set indexes "0 1 4 2 3 9 5 6 7 8 18 10 11 12 13 14 15 16 17 19"
set mol2sel [ atomselect $mol2id "index $indexes" ]

set transformation_matrix [measure fit $mol2sel $templateSel]
set mol2fullSel [ atomselect $mol2id "all" ]
$mol2fullSel move $transformation_matrix

animate write mol2 "keto_fitted.mol2" beg 0 end 0 skip 1 $mo-
l2id
exit
```

The script can be executed by typing the following commands into the terminal: vmd -dispdev text -e fit.tcl. The first lines of the script load files into vmd (**keto.mol2** and **anb.xyz**—this file contains only a subset of atoms from the original molecule). After reading the files, the next instructions create two selections—one is the template molecule (anb.xyz) and the second one describes the atoms from keto.mol2 that can be used to calculate RMSD. After fitting, the final coordinates are written in **keto_fitted.mol2**.

2.2 Parametrization of Nonstandard Amino Acid

In this section, we discuss the preparation of the nonstandard amino acids. The procedure is explained using tyrosyl anion as an example. In this type of situation, we once again start with geometry optimization at the density functional theory (DFT) level. To reproduce the electronic characteristic of the amino acid with good quality, we have to prepare a capped version of tyrosyl anion as shown in Fig. 4a. We have to remember that after deprotonation the charge of the molecule is −1 (tyrosyl anion).

After geometry minimization, we generate gesp file in the same way as in Subheading 2.1 (all files are located in SI in 2.2.TYM_-parametrization folder). The next step is to fit charges of atoms to the calculated potential with the RESP program. To do that, we have to take care of two aspects:

1. A sum of charges of the non-cap atoms must be equal to −1. To satisfy this requirement, the charges of the capped atoms must be forced to specific values which add up to zero (Fig. 4b).

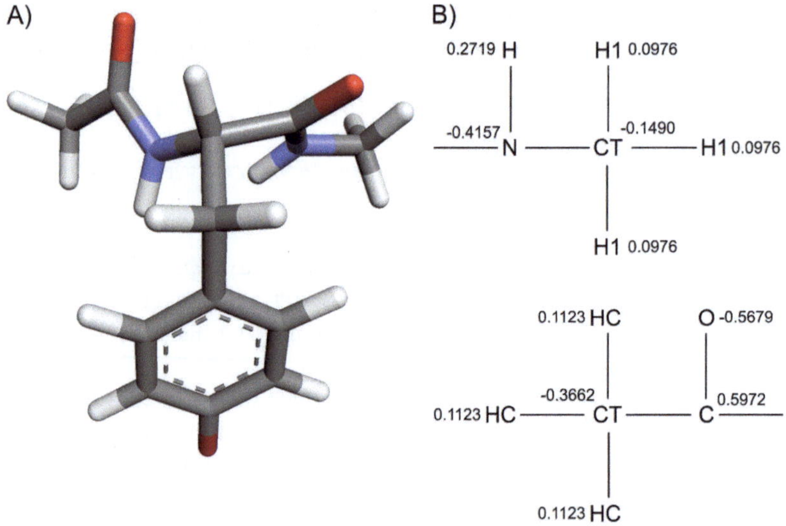

Fig. 4 (**a**) The capped version of tyrosyl anion and (**b**) the point charges of the caps used in the RESP procedure

2. Charges of equivalent atoms (e.g., at H atoms at the CB or equivalent H atoms in the atomic ring) must be forced by resp procedure to be identical.

The simplest way to ensure that those constraints are respected is to generate initial input files for resp using Antechamber (as in Subheading 2.1, **point 3**) and then modify it (examples of Antechamber outputs are located in 2.2.TYM_parametrization/antechamber). We use two of the files generated by antechamber: **ANTECHAMBER.ESP** (electrostatic potentials formatted for resp program) and **ANTECHAMBER_RESP1.IN** (resp input file). In the cntrl section of the file **ANTECHAMBER_RESP1.IN** we add iqopt = 2 option to let resp read the initial charges from an external file. The next step is to freeze the charges of cap atoms. The next section of resp input file looks as follows:

```
 1.0
Resp charges for organic molecule
-1 32
6 -1
6 -1
8 -1
1 -1
1 -1
1 -1
7 0
6 0
6 0
8 0
```

```
6 0
6 0
6 0
6 13
6 0
6 15
6 0
8 0
1 0
1 0
1 0
1 21
1 0
1 23
1 0
1 25
7 -1
6 -1
1 -1
1 -1
1 -1
1 -1
```

The first number (1.0) is the weighting factor for multiconformational RESP (in our example, there is only one conformation). The next line is a comment. Then there is a pair of numbers: molecule charge and number of atoms (−1 32). In the next lines are listed atomic numbers of each atom in the model and integer flag parameters. These parameters determine what resp does with the charge of a specific atom. 0 means normal fitting procedure, −1 means that the charge is not changed and a positive *N* value means that the charge of this atom must be equal to the charge of the *n*th atom of this molecule. In our example, there are five pairs of such equivalent atoms (four of them were already detected by antechamber): 13–14, 15–16, 21–22, 23–24, and 25–26. The modified resp input is located in 2.2.TYM_parametrization/resp./**resp.IN**.

The external file (**tym.qin**) with initial charges of the atoms used for capping the tyrosyl ions (Fig. 4b) looks as follows:

```
-0.366200 0.597200 -0.567900 0.112300 0.112300 0.112300 0.000000 0.000000
0.000000 0.000000 0.000000 0.000000 0.000000 0.000000 0.000000 0.000000
0.000000 0.000000 0.000000 0.000000 0.000000 0.000000 0.000000 0.000000
0.000000 0.000000 -0.415700 -0.149000 0.271900 0.097600 0.097600 0.097600
```

The charges are provided in the same order as the atoms in our model. The 0.000000 charges are replaced by the RESP procedure with new partial charges fitted to the model electrostatic potential

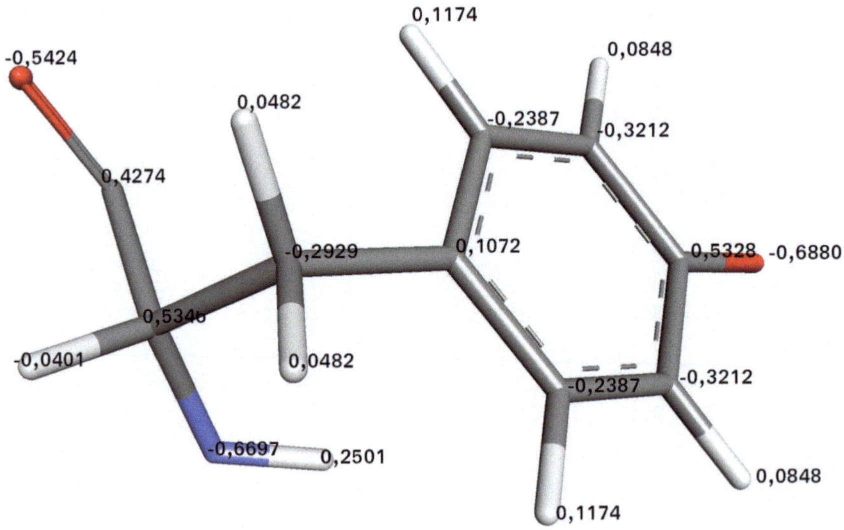

Fig. 5 The MM partial charges fitted to the tyrosyl ion residue by RESP procedure

described in the gesp file, while the charges already provided are fixed by −1 flag (*see* above). Unfortunately, RESP program is format sensitive, and you should always make sure that the input data has been correctly loaded.

Now we can perform resp fitting using command:

```
resp -O -i resp.IN -o tym_resp.out -p tym.punch -t tym_resp.chg
-q tym.qin -e tym.ESP -s tym_resp.esout
```

Fitted charges are written in 2.2.TYM_parametrization/resp./**tym_resp.chg** and are visualized on Fig. 5.

Finally to obtain the mol2 file of the amino acid, we need to replace the last column (charges) of the file 2.2.TYM_parametrization/antechamber/**keto.mol2** generated by Antechamber with new charges generated by RESP (2.2.TYM_parametrization/resp/**keto_new_charge.mol2**).

2.3 Parameterization of the Metal Center

The metal ions have an important role in protein structure and function. These metal ions are usually coordinated with amino acid residues. To study the protein structure and the functional relationship of these metalloproteins using MD simulation, one needs the force field parameters describing the interactions of the metal with other atoms of the system. But the force field parameters are available only for standard amino acid residues and standard organic molecules. Coordination of metal ions with amino acid residues varies from protein to protein, and these coordinated amino acid residues are considered as nonstandard residues during MD simulation if a so-called bonded model is used. The force fields for these metal centers need to be generated to perform an MD simulation. To illustrate such parameterization, here we use thebaine 6-Odemethylase (T6ODM), a non-heme 2-oxoglutarate/Fe(II)-dependent dioxygenase that catalyzes regiospecific 6-O-demethylation of morphine precursors thebaine (i.e., (5α)-3,6-dimethoxy-17-methyl-6,7,8,14-tetradehydro-4,5-epoxymorphinan) and oripavine ((5α)-6-methoxy-17-methyl-6,7,8,14-tetradehydro-4,5-epoxymorphinan-3-ol) in morphine biosynthesis pathway. The active site of T6ODM (PDB ID: 5O7Y) hosts a Fe ion coordinated with HIS238, ASP240, HIS295, succinate, and an oxo ligand. Although there are many methods for modeling the metal center, here we are discussing the bonded model with the Seminario method to extract force constant from the Hessian matrix (matrix of second-order partial derivatives of the energy with respect to the atomic coordinates) [3].

2.3.1 Protocol in Brief to Parametrize the Metal Center by the Seminario Method

1. Generate the metal center small model (Fig. 6) from the PDB file.
2. Optimize the geometry of the small model (equilibrium bond lengths and valence angles made by the metal are used as reference values in the force field).
3. Calculate harmonic frequencies for the optimized small model to get the Hessian matrix.

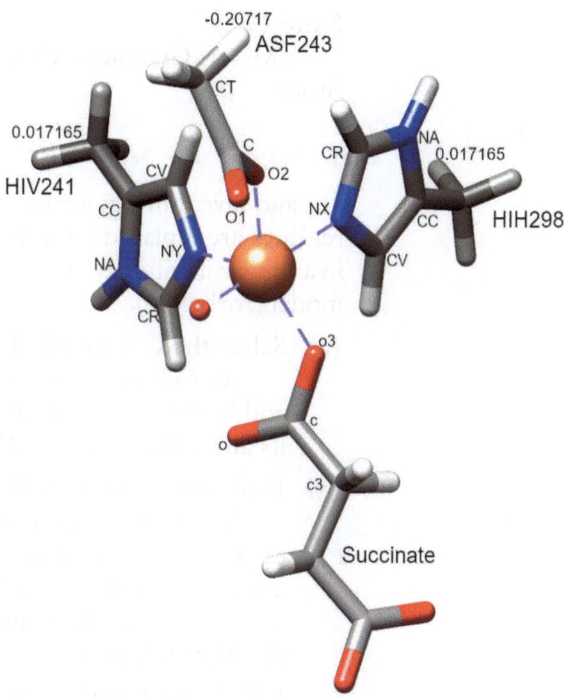

Fig. 6 Optimized T6ODM metal center with labeled point charges on H replacing CA atoms and atom types for atoms considered to make AS. frcmod parameter file

4. From the formatted checkpoint file containing the Hessian extract the missing force constants needed to prepare the **AS.frcmod** parameter file.
5. Calculate the electrostatic potential (ESP) for the optimized small model to get a **sm.gesp** file containing the ESP values computed for an array of points around the model (*see* 1.3. Fe_parametrization/gesp).
6. Fit the atomic partial charges of atoms with the RESP procedure of AMBER using the **sm.gesp** as an input file.
7. Generate prep files (containing topology information, atom types and atomic charges) for nonstandard residues.

2.3.2 Protocol in Detail

1. Prepare the **sm.xyz** coordinate file for Fe center (small model):

   In our PDB (PDBID: 5O7Y) file, there were missing residues from 38 to 40. They were regenerated by the loop modeling procedure available in MODELLER. The oxygen atom of 1,2-ethanediol coordinated with the metal site was replaced by the oxo ligand, the rest of the atoms of 1,2-ethanediol were removed, and Ni(II) was replaced with Fe(IV).

   Add hydrogen atoms to the PDB file using the H++ web server (http://biophysics.cs.vt.edu/).

   Then load the modified PDB file *t6odm.pdb* in VMD, extract the Cartesian coordinates of the metal center (*see* **Note 1**).

   VMD: Graphics→Graphical Representation→Selected Atoms

```
resid 241 243 298 368 369 370 and (sidechain or name CA)
```

   and save it in xyz format (*sm.xyz*). CA atoms of amino acid residues are replaced with hydrogen atoms using XYZViewer, in a similar manner as shown for the construction of the cluster model (*see* below).

   (a) Select the C atom which is needed to be replaced with H

   XYZViewer: File→Load structure→select *sm.xyz* file and load it→select the C by clicking on it, also note down its atom number, e.g., 20.

   (b) Replace C atom with H atom

   XYZViewer: Edit→Show matrix→"Matrix Editor" opens→scroll down to atom number 20 and click on C atom→"Periodic Table" opens, click on H atom→selected the C atom that is replaced with H→close the Matrix Editor.

   (c) Edit bond length between replaced H and C atom to which it is connected

XYZViewer: Edit→Edit in internal coords→a single atom→select the atom to edit, and three more atoms to define its internal coordinates→"Set coordinate window" opens→set the "Bond length" 1 Å→click on OK button.

2. Geometry optimization of a small model

For geometry optimization, we need a Gaussian input file.

```
#sm-opt.com

%chk=sm-opt.chk
%mem=6GB
# ub3lyp/lanl2dz 5d scf=(xqc,maxcycle=350) nosymm
EmpiricalDispersion=GD3BJ Opt=modredundant
----a blank line----
sm optimization
-----a blank line-----
0 5
XYZ coordinates
-----a blank line-----
X 1 F
-----two blank lines-----
```

%chk=sm-opt.chk: This line is to specify the name of the checkpoint file (*sm-opt.chk*), the results of the calculation are written in this file during calculation and it can be further used as a starting point of the second calculation.

%mem=6GB: Maximum amount of memory to be used for calculation.

 # ub3lyp/lanl2dz: We are using the *uB3LYP/lanl2dz* [4] level of theory. A different level of theory can be used, but the same level of theory has to be used for geometry optimization and force constant (Hessian) calculations. *5d* is for using five d-type orbitals (as opposed to six Cartesian d functions) *SCF=(xqc,maxcycle=350)*, for SCF procedure with a maximum of 350 iterations, *nosymm* to prevent Gaussian from trying to use symmetry during the calculation. *EmpiricalDispersion=GD3BJ* [5], to enable empirical dispersion (D3) with the Becke–Johnson dumping. *Opt=modredundant*, to freeze the Cartesian coordinates of hydrogen atoms that replace CA atoms of amino acid residues to prevent any dramatic changes in the overall geometry of the small model during optimization. Keywords can be written in more than one line. This section is terminated by a blank line.

sm optimization: Title section—a brief description of calculation, this is also terminated by a blank line.

0 5 and XYZ coordinates: Molecular specification section to specify charge (0) and spin multiplicity (quintet) of the model followed by xyz Cartesian coordinates.

X 1 F: Here we can specify the atom(s) with frozen coordinates by the ***modredundant*** keyword.

Command to run Gaussian:

```
g16 sm_opt.com
```

After geometry optimization, a small model should be checked for the integrity of the complex (i.e., if any of the coordination bonds did not get broken and the geometry of the model is resembling that in the crystal structure). Sometimes the accuracy of a metal site geometry in a protein is poor, e.g., due to ambiguity of electron density or lack of reference structures, so some differences between the experimental and computational structures may occur. Also one has to remember that the complex is optimized without explicit constraints imposed by the protein. Sometimes the enzymes strain the coordination of the metal active site to influence its electronic structure [6].

3. Frequency calculation for optimized small model

   Frequency calculation Gaussian input file (1.3.Fe_parametrization/optimization/**sm-opt.fq.com**):

```
#sm-opt.fq.com

%oldchk=sm-opt.chk
%chk=sm-opt.fq.chk
%mem=6GB
# ub3lyp/lanl2dz 5d scf=(xqc,maxcycle=350) nosymm Guess=Read
Geom=AllCheckpoint
EmpiricalDispersion=GD3BJ Freq
```

We use keyword *Guess=Read* to take orbitals from the previous optimization checkpoint file, *sm-opt.chk,* as an initial guess, and *Geom=AllCheckpoint* to retrieve from it also molecular geometry, charge, and multiplicity of the model. *Freq* is used to tell Gaussian that this is input for frequency calculation. After finishing the frequency calculation, we format the checkpoint file *sm-opt.fq.chk*:

```
formchk sm-opt.fq.chk sm-opt.fq.fchk
```

which is suitable for various visualizing programs, e.g., XYZViewer. From this formatted checkpoint, bond and angle values as well as force constant are extracted to construct the *frcmod* parameter file.

4. Make *frcmod* parameter file

   (a) Extract bond force constant

XYZViewer: File→Load structure→select *sm-opt.fq.fchk* and load it→Calculate→Forces from hessian (beta)→Forces→click the two atoms for which bond length and force constant has to be measured→"Force calculation" window opens displaying values of bond length and its force constant for that particular bond.

As an alternative to the XYZViewer program, to extract force constants, one can use Hess2FF software developed by Ulf Ryde [7].

Bonds to be measured for a metal center (Fig. 6):

- Between metal and atoms coordinated to a metal.
- (Optionally) between coordinated atom and atom connected with it.

(b) Extract angle force constant

XYZViewer: File→Load structure→select the formatted checkpoint file *sm-opt.fq.fchk* and load it→Calculate→Forces from hessian(beta)→Angle force→click the three atoms for which angle force constant has to be measured→"Force calculation" window opens displaying values of angle and its force constant for that particular angle.

Angles to be measured for the metal center:

- Between three atoms one must be metal.
- (Optionally) between three atoms, one of them must be an atom coordinated with metal.

Atom types that need to be provided in the *AS.frcmod* file are labeled in Fig. 6. Now make two columns in the Excel sheet: r for bond length and k for force constant. After measuring values for all the bonds, divide column k by 2 (this is due to the difference in assumed formula, i.e., it is $K(r\text{-}r_{eq})^2$ for Amber and $1/2\ k(r\text{-}r_{eq})^2$ for XYZViewer).

Construct the **AS.frcmod** file for the metal center which looks like below:

```
MASS
NX 14.01 0.530
FE 55.85

BOND
FE-ox 365.0 1.642
NY-FE 61.0 2.110

ANGL
NY-FE-o3 41.0 90.0
FE-NX-CV 107.0 125.0
```

```
DIHE
ox-FE-NY-CV 3 0.00 0.0 3.0
O2-FE-o3-c 3 0.00 0.0 3.0

NONB
NX 1.8240 0.1700
FE 1.409 0.0130
```

We are not measuring dihedral parameters, rather when we use our ***AS.frcmod*** file (without dihedral parameter) to generate topology and initial coordinate files by the leap program. The output of the first leap run contains a list of the missing dihedral angle parameters. So, from there we take those dihedral angles specifications and add them to our frcmod file while keeping the values of their barrier heights set to zero, as in the above ***AS.frcmod*** file.

5. ESP (electrostatic potential) calculation using Gaussian (1.3. Fe_parametrization/gesp/*sm-opt.resp.com*)

```
#sm-opt.resp.com

%oldchk=sm-opt.chk
%chk=sm-opt.resp.chk
%mem=6GB
# ub3lyp/lanl2dz 5d scf=(xqc,maxcycle=350) nosymm Guess=Read
EmpiricalDispersion=GD3BJ Pop=(MK,ReadRadii) IOp(6/33=2) iop
(6/41=10) iop(6/50=1)
----molecular specification, and a blank line-----
Fe 1.409
-----a blank line-----
sm.gesp
-----a blank line-----
sm.gesp
-----two blank line-----
```

pop=(MK, ReadRadii): To calculate atomic charges according to Merz-Singh-Kollman scheme, ReadRadii to read atomic radius for Fe from the input file.

iop(6/33=2): To write coordinates and ESP potential values.

iop(6/41=10): Specify the number of layers at which the ESP potential is calculated, i.e., 10.

iop(6/50=1): To write Gaussian output file *sm.gesp*, which is further used for charge fitting by Antechamber.

After *xyz* Cartesian coordinates, we specified van der Waals (vdW) radii of Fe and the name of the Gaussian *sm.gesp* output file to write ESP data.

6. RESP charge fitting using Antechamber (*see* /1.3.Fe_parametrization/resp for files)

   Run the Antechamber [8] using *.gesp* file and generate input file for RESP fitting.

```
antechamber -i sm.gesp -fi gesp -o sm.mol2 -fo mol2 -c resp -j 5
```

   This step also generates the *quot* (point charges for all the atoms) file. *quot* file is further modified to the *qin* file by making charges of all the atoms zero except for H atoms replacing CA. Since here we are generating a prep file for AMBER MD simulation, we need to make sure that the total charge of the system is an integer. For this purpose, we sum the backbone atoms' (N, H, CA, HA, C, O) charges of respective amino acid residues (in the metal center), which are available in the AMBER prep file for standard residues, and ascribe these total charges to H (Fig. 6) atoms replacing the CA. This *qin* file is further used by RESP in the next step of charge fitting.

   A two-step charge fitting procedure by RESP is the next step.

   Step 1: All charges of atoms are allowed to vary except those for hydrogen atoms replacing the CA atoms. Charges of these atoms (Fig. 6) are read from the qin file.

```
resp -O -i sm_RESP1.IN -o sm_RESP1.out -e ANTECHAMBER.ESP -p
sm_RESP1.punch -q qin -t sm_RESP1.qout
```

   Step 2: Final fit with constraints on charges of the degenerate atoms. For example, one of the hydrogen atom of the methyl group in an amino acid model is a hydrogen atom replacing alpha carbon and the other two hydrogen atoms, which are bound to Cβ, are chemically equivalent (degenerate), and thus, these two have the same charge.

```
resp -O -i sm_RESP2.IN -o sm_RESP2.out -e ANTECHAMBER.ESP -p
sm_RESP2.punch -q sm_RESP1.qout -t sm_RESP2.qout
```

7. Generate prep files (1.3.Fe_parametrization/paramerization)

   From 1.3.Fe_parametrization/resp./**sm_RESP2.qout**, point charges of Fe, oxo, and succinate and side chain atoms of HIS241, ASP243, HIS298 are extracted to construct their respective *prep* files. First, generate prep files for respective nonstandard amino acids from AMBER prep file *all_amino03.in* and then point charges (last column) of side chain atoms are replaced with point charges obtained after RESP fitting. Below is a prep file for ASP243:

```
#ASF.in
0 0 2
```

```
ASF
asf.db4
 ASF INT 1
 CORR OMIT DU BEG
 0.00000
 1 DUMM DU M 0 -1 -2 0.000 0.000 0.000 0.00000
 2 DUMM DU M 1 0 -1 1.449 0.000 0.000 0.00000
 3 DUMM DU M 2 1 0 1.522 111.100 0.000 0.00000
 4 N N M 3 2 1 1.335 116.600 180.000 -0.558201
 5 H H E 4 3 2 1.010 119.800 0.000 0.319676
 6 CA CT M 4 3 2 1.449 121.900 180.000 0.007225
 7 HA H1 E 6 4 3 1.090 109.500 300.000 0.082375
 8 CB CT 3 6 4 3 1.525 111.100 60.000 0.718039
 9 HB2 HC E 8 6 4 1.090 109.500 300.000 -0.140144
 10 HB3 HC E 8 6 4 1.090 109.500 60.000 -0.140144
 11 CG C B 8 6 4 1.527 109.470 180.000 0.281945
 12 OD1 O2 E 11 8 6 1.260 117.200 90.000 -0.451834
 13 OD2 O2 E 11 8 6 1.260 117.200 270.000 -0.664639
 14 C C M 6 4 3 1.522 111.100 180.000 0.443199
 15 O O E 14 6 4 1.229 120.500 0.000 -0.501445

IMPROPER
 -M CA N H
 CA +M C O
 CB OD1 CG OD2

DONE
STOP
```

For Fe, oxo, and succinate, we can use antechamber to generate mol2 files then prep files from mol2 and further replace the point charges with the above-calculated charges. Here we need to construct the two separate prep files for HIS241 and HIS298 because the coordination orientation of both residues with Fe is unique, HIS241 is in an axial, whereas HIS298 is in an equatorial position, so there is a need to distinguish them as their nitrogen atoms coordinating Fe make valence angles of very different values, e.g., NX-Fe-oxo—178.7° and NY-Fe-oxo—93.8°.

3 MD Simulation

Before QM:MM calculation, we usually need to do a MD simulation for the enzyme-substrate complex using the AMBER MD simulation package.

Protocol in brief for performing an MD simulation using AMBER:

1. Generate a prmtop file (description of molecular topology and force field parameters) and an inpcrd file (atom coordinate, velocity information, and periodic box dimensions).
2. Optimize the geometry of the model at 0 K (energy minimization).
3. Heat the system and equilibrate its density.
4. Perform production run.

3.1 Protein Preparation

Generate the prmtop and inpcrd files

Editing in the PDB file

Calculate the pK_a of amino acid residues using PDB2PQR (http://server.poissonboltzmann.org/pdb2pqr), then change the name of HIS residues to HID (delta protonation), HIE (epsilon protonation), HIP (positively charged HIS) in the PDB file accordingly. Since HIS241, ASP243, HIS298 coordinate metal center and are nonstandard residues, thus, we rename them to HIV241, ASF243, and HIH298, respectively. This PDB file has also previously docked two molecules of substrate thebaine. Since we already generated prep and parameter files for the metal center, we generate these files for thebaine using Antechamber, as is usually done for organic molecules (*see* Subheading 2.1).

Run leap program to generate *prmtop* and *inpcrd*

```
tleap -f leap.in (files in 3.MD-simulation/3.1 leap)
#leap.in

source leaprc.protein.ff03.r1
source leaprc.water.tip3p
loadamberparams AS.frcmod
loadAmberPrep HIV.in
loadAmberPrep ASF.in
loadAmberPrep HIH.in
loadAmberPrep Fe.in
loadAmberPrep oxo.in
loadAmberPrep sin.in
loadAmberPrep thb.in
mol = loadpdb t6odm.pdb
bond mol.residue1.atom1 mol.residue2.atom2
solvateBox mol TIP3PBOX 10.0
addions mol Na+ 0
charge mol
saveAmberParm mol t6odm.prmtop t6odm.inpcrd
charge mol
quit
```

To create the *t6odm.prmtop* and *t6odm.inpcrd* files, we use ff03 force field [9, 10] for protein (other protein force fields can also be used), and above-generated force field parameters (*AS.frcmod)* and prep (HIV.*in*, *ASF.in*, *HIH.in*, *Fe.in*, *oxo.in*, *sin.in*, *thb.in*) files for nonstandard atom types and residues. Since we perform explicit water MD simulation, we add TIP3P water to solvate our system and also add counter ions to neutralize the net charge. Because we adopted the bonded model for the metal center, we form bonds between Fe and coordinated atoms of respective residues using *bond* command.

Periodic boundary: When we add solvent to solute (protein in our case), we make a finite box size in which some of the solvent molecules are present at the boundary between solute and solvent, some in the bulk solvent, and some at the edge of the box. During MD simulation, water molecules at the edge may move out of the box. To avoid such effect, due to finite box size, we make the system infinite by applying the periodic boundary condition, i.e., the unit cell is replicated in all three directions to fill the space with its images. As a result, if a molecule leaves the unit cell through one wall, it reappears on the opposite side. Thus, through periodic boundary conditions, we construct an infinite system consisting of images of a unit cell.

3.2 Geometry Minimization

There is a high chance that the PDB file we are using for MD simulation has some close contacts, nonoptimal residue conformation, the sub-optimal position of added protons. Also, the addition of an explicit solvent can create close contacts between the solute and solvent. Therefore, the geometry minimization is needed to deal with those problems—otherwise, MD simulation will be unstable due to van der Waals clashes. The minimization procedure we adopt consists of three steps (*see* 3.MD-simulation/3.2. MM for files):

#Step1: Minimization of water and Na^+ ions

```
#min1.in

minimization of water and Na+ ions
&cntrl
 imin=1,
 ntpr=50,
 ntb=1, cut=10.0,
 ntr=1,
 maxcyc=5000, ncyc=500,
 restraint_wt=500.0, restraintmask=':1-372'
 /
```

imin=1: To do minimization.

ntpr=50: Energy information is printed to AMBER output file at every ntpr steps.

ntb=1: To apply constant volume periodic boundary condition.

cut=10.0: Cutoff distance for non-bonded calculation (should not be <8).

ntr=1: Turn on positional restraints on mask atoms.

maxcyc=5000: Maximum number of minimization cycles.

ncyc=500: For the first 500 steps of maxcyc, perform for steepest descent minimization, then switch to the conjugate gradient.

restraint_wt=500.0: Weight (harmonic force constant) for positional restraint.

restraintmask=':1-372: Mask atom for applying restraint.

There are many algorithms for minimization. The commonly used are the steepest descent which quickly removes larger strains but converges slowly and the second method is the conjugate gradient, which converges faster than the steepest descent. Thus, a combination of both is used to remove larger strains in the initial phase (500 cycles) and then faster convergence in the second phase (next 4500 cycles).

```
pmemd.MPI -O -i min1.in -o min1.out -p t6odm.prmtop -c t6odm.
inpcrd -ref t6odm.inpcrd -r min1.rst
```

Since we are using positional restraint on mask atoms so we are giving *t6odm.inpcrd* as a reference structure (-ref) to which we can impose restraint.

#Step2: Minimization of water and Na^+ ions with decreased restraint weight using restart file *min1.rst* from #Step1

```
#min2.in
```

minimization of water and Na+ ions with decreased restraint weight

```
&cntrl
 imin=1,
 ntpr=50,
 ntb=1, cut=10.0,
 ntr=1,
 maxcyc=5000, ncyc=500,
 restraint_wt=10.0, restraintmask=':1-372'
/
pmemd.MPI -O -i min2.in -o min2.out -p t6odm.prmtop -c min1.rst
-ref min1.rst -r min2.rst
```

#Step3: Minimization of the whole system using restart file *min2.rst* from #Step2

```
#min3.in
```

minimization of whole system

```
&cntrl
 imin=1,
 ntpr=50,
 ntb=1, cut=10.0,
 ntr=1,
 maxcyc=10000, ncyc=500,
/
pmemd.MPI -O -i min3.in -o min3.out -p t6odm.prmtop -c min2.rst
-r min3.rst
```

3.3 Heating, Density Equilibration and Production

3.3.1 After Minimization, Heat the System from 0 to 300 K (3. MD-Simulation/2.3. MD)

```
#NVT.in
Heating for 100 ps
&cntrl
 imin=0
 ntx=1, irest=0,
 ntpr=500, ntwr=500, ntwx=5000, iwrap=1,
 ntc=2, ntf=2, ntb=1, cut=10.0,
 ntr=1,
 nstlim=50000, dt=0.002, nscm=1000,
 tempi=0.0, temp0=300.0, ntt=3, gamma_ln=0.05,
&end
restrain backbone of the solute
1.0
FIND
* * M *
SEARCH
RES 1 372
END
END
```

imin=0: To perform MD simulation.

ntx=1: Read coordinate from restart coordinate file.

irest=0: It is not a restart of the MD simulation, velocity in the input coordinate file are ignored.

ntpr=50: Energy information is printed to mdout file at every ntpr steps.

ntwr: At an every ntwr step, restart file is written.

ntwx: At an every ntwx step, coordinates are written to mdcrd file.

iwrap=1: Wrap the coordinate to primary box while writing the restart and mdcrd file.

ntc=2: Use SHAKE to constrain bonds involving hydrogen atoms for increased efficiency of MD simulation.

ntf=2: Force evaluation, should be used with ntc=2.

ntb=1: To apply constant volume periodic boundary condition.

cut=10.0: Distance cutoff [Å] for non-bonded interactions.

nstlim=50000: Number of MD steps to perform.

dt=0.002: Time step [ps].

nscm=1000: Remove the translational and rotational motion of the whole system (centre of mass) at every nscm steps.

tempi=0.0: Initial temperature of the system.

temp0=300.0: Target temperature [K] of the system after heating.

ntt=3: Turn on Langevin dynamics.

gamma_ln=0.05: Collision frequency [ps^{-1}] when ntt=3.

The Berendsen temperature coupling (ntt=1) is a weak coupling; it cannot ensure that temperature is even all over the system, which can lead to the hot solvent cold solute. Meanwhile, the Langevin thermostat (ntt=3) with collision frequency is more efficient for equilibrating the temperature and is thus recommended. There are different thermodynamic ensembles used during MD simulation. The isobaric-isothermal NPT ensemble is closer to the laboratory condition, but the heating from the low temperature at NPT for temperature equilibration is inaccurate. Thus, heating is performed using the canonical NVT ensemble.

```
pmemd.MPI -O -i NVT.in -o NVT.out -p t6odm.prmtop -c min3.rst
-ref min3.rst -r NVT.rst
```

3.3.2 Density Equilibration

Heating at NVT creates void space in the solvent and for a periodic boundary, NPT is the only way to equilibrate the density. Thus, it is necessary to bring the system to the final temperature, using NVT with ntt=3, and then switch to NPT, with ntt=3, for density equilibration.

```
#NPT.in
```

Equilibration for 1 ns

```
&cntrl
 imin=0
 ntx=5, irest=1,
 ntpr=500, ntwr=500, ntwx=5000, iwrap=1,
 ntc=2, ntf=2, ntb=2, pres0=1.0, ntp=1, taup=2.0, cut=10.0,
 nstlim=500000, dt=0.002, nscm=1000,
 tempi=300.0, temp0=300.0, ntt=3, gamma_ln=0.05,
&end
restrain backbone of the solute
1.0
FIND
* * M *
SEARCH
RES 1 372
END
END
```

ntx=5: Option to read coordinate, velocity, and box information from a restart file.

irest=1: Restart the simulation after reading coordinate and velocity from a previously saved restart file.

ntb=2: To apply constant pressure periodic boundary condition.

ntp=1: MD with isotropic position scaling.

pres0=1.0: Reference pressure [bar] when ntp>0.

taup=2.0: Pressure relaxation time [ps] when ntp>0.

```
pmemd.MPI -O -i NPT.in -o NPT.out -p t6odm.prmtop -c NVT.rst -r
NPT.rst
```

4. Production run

Once the temperature and pressure are equilibrated, a longer production run is performed for T = const., p = const. conditions (NPT ensemble), which are meant to mimic laboratory conditions.

```
#NPT.in
```

3.3.3 Production Run 5 ns

```
&cntrl
 imin=0
 ntx=5, irest=1,
 ntpr=5000, ntwr=5000, ntwx=5000, iwrap=1,
 ntc=2, ntf=2, ntb=2, pres0=1.0, ntp=1, taup=2.0, cut=10.0,
 ntr=1,
 nstlim=2500000, dt=0.002, nscm=1000,
 tempi=300.0, temp0=300.0, ntt=3, gamma_ln=0.05,
/
```

The production run is a long MD simulation, and to get an unbiased sampling, the whole production run should be broken into many small consecutive runs rather than attempted in a single long MD run. For example, if we have to run 100 ns, then instead of doing a single 100 ns run, it should be divided into, e.g., 20 consecutive runs of 5 ns each.

```
pmemd.MPI -O -i prod.in -o prod.out -p t6odm.prmtop -c NPT.rst
-r prod.rst
```

3.4 Analysis of MD Simulation

3.4.1 RMSD Analysis

Before we proceed with further analysis of the MD trajectory, we have to check if the trajectory has converged. One method to check the convergence is to look at the RMSD profile along the trajectory. Here, as a reference structure for calculating RMSD, we use the energy minimized structure **min3.pdb** obtained at the end of this section. Files to conduct RMDS analysis are available at 3.MD-simulation/3.4 Analysis of MD simulation/rmsd.

```
#rmsd.in
```

```
trajin prod-1.mdcrd
trajin prod-2.mdcrd
trajin prod-3.mdcrd
strip :Na+,Cl-,WAT
autoimage origin
reference min3.pdb
rms reference mass out t6odm.rms time 5.0 :1-367@N,CA,C

cpptraj -p t6odm.prmtop <rmsd.in> rmsd.out
```

Above input file uses three MD trajectory files to calculate the RMSD.

strip: Na^+,Cl^-,WAT: Strips all ions and water molecules so that process is faster.

autoimage origin: Centers the periodic boundary trajectory on the first molecule and re-image.

reference min3.pdb: Reference structure to calculate RMSD.

rms reference mass out t6odm.rms time 5.0 :1-367@N,CA,C: Calculate the mass-weighted RMSD of protein backbone atoms using reference structure and write to *t6odm.rms* output file.

Once we get the 3.MD-simulation/3.4 Analysis of MD simulation/rmsd **/t6odm.rms** file, by plotting it, we can check the convergence of the trajectory. Initially, there is a major fluctuation in RMSD, which means changes in the molecule's conformation, and then the RMSD starts to level off. This indicates that our trajectory has converged, i.e., there are no further major changes in conformation; now we can proceed with further analysis.

3.4.2 Cluster Analysis

After MD simulation, we need to retrieve the representative structure(s) of our protein from our MD trajectory. In this tutorial, this is achieved by clustering analysis. This procedure compares the geometries of selected atoms or residues (by calculating RMDS) and generates the clusters of MD frames of similar conformations. Structures belonging to one cluster are similar and vary from the other clusters. Differences in structures which is the basis of clustering are determined by coordinate RMSD. There are many algorithms for clustering, here we are using DBScan. All files are available in SI at 3.MD-simulation/3.4 Analysis of MD simulation/clustering.

Steps in cluster analysis:

1. Estimate the epsilon and minpoint parameters.
2. Perform actual clustering using the above-determined parameters.

Estimation of epsilon and minpoint parameters

To do the clustering analysis using DBScan, we need two parameters: epsilon (distance cutoff to form a cluster) and minpoint (minimum number of points to form a cluster). These parameters can be estimated from a so-called k-dist plot, which can be generated by CPPTRAJ.

```
#k-dist.in

trajin prod-1.mdcrd
trajin prod-2.mdcrd
trajin prod-3.mdcrd
autoimage origin
strip :WAT, Na+, Cl-
cluster c1 dbscan kdist 1 rms :368@FE:369@O:370@C1,O1,
O2:372@C19,O2,C6
cluster c2 dbscan kdist 2 rms :368@FE:369@O:370@C1,O1,
O2:372@C19,O2,C6
cluster c3 dbscan kdist 3 rms :368@FE:369@O:370@C1,O1,
O2:372@C19,O2,C6
cluster c4 dbscan kdist 4 rms :368@FE:369@O:370@C1,O1,
O2:372@C19,O2,C6
cluster c5 dbscan kdist 5 rms :368@FE:369@O:370@C1,O1,
O2:372@C19,O2,C6

cpptraj -p t6odm.prmtop <k-dist.in> k-dist.out
```

Cpptraj will make five kdist (**Kdist.1.dat**, **Kdist.2.dat**, **Kdist.3.dat**, **Kdist.4.dat**, and **Kdist.5.dat**) files from three MD trajectories based on five different values of kdist (Å) and RMSD of mask atoms. We considered these atoms as mask atoms because their mutual arrangement is pivotal for the progress of the enzymatic reaction, and we wanted to get a good starting point for the next step of QM:MM calculations.

By plotting these kdist, we can estimate epsilon and minpoint. Initially, the slope of the curve is steep and then becomes flat (increase in density). The distance around which the curve starts to flatten is taken as epsilon value. The kdist at which the shape of the curve does not change too much is considered as minpoint. Generally, after kdist = 4, the shape of the curves does not change too much. But it is suggested to do some clustering experiments with different kdist values. Further, clustering based on different kdist values starting from 1..., the number of clusters decreases with the increase in kdist value. Thus, it helps us to choose appropriate epsilon values for clustering so that we can get enough clusters to get a good representative structure for our MD trajectory.

Clustering

```
#cluster.in

trajin prod-1.mdcrd
trajin prod-2.mdcrd
autoimage origin
strip :WAT, Na+, Cl-

cluster C0
dbscan minpoints 4 epsilon 0.3 sievetoframe
rms :368@FE:369@O:370@C1,O1,O2:372@C19,O2,C6
sieve 10 random
out cnumvtime.dat sil Sil
summary summary.dat
info info.dat
cpopvtime cpopvtime.agr normframe
repout rep repfmt pdb
singlerepout singlerep.nc singlerepfmt netcdf
avgout Avg avgfmt restart
```

Clustering parameters

minpoint = Minimum number of points to form cluster = 2.

epsilon = Distance criteria to form a cluster = 0.3.

Distance matrix options

rms = Use RMSD of mask atom to calculate the distance between frames.

sieve 10 = Every 10 frames are considered for initial clustering then sieved frames added to those clusters, *random* keyword used to select random frames.

Output options

out cnumvtime.dat: This file contains information for cluster number vs. time. Noise frames are assigned to a value of −1.

summary summary.dat: Overall summary of clustering (number of clusters formed).

info info.dat: It contains all the clustering results. This file also has information on DBI and pSF values, which can be later used, e.g., to compare results obtained with different clustering algorithms. Lower DBI and higher pSF values indicate better clustering.

```
cpptraj -p t6odm.prmtop <cluster.in> cluster.out
```

Running the above command generates the following output files. The **summary.dat** (3.MD-simulation/3.4 Analysis of MD simulation/clustering) gives information about the number of clusters formed. Representative structures of each cluster are also generated, which we can use for our further studies like structure analysis, and as a starting structure for QM or QM:MM calculations. We can also use **cnumvtime.dat** (3.MD-simulation/3.4 Analysis of MD simulation/clustering) to once again check if the simulation converged. We should observe a stable population of the clusters for the phase of simulation where we have observed stable RMSD.

4 Cluster-Based QM Calculations

The first step in the investigation of the reaction pathways is the construction of the appropriate model representing the enzyme active site. Generally, there are two methods to approach this problem. The first is to use the whole protein with a significant portion of the solvent where only active site and residues with reagents are treated with QM level of theory while for the rest of the model, one applies simple molecular mechanics. This approach, known as QM:MM, is described in detail in Subheadings 6 and 7. In the other approach, one cuts from the whole protein-only residues of the active site and regents and applies the QM calculation for the description of all selected atoms. Such fragmentary models are known as "cluster models" (and the link to the statistical clustering analysis described in Subheading 3.4.2 is solely semantic). Such a cluster model of enzyme active site can be based on:

- Crystal structure of the enzyme, preferably with a substrate or substrate analog bound in the active site.
- Structure of enzyme–substrate complex derived from crystal structure that was relaxed during MD simulation, followed by minimization in MM.
- MD-relaxed structure of enzyme–substrate from the homology model.

Alternatively, a substrate can be substituted with a product, and in such a case, the reaction can be followed in reverse (i.e., starting from enzyme–product complex).

4.1 Construction of the Cluster Model

The construction of the cluster model should be done taking the following guidelines into account:

1. All reagents should be included in the model.
2. All residues that are known to be involved in the reaction are included in the model.
3. All solvent molecules that are known to interact via H-bonds with reagents or important protonable residues are included in the model.

4. In case the active site contains the metal ions, all its ligands are introduced into the model ensuring representative electronic surroundings of the metal complex.
5. The second shell of residues that form H-bonds or strong electrostatic interactions with primary residues can be included. Hydrophobic residues that surround the active site can also be included.
6. The residues are trimmed by cleavage at the nonpolarized C–C bond. The dangling bonds are saturated with H atoms.
7. The position of the capping H atoms and adjacent C atoms are frozen by Cartesian constraints.
8. If parts of the enzyme main chain have to be included (e.g., to model oxo anion hole), the cleavage site of the peptide backbone is end-capped by CH_3 groups yielding –NH–CH_3 or OC–CH_3. The position of C atoms and H atoms along the path of the main chain should be constrained.
9. The position of cofactors (e.g., NADH, FAD, heme) is constrained at the atoms that form covalent bonds with the protein, H-bonds, or other detectable interactions inside the protein (e.g., salt-bridges).
10. Constrains should allow conformational changes that would normally be accessible to residues. However, as the outer shell of protein is removed, sometimes it is necessary to introduce additional constraints preventing artificial changes of geometry during geometry optimization.
11. The size of the cluster model should be kept reasonably low (100–200 atoms) but can be extended if a lower level of theory is used (e.g., semi-empirical methods).

This procedure is demonstrated based on the active site of short-chain dehydrogenase/reductase (SDR) from *Aromatoleum aromaticum* (*S*)-1-phenylethanol dehydrogenase (Fig. 7a, *S*-PEDH) [12, 13]. The SDRs catalyze reversible NAD^+-dependent reaction of alcohols oxidation to ketones/aldehydes (pH optimum at neutral to slightly basic) or NADH-dependent reduction of ketones/aldehydes to alcohols (pH optimum in the range of 5–6).

The industrially interesting reaction, enantioselective reduction of ketones to chiral alcohols, starts with the binding of NADH and ketone. The carbonyl group of the ketone is bound by two H-bonds, formed with Ser (in *S*-PEDH Ser141) and Tyr (Tyr154). This ensures that the proper face of the carbonyl sp^2 atom is turned toward the NADH ring which transfers hydride during the reduction of the ketone to a secondary alcohol. However, to reduce ketone to alcohol, the enzyme also needs to protonate the carbonyl group. This process is achieved by the proton relay system, which consists of a H-bond network involving

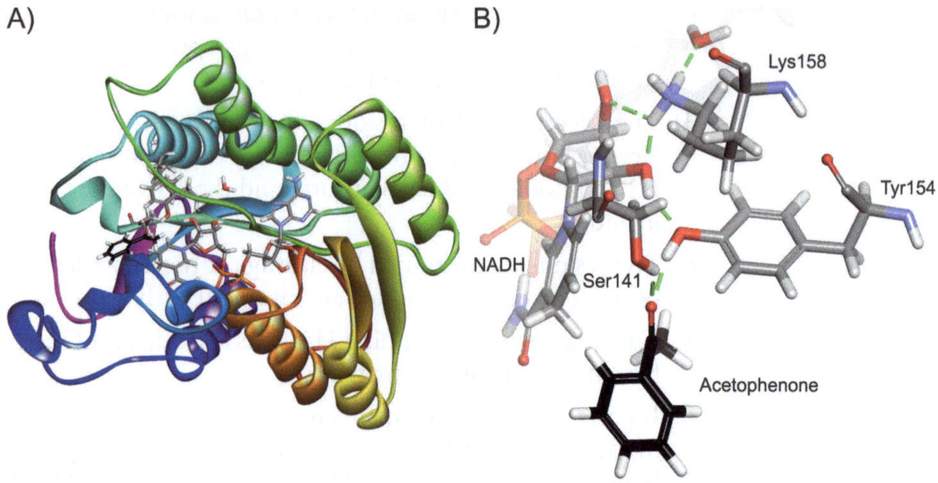

Fig. 7 (**a**) Structure of S-PEDH from *A. aromaticum* [11]; (**b**) active site of the enzyme with docked acetophenone; green dashed lines depict H-bonds (*see* 4.Cluster-QM/PEDH_acetophenon_H20.pdb)

Tyr154, alcohol group of the NADH ribose, the amine group of Lys (Lys158), and finally water molecule (Fig. 7b). To properly study this reaction, we need to include in the cluster model all these crucial elements:

- Ketone.
- Part of NADH, which contains at least the nicotinamide ring and ribose ring (the rest is not involved in the reaction and works as an anchor for cofactor binding).
- Residues involved in substrate binding and protonation (Ser 141 and Tyr154).
- Residues that transfer a proton to Tyr (ribose, Lys158).
- If we want to model the pH-dependence of the enzyme, the inclusion of water that binds to Lys may be considered. In this case, in acidic conditions (preferential for ketone reduction), the Lys residue is protonated ($-N\varepsilon H_3^+$) and in basic conditions, it is deprotonated ($-N\varepsilon H_2$).

The model is trimmed—in the case of amino acids, bonds between Cα and Cβ are replaced by C-H and their positions are constrained (Fig. 8, yellow). In the case of NADH, the C6–O bond is replaced by C–H, and again their coordinates are frozen. As a result, all residues can rotate along the Cα–Cβ axis, and movement of the ring is not constrained. In the end, we receive a very handy model that contains 94 or 95 atoms (depending on the protonation state of Lys). The 94-atom model has a charge of 0 and the 95-atom model a charge of +1.

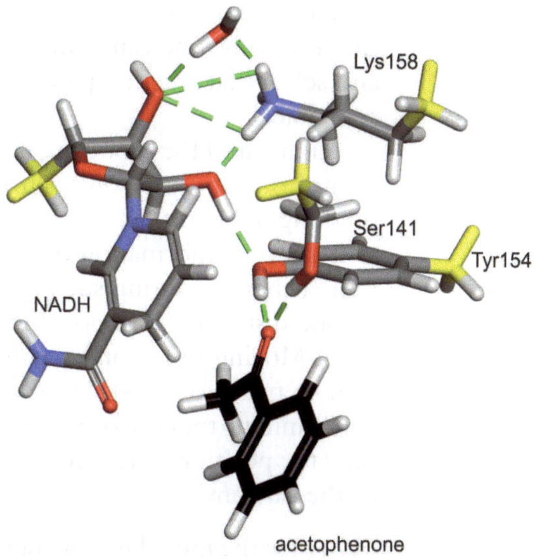

Fig. 8 The structure of the cluster model of the S-PEDH active site [13]. Yellow color indicates the position of the atom with constrained coordinates. The green dashed lines indicate the H-bond network. (4.Cluster-QM/)

4.2 Investigation of the Reaction Pathway with the Cluster Model

Before we describe the steps which lead to the modeling of the reaction mechanism, it is necessary to consider the chemistry of the hypothetical mechanism. It is always useful to draw the sequence of assumed steps. Also, one has to consider if the chosen methodology allows the investigation of the process. In this particular case, the reaction is relatively simple—it involves the transfer of a hydride from NADH to C atom of the keto group and a proton from Tyr to the oxygen atom of the keto group. As a result, we can safely use a spin-restricted approach as radicals are not produced during reactions (no homolytic C–H/O–H cleavage occurs). However, if the reaction involved homolytic C–H activation, one would have to use an unrestricted approach (with the possible breaking of orbital symmetry). Also if the enzyme contains a metal active site that can attain multiple possible spin states, one has to refer to experimental data on the subject or test different possibilities with theoretical calculations. It should be underlined here that type of selected theoretical method may influence the results. For example, it has been demonstrated that certain pure DFT functionals tend to predict the lower spin state of the metal complexes if compared to hybrid-type functionals which contain various proportions of the Hartree–Fock-based exchange part [11].

We can imagine the process of ketone reduction in two manners: hydride and proton can be transferred concomitantly or in two steps. The latter possibility of course can proceed in two ways: first, the hydride is transferred and then the proton or vice versa. It is very difficult to determine experimentally the order of events taking place in the enzyme's active site. Therefore, the theoretician usually has to consider all chemically viable scenarios and test all of them. Finally, one can compare the energy profiles of all scenarios and select that reaction pathway, which is thermodynamically most probable.

In most of the modeling programs, we can control the position of atoms that are involved in the reaction and move them along predesigned paths. This procedure is called a potential energy scan as it delivers information on how the energy and geometry of our model changes during such transfer (we get energy in the function of one distance). It is possible to move more than one atom at a time. Moving two atoms results in a 2D scan (energy in the function of two distances). However, in this simple case, to study the concomitant transfer of hydride and proton, one needs to control only the position of the hydride (i.e., the distance of H^- to C atom of the carbonyl).

1. After the model is constrained, we conduct a complete geometry optimization of the model at the selected level of theory. As the model involves lots of H-bond, it is good to use a basis set that enables density polarization also on the light atoms (e.g., 6-31g(d,p)) (*see* **Note 2**).

   Example of the root section in Gaussian:

```
# opt=modredundant freq b3lyp/6-31g(d,p) nosymm geom=connec-
tivity pop=regular
```

   The modredundant keyword introduces additional dihedrals associated with H-bonds (four heavy atoms) and between substrate and NADH. This makes the optimization procedure more stable and less prone to errors. The example of the input files is available in the supplement (4.Cluster-QM/**S-PED-H_ES_p-H_opt.gjf**)

2. We perform a vibration analysis of the obtained geometry (freq keyword). This provides two types of information: the number of imaginary frequencies and corrections to electronic energy derived from vibration calculations (zero-point energy, thermal energy, enthalpy, free energy corrections). If the geometry is properly optimized, we should not observe any imaginary frequencies that are not associated with the imposed constraints (i.e., fix points of our model). Additionally, in many programs, it is possible to remove those atoms from vibration analysis so, in the end, we do not see their contributions to thermodynamic corrections. In case we still observe a nonzero number of imaginary frequencies, the optimization should be repeated. If the problem persists, it is good to visualize this vibration mode and potentially change the conformation of the associated part of the model. Then the geometry optimization has to be repeated (*see* **Note 3**).

3. We select reaction coordinate—in this case, it is a distance between the H-atom of the nicotinamide ring and carbonyl C atom.
4. The distance (in this case 2.74 Å) is divided into steps. Some programs only provide even steps scan (e.g., Gaussian, Turbomole), other enable custom steps (e.g., Jaguar). Our goal is to catch the transition state so we need to have a very good sampling of the region where we expect TS to occur. In case we have no prior knowledge about the localization of the transition state, it is best to use even sampling. For H-atom transfer, the step should be no bigger than 0.1 Å. In our case, we scan a range of 2.7 down to 1.1 Å in 16 steps (the endpoint of the scan is standard C-H distance of alcohols, which is ~1.1 Å).

   In order to conduct such a scan with the Gaussian package, the following section is added to the modredundant section at the end of the file (S-PEDH_p-H_scan_TS.gjf):

```
B 77 67 S 16 -0.100000
```

   (B—bond, 77 67—numbers of atoms, S—scan, 16—number of steps, -0.1 step size in Å)
5. In each step (i.e., for a given distance of H---$C^{carbonyl}$), the program conducts geometry optimization and calculates the lowest energy which complies with imposed restrains (i.e., given H---$C^{carbonyl}$ distance).
6. The energy of the optimized geometries along the scanned pathways is analyzed. The geometry with the highest energy on the profile is selected as the closest to the transition state.
7. We conduct a vibrational analysis for the selected geometry (*see* 4.Cluster-QM/**S-PEDH_p-H_preTS_freq.gjf**). The result output file should contain an imaginary vibrational mode associated with our reaction coordinate (in this case transfer of the H^- from NADH to carbonyl). Visualization enables the detection of any coupled motions. In this particular example, the hydride transfer turns out to be associated with a shift of the proton from Tyr to the oxygen atom of the ketone, proton from OH group of ribose to Tyr as well as slight proton shift from the amine group of Lys to OH group of ribose (*see* animated gif: 4.Cluster-QM/**pre_TS_movie.gif**).
8. We conduct optimization of the transition state staring from the selected geometry (*see* 4.Cluster-QM/**S-PEDH_p-H_TS_opt.gjf**). We can read the Hessian needed for the optimization from the already conducted vibration job (guess = read of the checkpoint file from 4.Cluster-QM/**S-PEDH_p-H_preTS_freq.gjf**). It is also possible to calculate it once again (*see* **Note 4**).

9. If the TS optimization finishes with success (we inspect the geometry and the H^- is indeed localized somewhere midway between C atoms, it did not slide to the reactant or product atom), we conduct vibration analysis for the final geometry. In the studied case, the C–H–$C^{carbonyl}$ distances are 1.37 and 1.3 Å, and the frequency of the imaginary vibration is i787 cm^{-1} with high intensity (4200). As for preTS analysis, the hydride transfer is coupled to the motions of the protons in the proton relay system (*see* TS_movie.gif). The TS should not contain any other imaginary frequencies. If they were present and were not caused by our constrains, we had localized higher-order saddle point instead of the transition state.
10. In case the TS optimization was not successful, we should repeat the scan in the vicinity of the maximum of our profile. Perhaps a more detailed sampling is needed (e.g., 0.05 Å). Most probably such investigation will provide slightly different geometry with the highest energy. In such a case we should go back to point 7 and repeat TS optimization (*see* **Note 5**).
11. It is a good practice to check the structures of stationary points connected by TS. It may happen that the geometry we have found does not lie on a path connecting those two minima. The validation can be performed with intrinsic reaction coordinate (IRC) calculations [14]. This procedure requires an initial force constant matrix, which can be read from the checkpoint file from the vibrational analysis of TS (Gaussian keywords: IRC(RCFC)). During the calculations, the atoms are moved along the steepest descent path from the TS (in both directions). After IRC, we can optimize the obtained structures of product and substrate, and conduct vibrational analysis. They should be identical as the stationary points from which we initiated the scan (in our case the structure resulting from 4. Cluster-QM/**S-PEDH_ES_p-H_opt.gjf** calculations) (*see* **Note 6**).
12. Having geometries of all steps (in this case ES, TS, EP), we collect final geometries, energies, and vibration corrections.
13. It has been demonstrated that a good approximation of the geometry can be obtained with a lower basis set but in order to obtain a good description of the energy, it is important to use a bigger basis set than DZVP (in this case 6-311+g(2d,2p)). It is possible to conduct geometry optimization with a very big basis set, but it is not very practical and, in most cases, does not result in significantly better results. Despite the increasing power of the supercomputer centers, it is still practical to introduce those corrections as single-point calculations of energy (*see* **Note 7**).

14. If the geometry optimization was not conducted in a model solvent at this stage, the model solvent corrections can be introduced. In most modeling packages, several different model solvents can be selected (e.g., COSMO, PCM, CPCM, IEFPCM, SDM [15]). An example file (4.Cluster-QM/**S-PEDH_p-H_P_pcm.gjf**) uses the polarizable continuum model (PCM) with $\varepsilon = 4$ (*see* **Note 8**).
15. We construct the energy diagram which describes energies of the enzyme–substrate complex (ES), transition state (TS), and enzyme–product complex (EP) stationary states (Table 1). Of course, for a more complex pathway, we may localize serval consecutive TSs which are separated by respective intermediates (ES, TS1, I1, TS2, I2, . . ., EP). For each stationary point, we use electronic energy calculated at the highest available level (e.g., gas-phase with higher basis set) and add corrections for solvatation (provided that the optimization was done without model solvent) and selected correction from the vibration analysis, such as zero-point energy, thermal energy, enthalpy, or Gibbs free energy. In the case of *S*-PEDH modeling, the geometry optimization, vibration correction, and solvent corrections ($\varepsilon = 4$, Polarizable Continuum Model using the integral equation formalism variant IEFPCM) were calculated on the B3LYP [16]/6-31g(d,p) level of theory with a spin-restricted approach. Then the electronic energy was recalculated at the higher level B3LYP/6-311+g(2d,2p), and additionally, a D2 Grimme correction for dispersion forces [17] was introduced (*see* **Note 9**). The energies are referred to as energy of ES stationary point (by calculation of total energy difference between any of the stationary points and ES) and presented in chemical units (kJ/mol or kcal/mol). It is also a good practice to calculate the energy diagram of the pathway at various levels of theory (e.g., for energies calculated at the geometry optimization level and for each introduced correction) (*see* **Note 10**).

Based on the obtained energetics, we should first analyze if the obtained results are in the sensible range. Here we obtain reaction barriers in the range of 87–113 kJ/mol, which can be lowered to ~67–90 kJ/mol if dispersion corrections are included. So regardless of the selected type of "energy descriptors" (be it just electronic energy, ZPE, and solvent-corrected energy or ΔG), we obtain kinetically viable results (the experimental activation energies for these reactions are in the range of 60–80 kJ/mol, depending on the substrate). It is always necessary to validate if the obtained results are chemically sensible (barriers in the range of 250 kJ/mol are under ambient conditions almost impossible to cross). On the other hand, we cannot expect such a small and static model to reproduce the chemical accuracy of the dynamic enzyme.

Table 1
Example of energy profile calculated for acetophenone; hb –base set 6-311+g(2d,2p); solvent—PCM model $\varepsilon = 4$

Substrate		ΔE (kJ/mol)	$\Delta E_{solvent}$ (kJ/mol)	ΔE_{hb} (kJ/mol)	$\Delta E_{hb + solvent}$ (kJ/mol)	$\Delta E_{ZPE + solvent}$ (kJ/mol)	$\Delta H_{+ solvent}$ (kJ/mol)	$\Delta G_{+ solvent}$ (kJ/mol)	ΔvdW corrections (kJ/mol)
Acetophenone	ES	0.00	0.00	0.00	0.00	0.00	0.00	0.00	0
	TS	104.32	101.04	113.99	110.71	92.31	87.59	103.89	−22.65
	EP	12.47	12.08	20.19	19.79	15.23	13.58	16.01	−12.32

4.3 Analysis of Different Reaction Pathways

When modeling the catalytic reaction pathway, we should **always** consider other hypotheses. Just finding one which leads to product formation and yields kinetically accessible barriers does not prove its correctness. As mentioned above, the modeling only provides the answer for the most probable proposed scenarios. In the case of our example, we can test two plausible alternate scenarios:

- Protonation before hydride transfer.
- Hydride transfer before protonation.

In these cases, the reaction should proceed in two steps (two transition states) with one intermediate. In the first scenario, we should follow the following procedure:

1. Scan the shift of the proton from Tyr-OH to oxo group of the ketone, conduct vibrational analysis for the structure with the highest energy, run the TS optimization.
2. Optimize the structure of the intermediate product (protonated ketone).
3. Scan the shift of the hydride from NADH to carbonyl C atom, localize TS for the maximum.
4. Optimize the geometry of enzyme–product.

The second scenario can also be evaluated assuming that **steps 1** and **3** are inverted. Of course, to get the same level of energy, all the corrections applied for the initial mechanism should be calculated and added. Finally, the pathway with the lowest barriers is selected as the most probable.

Sometimes it is possible to exclude one of such alternate mechanisms by optimization of the intermediate. If the intermediate is not stable, the geometry will optimize to the reactant state. This is what was observed in the case of the ketone protonation without the transfer of hydride atom—the proton shifted back to the tyrosyl ion. The geometry was obtained from TS of the concomitant scenario by transferring back the hydride to NADH (at this stage the proton from Tyr was already transferred to ketone—*see* below).

It should be underlined that one can get different advancements of the proton/hydride shifts even during concomitant scenario which was scanned only along the NADH-CO reaction coordinate. For example, for acetophenone and p-hydroxyacetophenone in TS, the proton shift was very advanced with C=O–H distances of 1.09 and 1.07 Å. Meanwhile, for a substrate such as p-nitroacetophenone or p-acetylopyridine, these distances were 1.35 and 1.4 Å while the H–O–Tyr distances were 1.09 and 1.07 Å (i.e., proton still attached to Tyr). This indicates that for the two former cases keto groups were already protonated before TS of hydride, while for the two latter cases the protons were still part of the tyrosine. However, both H atoms were shifted in one step without a discernible intermediate.

5 Correlation of Modeling with Kinetic Experiment

If our mechanistic hypothesis is correct, we can expect that the results of the calculation will correlate with the experiment. It should be underlined here that the calculations can be based on any method described in this chapter, that is, QM-only, QM:MM, or QMMM MD calculation and the selection of the particular method should be guided by the desired level of accuracy as well as computational costs. However, before we go into theory vs. experiment calculations, there are several points which should be taken into account before such a correlation is tried:

- We need several enzyme substrates with different structures and exhibiting different electronic properties in terms of the studied reaction.
- We should have a strong basis to assume that despite the structural diversity of the substrates the catalysis proceeds along the same reaction mechanism.
- It is possible to measure k_{cat}, i.e., the observed kinetics is measured at enzyme saturation by a studies substrate (no influence of different K_m of substrates) and is not modulated by other reagents (like co-substrates, reagents, etc.) or effects (substrate inhibition, substrate-dependent allosteric effect).
- The reaction under study is sensitive to the type of substrate (otherwise we can expect that our kinetic assay is limited by other factors not associated with the modeled process).
- We can collect the experimental values of k_{cat} with an error lower than the differences in k_{cat} for different substrates.

If those conditions are met, one can proceed with the calculation, which requires the modeling of the reaction pathway for each of the substrate. As a result, we get several "kinetics descriptors" (*see* 5.Correlation/correlation_descriptors.xlsx) that can be correlated with ln k_{cat} obtained from the experiment.

For each substrate, we have to identify the rate-limiting barrier, which contributes the most to the observed kinetics. In the presented example, we had only one TS, and this simplifies the problem. We repeated the described procedure for all substrates. Sometimes, for similar substrates we do not need scans—we can directly modify the geometry at TS and leave the program to perform the optimization. Finally, we do the correlation analysis (calculate the linear correlation) between each of the selected descriptors associated with the limiting barrier and ln k_{cat}. If our mechanistic hypothesis and calculation scheme are correct (as well as the experimental assay which delivered k_{cat}) we should obtain a linear correlation between some of the energy descriptors (Fig. 9, *see* also [18] for another example of successful correlation with experiment). It is advisable to include each of the experimental results as an individual case to account for the experimental error of k_{cat} determination. Unfortunately, we cannot account for the error of modeling because we have no clear way to estimate the level of approximation introduced by the use of the cluster model of the enzyme.

We can also correlate the calculated reaction energies with the experiment. In this case, we use the energy difference between ES and EP stationary state. For the experiment, we can either use the enthalpy of the reaction (measured using calorimetric method) or the equilibrium constants established in the experiment (in this case the enzyme catalyzes reversible process so it is possible to determine reaction equilibrium). A similar correlation procedure is used to obtain correlations. The example of such a correlation is presented in Fig. 10.

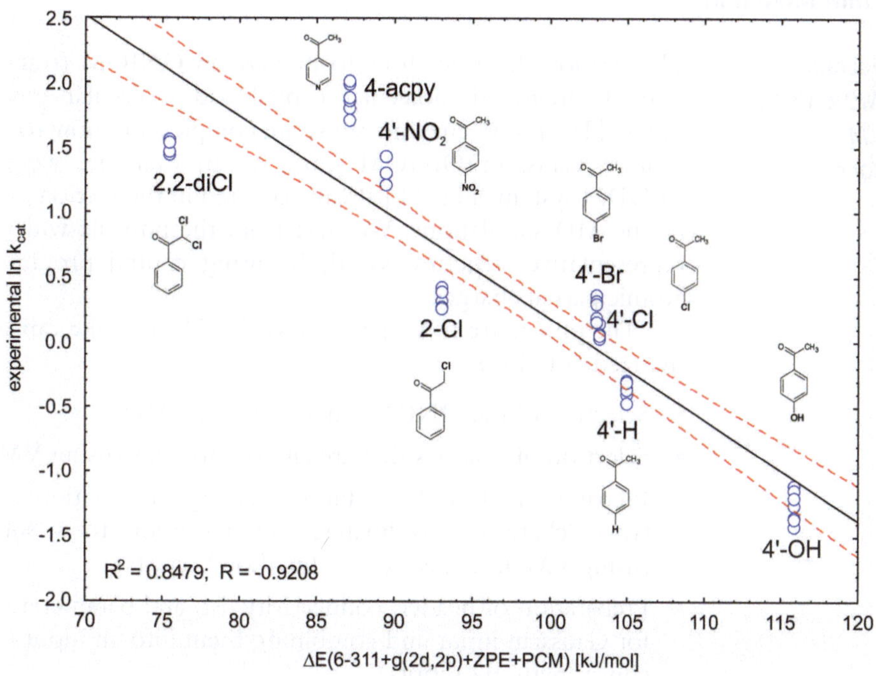

Fig. 9 Correlation analysis—scatter plot linking energy barrier of the TS with experimental ln kcat; the labels refer to substituents of acetophenone (p-H acetophenone). The correlation explains 84.8% of the observed variability [13]

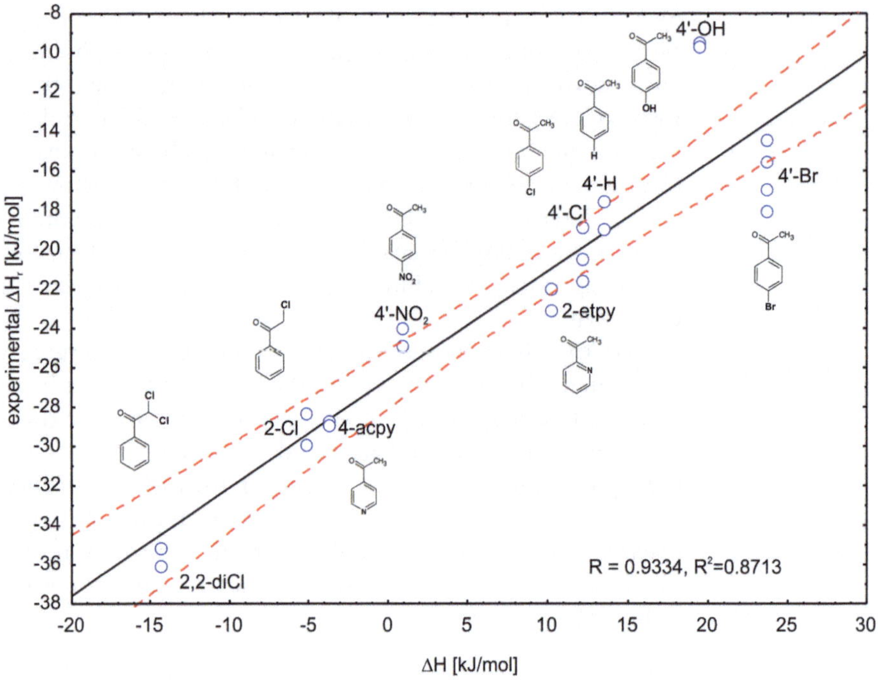

Fig. 10 Correlation plot between theoretical and experimental ΔH_r of reaction [13]

6 QM:MM Modeling

6.1 Preparation of the ONIOM Model for QM:MM Calculations

This section describes how to prepare an ONIOM (our own *n*-layered integrated molecular orbital and molecular mechanics) model [19] for an enzyme–substrate complex and how to run and analyze Gaussian ONIOM jobs. As an example, we use the T6ODM system (Fig. 11) already covered in the section dedicated to the MD simulations. We start from the pdb file with an MD representative structure saved, by using cpptraj program, with atomic partial charges.

The procedure to prepare the ONIOM input file can be summarized as follows:

- Selection of the ONIOM model (using VMD).
- Selection of residues that are allowed to relax (using VMD).
- Preparation of an *.atq file containing information on atom types, charges, coordinates, and allocation to a subsystem (using VMD, text editor, and python2 scripts).
- Preparation of header, connectivity list, and parameter sections for Gaussian input and combining them into an input file (text editor, python2 scripts).

Prerequisites:

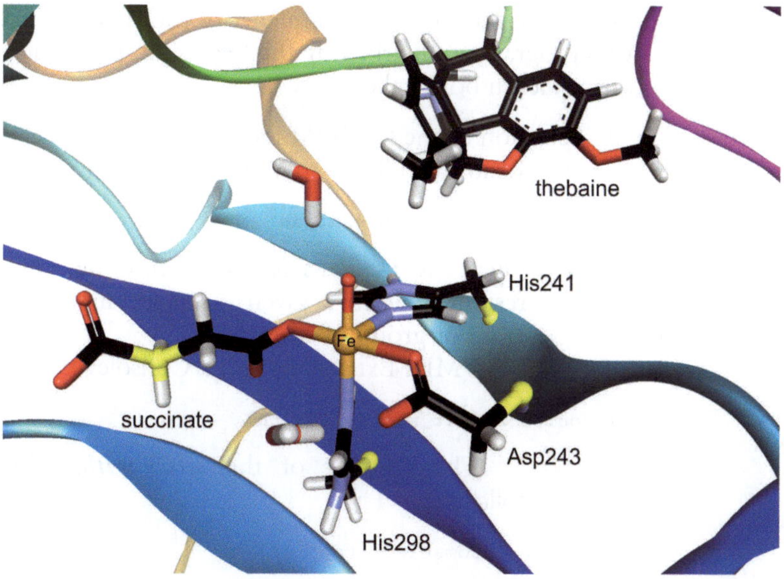

Fig. 11 QM subsystem (shown as sticks) and link atom hosts (shown as yellow balls) for T6ODM ONIOM setup

- VMD, some text editor, python2 interpreter, Linux console, Gaussian.
- Pdb file with atomic charges saved in the occupancy field of the file.
- Tcl, python2, and awk scripts accompanying this section.

6.1.1 Selection of the ONIOM Model and Optimization Zone

The pdb file saved with the atomic partial charges by cpptraj (**t6odm-thb-thb-sin.pdb**) contains coordinates for the whole MD simulation box (protein, substrate, water molecules, counter-ions—SI *see* files 5.QMMM/1_step_select_opt_fixed_zones/). As the ONIOM calculations are static (geometry optimization, $T = 0$ K) residues that are further than some cutoff distance from the active site usually have to be constrained in order to avoid the local minima artifacts (some random geometry change far from the active site that is not coupled to the reaction coordinate but may affect the calculated energy). Since such remote from the active site and static (frozen) residues have a negligible effect on the computed reaction energy profile, there is no need to include the whole simulation box in the ONIOM model. Hence, in this step, we select as the ONIOM model the whole protein and substrate plus those solvent molecules that have at least one atom within a range of 20 Å from the active site iron. Then, the optimization zone is set, and it includes all residues with at least one atom within 15 Å from the iron.

In what follows, commands executed in the Linux shell begin with the ">" sign, commands executed in the Tcl shell of VMD begin with the "%" sign, a sequence of choices made from the menu of a program are marked by the "➞" sign, file names are in bold, script files in bold italic.

1. Selection of the ONIOM model.
   (a) Load the pdb file to VMD:

```
> vmd t6odm-thb-thb-sin.pdb
```

   (b) Open the Tk Console and execute the script that selects residues within a given distance range from a selected atom or group of atoms:
   VMD: Extensions ➞ Tk Console

```
% source select_within_20A.tcl
```

   The content of the ***select_within_20A.tcl*** is the following:

```
set sel_0 [atomselect top "all"]
$sel_0 set occupancy -1
# here we provide the radius (20 A) and its origin (atom of
type FE):
set sel_1 [atomselect top "within 20 of type FE"]
set res_id [$sel_1 get residue]
set sel_2 [atomselect top "residue $res_id"]
$sel_2 num
$sel_2 set occupancy 0
# here we provide the name of the pdb fill we want to save:
$sel_2 writepdb resids_within_20A_of_FE.pdb
```

   # residues with at least one atom within a specified radius.
   # from the chosen atom (or residue).
   # will have occupancy set to 0.
   # all other residues will have occupancy set to −1.
   As a result, we get a file: **resids_within_20A_of_FE.pdb**.

   (c) Load the **resids_within_20A_of_FE.pdb** file to VMD:

```
> vmd resids_within_20A_of_FE.pdb
```

   (d) Visualize the zone within 20 Å of the selected atom:
   VMD: Graphics ➞ Representations ➞ Create Rep ➞ Selected Atoms: Occupancy 0.0000
   Coloring Method: ColorID 0

   (e) Change the selected atoms to include the whole protein plus other molecules with atoms within the above selected (20 Å) range:
   protein or occupancy 0.0000

   (f) Save the last selection to a pdb file:

VMD: File ➞ Save Coordinates ➞ Selected atoms: protein or occupancy 0.0000

File type: pdb [Save ↵]

Filename: .../**t6odm_within_20A.pdb** [OK ↵]

As a result, we get **t6odm_within_20A.pdb** file with all residues and molecules we want to have in the ONIOM model.

2. Selection of residues that are allowed to relax their geometry
   (a) Load the **t6odm_within_20A.pdb** file to vmd and run the ***select_within_15A.tcl*** (5.QMMM/1_step_select_opt_fixed_zones/)

   script to select all residues with atoms within 15 Å from Fe:

```
> vmd t6odm_within_20A.pdb
```

VMD: Extensions ➞ Tk Console

```
% source select_within_15A.tcl
```

(the content of this script is analogous to that of ***select_within_20A.tcl***, only the radius is set to 15 and the name of the saved file is changed to **resids_within_15A_of_FE.pdb**)

As a result, we get **resids_within_15A_of_FE.pdb** file in which all residues having an atom within 15 Å of Fe have occupancy set to 0.00, whereas all other residues have occupancy set to −1.00. These occupancy values are used later on to distinguish between atoms allowed to relax (value of 0) and atoms with frozen coordinates (value of −1).

   (b) Visualize this pdb file in VMD, check if all residues you want to have in the optimization zone are in the group with occupancy set to 0.00:

```
> vmd resids_within_15A_of_FE.pdb
```

VMD: Graphics ➞ Representations ➞ Create Rep ➞ Selected Atoms: occupancy 0.0000

Coloring method: ColorID

(*see* **Note 11**)

*6.1.2 Preparation of the *.atq file (5.QMMM/2_step_prepare_atq_file)*

The atq file is an auxiliary text file that has a simple format consisting of nine columns (*see* below). The information content of an atq file is exactly the same as of the geometry section of an ONIOM Gaussian input, but the atomic symbol, AMBER atom type, and atomic partial charge are in separate columns (1–3). Column 4 contains either "0" (relaxed atom) or "−1" (frozen atom), columns 5–7 contain x,y,z coordinates (in Å). Column 8 contains either "L" if an atom is in the MM subsystem or "H" if an atom is in the QM subsystem. For MM atoms that are replaced by H-link atoms during QM calculations (so-called link atom hosts), an extra entry in column 9 is needed. It consists of an atomic symbol ("H"), atom type, and partial charge of the link atom, all separated by "-" and the number of an atom from the QM subsystem to which this link atom is bonded. For historical reasons (some bug in a previous version of Gaussian) the charge of the link atom is set to a tiny, but nonzero value.

Excerpt from an example atq file (column numbers are not included in the atq file):

```
1 2 3 4 5 6 7 8 9
....
N N -0.558200 0 1.21400 6.53200 -2.16800 L
H H 0.319700 0 2.21600 6.63900 -2.23100 L
C CT 0.007200 0 0.61200 5.73400 -1.12800 L H-HC-0.000001 3807
H H1 0.082400 0 -0.31200 5.29700 -1.50600 L
C CT 0.718000 0 1.46500 4.53300 -0.75200 H
H HC -0.140100 0 2.51300 4.77400 -0.93000 H
H HC -0.140100 0 1.39100 4.37100 0.32400 H
C C 0.281900 0 0.98800 3.25600 -1.45300 H
O O2 -0.451800 0 0.32200 3.27500 -2.49300 H
O O2 -0.664600 0 1.44800 2.18700 -0.98900 H
C C 0.443200 0 0.35400 6.57500 0.17200 L
O O -0.501400 0 1.10100 7.48600 0.55200 L
...
```

In this step, we prepare an atq file using the atomic coordinates and fixed/relaxed zone information stored in the pdb file saved at the end of the previous step.

1. Selection of QM subsystem and link atom hosts (atoms that are replaced by H link atoms).
   (a) Load **resids_within_15A_of_FE.pdb** into VMD and make two separate representations for: fragments / atoms that make QM subsystem and link atom hosts

      VMD: Graphics ➞ Representations ➞ Create Rep (Drawing method: licorice):

      (resid 372 439 9791) or (resname FEO OXO) or (resname HIV HIH ASF and sidechain) or (index 5743 to 5748)

      VMD: Graphics ➞ Representations ➞ Create Rep (Drawing method: CPK):

      (index 3776 3804 4664 5749)

      The QM subsystem and link atom hosts are shown in Fig. 11.

   (b) Check if the two representations contain all the atoms that you would like to include in the QM subsystem. Once done, save the visualization state for your later reference (can be loaded into VMD to easily reproduce the visualization state):

VMD: File ➞ Save Visualization State ➞ Filename: .../**t6odm_hli.vmd** [OK ↵]

(c) Write down atom numbers for atoms from the QM layer that are bound to H-link atoms (3779, 3807, 4667, 5747)

VMD: hit 0 (zero), left-click on a given atom, read out the "index" value from the command line window, as VMD counts index from 0, add 1 to the value)

(d) With text editor, open the script file ***oniom_levels.tcl*** and from the VMD/Graphical Representations/Selected Atoms window copy the atom masks into the "..." fragments of sel_h and sel_i lines, and in the last line modify the name of the pdb output file to be saved by the script. Save the file.

After editing the content of the ***oniom_levels.tcl*** file is:

```
set sel_0 [atomselect top "all"]
$sel_0 set beta 0
# here provide info about the QM subsystem:
set sel_h [atomselect top "(resid 372 439 9791) or (resname FEO
OXO) or (resname HIV HIH ASF and sidechain) or (index 5743 to
5748)"]
$sel_h num
$sel_h set beta 2
# here provide info about link atom hosts:
set sel_i [atomselect top "(index 3776 3804 4664 5749)"]
$sel_i num
$sel_i set beta 1
# here give the name of the pdb file to be saved:
$sel_0 writepdb t6odm_oniom_hli.pdb

# saves a pdb file with info which atoms belong to which
subsystem
# beta values are set to: 2 - atoms from the QM subsystem
# 1 - link atom hosts
# 0 - other atoms from the MM subsystem
```

(e) Open the VDM TK Console and execute the ***oniom_levels.tcl*** script:

```
% source oniom_levels.tcl
```

This generates the **t6odm_oniom_hli.pdb** file, that in the beta factor column stores the information on the subsystem assignment.

(f) Make a copy of the **t6odm_oniom_hli.pdb** file and edit it using a text editor:

```
> cp t6odm_oniom_hli.pdb t6odm_oniom_hli_noX.pdb
```

In the **t6odm_oniom_hli_noX.pdb** file:

- Replace the fifth column ("X") with a space " ".
- Remove the first line of the file, which is not starting with "ATOM".
- Remove the last line, containing "END".
- Revert the nonstandard amino acid names to standard ones: HIH ➞ HIS, HIV ➞ HIS, ASF ➞ ASP.
- From the original pdb file (**t6odm-thb-thb-sin.pdb**), copy the column with atomic partial charges and place them in the last column of the **t6odm_oniom_hli_noX.pdb** file.

(g) With the use of an awk script ***vmd_hli_pdb_plus_charge_to_atq.awk*** generate the atq file using the **t6odm_oniom_hli_noX.pdb** file as input:

```
> awk -f vmd_hli_pdb_plus_charge_to_atq.awk t6odm_oniom_hli_noX.pdb > t6odm_oniom.1st.atq
```

(*see* **Note 12**)

(h) Make a copy of this file and edit the copy:

```
> cp t6odm_oniom.1st.atq t6odm_oniom.2nd.atq
```

Edit the **t6odm_oniom.2nd.atq** file, particularly pay attention to lines beginning with "ATN!", as these contain atoms for which the awk script did not have atom types defined:

- Insert appropriate atom types and format the lines that were marked with "ATN!".
- For atoms coordinated to the metal, which have other than the standard type, change their type consistently with preceding MD simulations (His241:NE2 (atom # 3787) type: NY, His298:NE2 (atom # 4675) type: NX).
- Change the type of the nitrogen atom in the N-terminal amino group to N3 (protonated N-terminal), if this is the case for your system.
- Change the type of oxygen from the C-terminal group from "O" to "O2".
- Change the entries of the last column:
  0 ➞ L
  2 ➞ H

1 → L H-HC-0.000001 XXXX (where XXXX is the number of the QM subsystem atom that is bound to the H-link atom replacing this L-atom).

(i) Check the charges of your system:

```
> ./atq_sum_charges.py t6odm_oniom.2nd.atq
```

The output of this is:
From the atq file I have read 6857 lines
found 87 atoms H, and 4 LAH atoms and 6766 remaining MM atoms
Charge of the H-layer: 1.1938
Charge of LAHs: 0.5297
Charge of other MM atoms: -17.7476
TOTAL charge of the system: −16.0241

(j) Check how the QM-part + link atom hosts look like:

```
> ./xyzn_z_atq.py t6odm_oniom.2nd.atq t6odm_oniom.qm_lah.xyz
```

Visualize the **t6odm_oniom.qm_lah.xyz** file with e.g. VMD.

6.1.3 Preparation of the ONIOM Input File (5.QMMM/3_step_prepare_oniom_input)

In this step, we first use the atq file generated previously and prepare the geometry block of the Gaussian ONIOM input. Then, we prepare a block with a connectivity list and a header section of the input and finally combine all of them into the whole input file.

1. Generate the geometry block of the ONIOM input:

```
> awk -f atq_to_input.awk t6odm_oniom.2nd.atq > t6odm_oniom.2nd.atq.for_inp
```

2. Generate the connectivity list:

```
> ./connec_list_for_atq.py t6odm_oniom.2nd.atq t6odm_oniom.2nd.atq.connect_1st
> cp t6odm_oniom.2nd.atq.connect_1st t6odm_oniom.2nd.atq.connect
```

Using a text editor, remove (explicit) bonds between Na+ (atom # 5846) and all other atoms from the **t6odm_oniom.2nd.atq.connect** file. These explicit bonds are removed since in AMBER force field interactions with the Na^+ ion are described by nonbonded terms only.

In the connectivity list, each line provides bonding information for a given atom (its number is provided at the beginning of the line); however, information on bonds made by this atom with atoms of lower indexes is provided in lines for these previous atoms and it is not repeated. Thus, to remove all bonds between atom # 5846, search all occurrences of

"5846" in the file and remove all of them except for the line beginning with "5846," which should be truncated to just the "5846" number. For example:

"4491 4492 1.0 4493 1.0 5846 1.0" should be changed to: "4491 4492 1.0 4493 1.0"

"4492 5846 1.0" ➡ "4492"

"4493 5846 1.0" ➡ "4493"

"5846 6012 1.0 6060 1.0 6492 1.0 6501 1.0" ➡ "5846"

3. Prepare the parameters section of the Gaussian input:

   The file **t6odm_oniom.params** has to be prepared manually using information that is available in Amber parameter and frcmod files that were used to prepare the *prmtop file used for MD simulations. For example, van der Waals parameters for the Na^+ ion ("VDW Na+ 1.369 0.0874393") are taken from the **frcmod.ionsjc_tip3p** file that is included in the AmberTools files. For the format of this section, *see* the Gaussian web page.

   Excerpt from the **t6odm_oniom.params** file with two example lines per VDW, bond, and angle sections:

```
VDW Na+ 1.369 0.0874393
VDW NX 1.8240 0.1700
...
HrmStr1 FE ox 365.0 1.642
HrmStr1 NY FE 61.0 2.110
...
HrmBnd1 CC CV NX 70.0000 120.0000
HrmBnd1 CC CV NY 70.0 120.0
...
```

4. Prepare a file with the head section of the input file: t6odm_oniom.head:

```
%chk=
%mem=24GB
# ONIOM(UB3LYP/def2SVP EmpiricalDispersion=GD3BJ:Amber=Soft-
First) Geom=Connectivity
5d scf=(xqc,maxcycle=350) nosymm

Comment here

-16 5 1 5 1 5
```

   The total charge of the system is −16, the total charge of the QM subsystem is +1, multiplicity in both cases is 5—this information is provided in the last line of this file.

5. Join the files (concatenate them) inserting blank lines between the input sections:

```
> cat t6odm_oniom.head t6odm_oniom.2nd.atq.for_inp <(echo)
t6odm_oniom.2nd.atq.connect <(echo) t6odm_oniom.params
<(echo) > t6odm_oniom.com
```

Edit the **t6odm_oniom.com** file—e.g., insert the appropriate name for the checkpoint file, modify the keyword list, insert a comment.

6. Run Gaussian for the t6odm_oniom.com input.

6.2 How to Analyze the Log File and Recompute Atomic Charges for the QM Subsystem

1. Extracting geometry of the whole system out of the Gaussian output (5.QMMM/4_step_analyze_result_recomput_charges).

   In the ***oniom_log_to_xyz2.awk*** script file, set the ***Ntot*** variable to the number of atoms in your system (here 6857); then the script can be used to extract geometry out of the Gaussian output:

```
>awk -f oniom_log_to_xyz2.awk t6odm_oniom.log > t6odm_oniom.xyz
```

   The last geometry can then be easily extracted with the tail command by providing as a parameter the number of lines by 2 greater than the number of atoms in the system:

```
>tail -6859 t6odm_oniom.xyz > t6odm_oniom.last.xyz
```

2. Visualizing the geometry changes between the starting and optimized structures.
   (a) Load the **t6odm_oniom_hli.pdb** and **t6odm_oniom.last.xyz** into VMD as molecules with ID 0 and 1, respectively, then execute the ***rmsd_b.tcl*** script:
   VMD: Extensions ➡ Tk Console:

```
% source rmsd_b.tcl
```

   (b) Color-code the atoms accordingly with their displacements:
   VMD: Graphics ➡ Representations:
   Selected Molecule: 1: t6odm_oniom.last.xyz
   Coloring Method: Beta
   Atoms are color-coded according to their displacement, which helps to identify regions of the system that changed their coordinates the most. The colors of the scale, its *Midpoint* and *Offset* can be changed at:
   VMD: Graphics ➡ Colors ➡ Color Scale

3. Recomputing the atomic charges of the QM part of the system and preparing a new atq file.
   (a) Prepare an atq file with coordinates of the ONIOM optimized system:

```
>all_geo_from_xyz_to_atq.py t6odm_oniom.last.xyz t6odm_oniom.2nd.atq t6odm_oniom.3rd.atq
```

   (b) Using VMD and a text editor, prepare the **t6odm_oniom.omit_q** file with atom indexes (one per line, index counted from 0, as in VMD) for atoms whose charges should not be included, in the set of atomic charges polarizing the QM system, when calculating new RESP charges of the QM subsystem. In our case, and consistently with the default electronic embedding scheme in Gaussian, these are atoms that are located one bond away from the link atom hosts.

The content of the **t6odm_oniom.omit_q** file is:

```
3774
3777
3789
3802
3805
3812
4662
4665
4677
5750
5751
5752
```

(c) Prepare the **t6odm_oniom.3rd.g_charge** file with coordinates and charges of atoms from the MM subsystem (without LAH:s and those specified in the **t6odm_oniom.omit_q** file):

```
>atq_omit_to_charge.py t6odm_oniom.3rd.atq t6odm_oniom.omit_q t6odm_oniom.3rd.g_charge
```

The first few lines of the **t6odm_oniom.3rd.g_charge** file look like this:

```
-25.39    -3.428   -10.086   -0.664400
-26.068   -3.076   -10.746   0.471500
-24.714   -2.681   -10.157   0.471500
-25.841   -3.482   -9.184    0.471500
-24.854   -4.725   -10.531   0.140300
...
```

(d) Extract the geometry of the QM subsystem (with LAH: s replaced by H atoms).

```
> qm_plus_H_from_xyz.py t6odm_oniom.last.xyz t6odm_oniom.3rd.atq t6odm_oniom.last.qm_H.xyz
```

(e) Prepare files with header and tail for Gaussian ESP calculations in the presence of MM charges:

Content of the gaussian_esp_q.head:

```
%chk=esp_q.chk
%Nproc=24
```

```
%Mem=24GB
# UB3LYP/def2SVP 5d scf=(xqc,maxcycle=350) charge nosymm
Pop=(MK,ReadRadii) iop(6/33=2) iop(6/42=6) iop(6/50=1)
```

SP with MM charges (like in ONIOM-EE) before RESP.

```
1 5
```

gaussian_esp_q.tail:

```
Fe 1.456
gaussian_esp_q.gesp
gaussian_esp_q.gesp
```

(f) Join the files to assemble the Gaussian input file:

```
> cat gaussian_esp_q.head t6odm_oniom.last.qm_H.xyz <(echo)
t6odm_oniom.3rd.g_charge <(echo) gaussian_esp_q.tail > gaus-
sian_esp_q.com
```

From the thus generated **gaussian_esp_q.com** file, remove two lines after charge and multiplicity, which contain number of atoms and comment from the xyz-file.

(g) Run Gaussian for the **gaussian_esp_q.com** input.

(h) Calculate the sum of charges for MM atoms of residues that are split between QM and MM subsystems (from the **t6odm-thb-thb-sin.pdb** file):

```
HIH and HIV: N, H, CA, HA, C, O: 0.0172
ASF: N, H, CA, HA, C, O: -0.2071
SIN: C3, H3, H4, C4, O3, O4: -1.0562
```

(*see* **Note 13**)

(i) Fit RESP charges with charges on H-link atoms fixed to values determined from the above sums: HIH and HIV (atoms # 1 and 20 in the **t6odm_oniom.last.qm_H.xyz** file): 0.0172; ASF (atom # 13): −0.2071; SIN (atom # 40): −0.0562. RESP charge fitting procedure is described in detail in the section on deriving parameters for MD, yet here the symmetry constraints between the atoms are removed completely.

(j) Generate a new atq file (**t6odm_oniom.4th.atq**) with the atomic charges of the QM subsystem as determined in the RESP procedure; **resp_2.punch** is the output of the second passage of the RESP program.

```
>add_resp_for_hlayer_to_atq_file.py resp_2.punch t6odm_o-
niom.3rd.atq t6odm_oniom.4th.atq
```

The resulting **t6odm_oniom.4th.atq** file contains optimized geometry of the system and atomic charges of the QM subsystem as determined with the RESP procedure in the presence of the MM point charges. It should now be used to prepare a new Gaussian input to run geometry re-optimization with the new set of atomic charges on the QM subsystem. In our experience, such a two-step optimization procedure, with an intervening RESP charge fitting, yields geometries and energies close to those that would be obtained with an electronic embedding scheme (which is usually far more costly and difficult to converge).

7 QMMM MD Simulation

QM cluster and QMMM modeling approaches do not take dynamic effects into account. The results obtained using these methods strongly depend on the initial geometry of the selected model (local minima problem). Using QMMM MD simulation, it is possible to obtain a Free Energy profile of reaction. One of the possibilities is to use an umbrella sampling approach to obtain the potential of mean force (PMF). We present how to generate such a profile at AMBER:AM1 level of theory using the AMBER software package.

7.1 Preparation of QMMM MD Calculations

In the example, we explain how to use the umbrella sampling method to study a reaction catalysed by 3-ketosteroid dehydrogenase from *R. erythropolis* (PDB code 4c3y). The mechanism proposed for this enzyme consists of two steps, but we discuss only the first one (analysis of the second step would be analogous to the first). In this process, a proton is transferred from 3-ketosteroid (in our example: dihydrotestosterone, DHT) to tyrosyl anion.

Procedure:

1. Make sure that the QM method selected for QMMM MD simulation can describe the analyzed process. In our example, we use the semiempirical method AM1. A few semiempirical methods are supported natively by AMBER. It is also possible to use ab inito or DFT methods with AMBER and external computational chemistry programs (there are interfaces for Gaussian, ADF, GAMESS-US, NWChem, Q-Chem, Tera-Chem, MRCC) [20]. Before MD simulation, we have modeled full reaction path using AMBER:AM1 with QMMM approach—results were comparable with a higher level of theory (AMBER B3LYP)—at least in terms of geometry and chemistry of the predicted mechanism. Barriers obtained with AM1 were significantly overestimated.
2. We start with a structure obtained from MD simulation (after cluster analysis and cooling).

3. To set up QMMM calculations in AMBER, we have to modify the input file setting ifqnt flag in &ctrl section to 1. Only then AMBER is be looking for &qmmm section. Then we have to specify the &qmmm namelist:

```
&qmmm
 qmmask="@4542-4555,7294,7295-7344,7384,7385,7386,7402-7428",
 qmcharge=-1,
 qm_theory='AM1'
 &end
```

4. Selection of the QM region—recommendations are the same as for the simple QMMM case. In this example, we selected three residues to QM layer: TYR318, a fragment of FAD cofactor, and dihydrotestosterone. To qmmask parameter, we can assign any valid atom mask, but we have to bear in mind that it cannot be longer than 256 characters (like any other atom mask in AMBER input).
5. Selection of flexible region—it is not necessary to make all atoms of the model flexible. In our example, only residues within 20 Å from dihydrotestosterone are allowed to move during QMMM MD simulation. To select flexible atoms in the AMBER input file, we can define atom mask or group input. For this purpose, using an atom mask is impractical because of its length limit to 256 characters. It is possible to use a distance-based atom mask, but in such a situation, number of frozen residues is dependent on initial coordinates. That is why we use tcl script (SI flexibleSelection.tcl in 6. QMMM_MD directory) to generate group input for AMBER.
6. Now we are ready to start the first QMMM geometry minimization of our model (all input and output files are located in 6. QMMM_MD/minimization in SI).
7. Selection of reaction coordinate—because we want to analyze the transfer of H^+ ion as the reaction coordinate, we choose the distance differences defined as in Fig. 12. If we chose only one distance to control, we would not be sure that after breaking/forming the bond, one of the reactants would not move away from the active site during MD simulation. We must be sure that we get geometries that match each stage of the reaction. To do this, we have to run a series of independent MD simulations with different values of ξ. To ensure that our coordinate is close to the desired value during MD simulation, we add parabolic potential (i.e., 500 kcal/Å^2 mol force constant) which holds H^+ close to the desired ξ. We specify this restraint in distDiff.rst file:

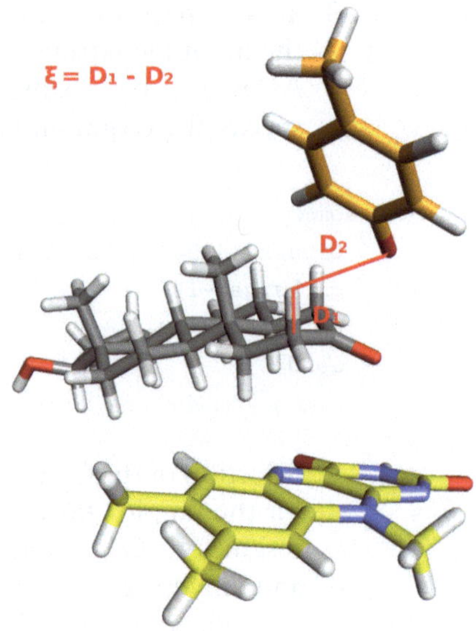

Fig. 12 QM-part of the QMMM MD job with a definition of the studied reaction coordinate ξ

```
 &rst
 iat=7310,7337,7337,4551, r1=-20, r2=-1.6 , r3=-1.6 , r4=20 ,
rk2=500,
 rk3=500,rstwt=1.0,-1.0,
 &end
```

(*see* **Note 14**).

7.2 Scanning Atom Transfer with QMMM MD Simulation

1. The next step is to generate the coordinates of the initial geometries that are used in the scan. To do that, we need to run the python script umbrellaSetup.py. This script creates 70 directories (we use 70 windows in our QMMM MD simulation dividing ξ into 70 steps) and slurm script AMBERScan.slurm to run a simulation. In every directory, distDiff.rst file is generated with the appropriate value of reaction coordinate, also some AMBER scripts are copied. Script AMBERScan.slurm performs sequential minimization series. The result coordinates from the first calculation is the input geometry for the next one, etc.
2. The geometries minimized in the previous step are ow used as initial geometries for series of QMMM MD simulations. At each simulation, we add restraint on a reaction coordinate. Each simulation starts with 5 ps of relaxation MD (md_eq.in), then 20 ps of MD simulation (md.in) is performed. For practical reasons, we do not use the heating procedure as in the regular MD simulation. Instead, in the relaxation phase, we set tempi=303 and as the input file contains no velocities, AMBER

assigns them from a Maxwellian distribution. In the MD phase, distDiffDump.dat file is created—it contains values of reaction coordinate of every hundredth frame. To run all these simulations, one has to execute runPMF.slurm script, which was also generated by umbrellaSetup.py. This script works with slurm system.

7.3 Analysis of QMMM MD Simulation

1. The next step is to verify if the values of the reaction coordinate from two consecutive windows overlap. If this condition is not met, it means that we have to decrease the scan step when generating the geometry for QMMM MD simulations or we have to decrease the value of force constant, to ensure that we have continuously sampled the reaction coordinate. To verify this, we need to run the script overlapTest.py. It is recommended to visualize the obtained data on some plots. If we run the two scripts: pmfHist.py and pmfPlots.py, we will create two files: hist.png and plots.png. The first file (Fig. 13a) contains histograms from all windows—as we can see there is no gap between any pair of histograms. The second file contains a plot of the reaction coordinate from all windows (Fig. 13b).
2. To generate a PMF profile from output files from our QMMM MD simulations, we use WHAM (Weighted Histogram Analysis Method). Unfortunately, AMBER does not provide this functionality, and we have to use an external code. We used the version developed by the Grossfield lab [21] (*input—download the wham file from the webpage*). The input file for this program was generated by umbrellaSetup.py script. All we need to do now is to run the following command:

```
wham -2.2 2.2 600 0.000001 303 0 whamInput.dat pmf.dat
```

The first two parameters define the range of ξ for the histogram analysis (-2.2 to 2.2), the next one is the number of bins in the histogram. Then we defined convergence tolerance, temperature (303 K). The value of the next argument is zero because we use aperiodic coordinates. The last two arguments are the input file and the result file (*see* **Note 15**).

Please note that the force constant in WHAM input is twice as big as the one we wrote in AMBER files. This is because wham software multiply entered force constant by 0.5 during calculations.

As a result, we obtain PMF which represents an averaged free energy profile of the proton transfer (Fig. 14).

In addition to the PMF, the above simulation provides a whole range of structures that can be used for other calculations. For example, these geometries can be used for the optimization of the transition states along multiple pathways (i.e., for different

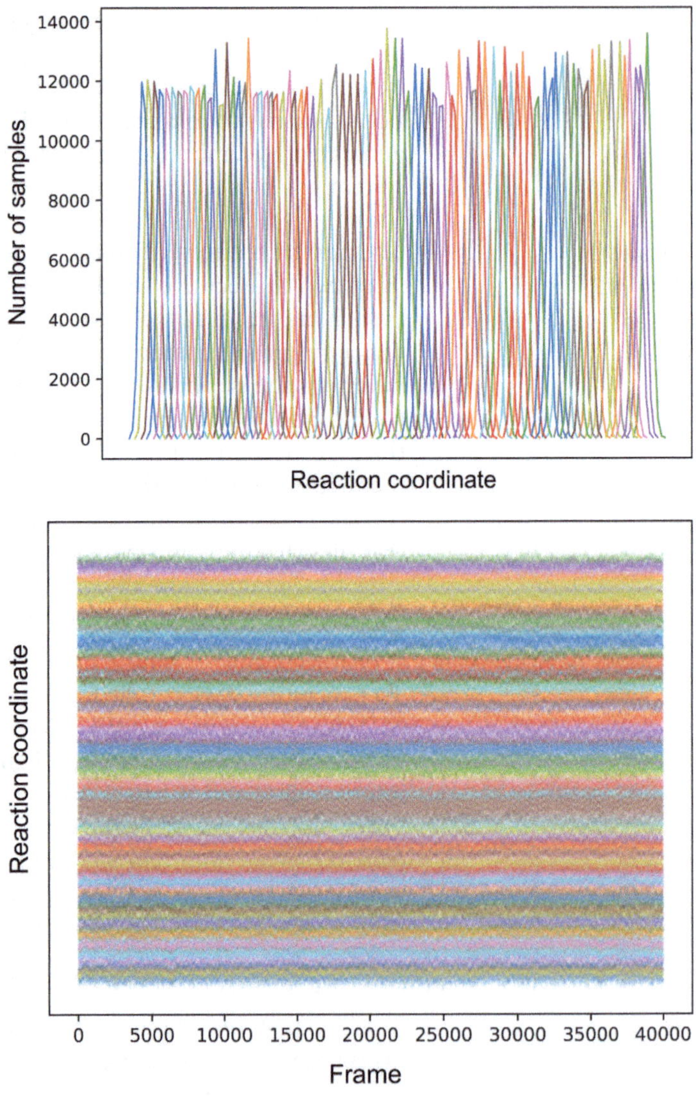

Fig. 13 (**a**) Contours of the histogram from umbrella sampling simulation. (**b**) Reaction coordinate over time for all windows used in umbrella sampling simulation

conformations of the proteins), providing information on the influence of enzyme confirmation on the catalytic process.

8 Prediction of Kinetic Isotope Effects

In this section, we demonstrate how we can estimate isotope effects from modeling. This section is general and is not associated with any specific software. The calculations presented here can be carried out using any QM software, which supports the second derivatives

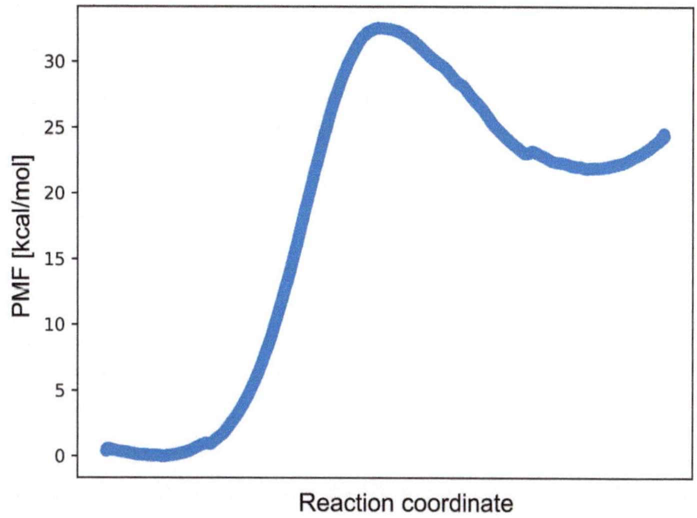

Fig. 14 An averaged free energy profile of the proton transfer

of energy. The procedure is the same for QM cluster or QMMM approach.

First of all, we have to specify what are the isotope effects, and what exactly we can estimate from the calculation. In chemistry, we usually consider kinetic isotope effects (KIE).

The formal definition of KIE is as follows:

$$\mathrm{KIE} = \frac{k_{\mathrm{L}}}{k_{\mathrm{H}}} \tag{1}$$

where k_{L} is the rate constant for the reaction involving reactant substituted with light isotopes and k_{H} is the rate constant measured for a reaction where at least one atom of the substrate was substituted by heavy isotope.

The detailed theory of KIE calculations is very complex (*see* [22, 23] for a comprehensive review). In this section, we present only the basics. Several factors affect KIE value. The first and most important one is the dependency of the free energy activation barrier ($\Delta G_{\mathrm{act}}^{\mathrm{QC}}$) to isotopic substitution.

$$\mathrm{KIE}\left(\Delta G_{\mathrm{act}}^{\mathrm{QC}}\right) = \mathrm{e}^{\frac{\Delta G_{\mathrm{act}}^{\mathrm{QC,H}}(T,x) - \Delta G_{\mathrm{act}}^{\mathrm{QC,D}}(T,x)}{RT}} \tag{2}$$

To explain how to calculate this property, one more time we analyse reaction catalyzed by 3-ketosteroid dehydrogenase from *R. erythropolis*. This time we use the cluster model of this enzyme active site (all input and output files are located in SI in the KIE directory).

To calculate the KIE, we have to:

1. Obtain and optimized structure of reactants and transition state (as described in Subheading 4.2).

2. Calculate the second derivatives of energy for those structures (i.e., frequency calculation) at the desired temperature.
3. Collect thermal corrections to free energy calculated directly from these derivatives. Gaussian software package prints in output from freq job (the thermal correction to Gibbs free energy) and those values can be used to calculate KIE.
4. Calculate the thermal correction to the free energy for the heavy isotopologue—this can be achieved by reading the checkpoint from the freq calculations (freq = readfc) and providing data on the mass of the substituted atom (here substitution of protium to deuterium).

```
H(Iso=2) 0 x y z
```

In this example, we substitute the hydrogen atom, which is transferred from ketosteroid to tyrosyl anion, with deuterium. We can verify that Gaussian correctly set the mass of the selected atom by investigating the output file.

```
Atom 198 has atomic number 1 and mass 1.00783
Atom 199 has atomic number 1 and mass 2.01410
Atom 200 has atomic number 1 and mass 1.00783
```

We collect thermal correction to Gibbs free energy from all output files (Table 2).

Now we can substitute these values into Eq. (2) (after recalculation of units—Hartree to J/mol). Finally, we obtain an estimation of KIE as 4.65. Please note that it does not matter if we use only thermal correction to Gibbs free energy or the sum of the electronic energy and this correction—electronic energy does not depend on the mass of atoms so it does not influence the value of KIE.

There are also external programs that can read the second derivatives from Gaussian outputs and calculate KIE from frequencies (Kinisot [24], ISOEFF [25]).

Another factor that can be important for the KIE calculation is tunneling, i.e., quantum effect associated with traversing the energy barrier even if the atom has too low energy to go over the barrier. The easiest way to estimate the tunneling coefficient is to use Wigner correction [26]:

$$\chi(T) = 1 + \frac{1}{24}\left(h\ \mathrm{Im}\left(\frac{v)}{k_b T}\right)^2 \tag{3}$$

As one can see, the tunneling coefficient depends only on the temperature and imaginary frequency of transition state. To calculate tunneling contribution to KIE, we have to calculate the tunneling coefficient for nonsubstituted and isotopically substituted

Table 2
Thermal correction to Gibbs free energy calculated for light (nondeuterated) and heavy (deuterated) substrate [Hartree]

Stationary point	Nondeuterated isotopologue	Deuterated isotopologue
ES	1.66398	1.660581
TS	1.664305	1.66238

reactants. The values of the imaginary frequencies (provided in cm^{-1}) associated with the transition state can be retrieved from the freq output job (calculated for light and heavy reactant). In the case of results calculated in Gaussian, one can use also formatted checkpoint and Gaussview to obtain imaginary frequency for the heavy isotopologue.

Then we can calculate:

$$\mathrm{KIE}(\kappa) = \frac{\kappa^{\mathrm{L}}}{\kappa^{\mathrm{H}}} \quad (4)$$

Finally, we can estimate KIE as:

$$\mathrm{KIE} = \mathrm{KIE}(\Delta G)\mathrm{KIE}(\kappa) \quad (5)$$

In our example, the values of the imaginary frequencies of TS are presented in Table 3.

When we substitute these values into the Wigner equation, we obtain:

$$\mathrm{KIE}(\kappa) = 1.36 \quad (6)$$

Finally, if we consider tunneling, we can estimate combined KIE as:

$$\mathrm{KIE} = \mathrm{KIE}(\Delta G)\mathrm{KIE}(\kappa) = 6.32 \quad (7)$$

This estimation yields the so-called intrinsic KIE associated with a particular barrier. In case the effect is strong and we did this calculation for the rate-limiting step in our mechanism, we can expect correlation with the experiment. KIE is the best way to correlate the modeling with the experimentally observed kinetics. However, we have to bear in mind that enzymatic reactions are very often multi-step processes with binding and release of substrate/ products and sometimes conformational changes. Each of these processes can be or not sensitive to isotopic substitution. It is very often that if some other rate-limiting step occurs on the reaction pathway the experimentally determined KIE is of lower value than this predicted by theory [27].

Table 3
Imaginary frequencies calculated at TS for a nondeuterated and deuterated substrate

Substrate type	Imaginary frequency (cm^{-1})
Nondeuterated isotopologue	i1256.7172
Deuterated isotopologue	i938.6045

9 Notes

1. In 5O7Y PDB file, the residue number starts from −2, but we renumber PDB file with the residue number starting from 1. Thus, for the whole work, we are using the new numbering scheme, e.g., HIS238 is renumbered to HIS241.
2. At this stage, we should consider if the calculations will be conducted in vacuum and only corrected with model solvent (single point corrections) or the geometry optimization will be also conducted in a model solvent. The latter solution may be of importance if the solvent strongly penetrates the active site and charges are being created during reactions. In some cases, the model solvent can stabilize more charge-separated states of the amino acid (zwitterion not neutral). On the other hand, model solvents can also introduce their own biases. The former solution is definitely faster. Also, at this stage, one has to consider using the dispersion correction. This is especially important if there are many vdW interactions present in the active site of the model.
3. In the case we want to compare the results to a particular temperature at this step, we should provide the temperature for which the thermal energy, enthalpy, or free energy is to be calculated; also one can consider the introduction of the scaling factor appropriate for the functional and basis set selected for calculations.
4. In most packages, it is possible to initiate TS search without prior vibration analysis; however, if the search is not successful, it might be difficult to understand why the search failed. The vibration analysis provides information if the vibration mode is correct, i.e., along the expected direction of the reaction coordinate, and if its frequency and intensity are appropriate for a given type of atom transfer; in this case, the frequency of the initial vibration was i800 cm^{-1} and intensity was very high 4978.
5. In case we are still unable to localize TS, we have to consider that our mechanistic hypothesis is wrong and that there is no TS in this region. Perhaps the order of steps is different. In this particular example, the ketone could have to be initially protonated to enable the reduction of the carbonyl atom.

6. In the case of a several-stage reaction, it is worth to pay attention if the energies of the same intermediate, obtained from different IRCs (and optimized), are comparable. If they differ, then the reason for this should be found. It may be, for example, a change in the conformation of the fragment of the model.
7. The reliability of this approach can be tested by reoptimization of the obtained stationary points in higher basis set; in most cases, no significant geometry changes result from such reoptimization; if significant changes are obtained, this proves that the basis set used was too small to account properly for the chemistry of the studied process.
8. Each of these methods requires several additional parameters to be added, and one should refer to the manual for the program/method specific guidelines. However, for each of them, one has to define the dielectric permeability (ε), which will directly influence the extent to which the model solvent will screen electrostatic interactions between charged species in the model. There are no hard guidelines for the selection of ε value. Usually, values in the range of 4 (average experimental ε of the denaturated protein]) up to 40 are used [28]. In reality, this parameter changes continuously and depends on the degree of exposition of the active site to the water phase (which has ε 78 at 25°C), so in fact, the dielectric constant is not "constant" at all. We recommend using ε of 4 for reactions that are buried deep inside the enzyme and 20 or even higher for reactions taking place in shallow clefts of enzymes that are normally filled with water and if the reagents are in contact with water molecules. Finally one can test with fast single-point calculations the dependence of the energy profile on the value of ε. If big changes of energies are observed for solvent corrections calculated with different values of ε, it is a strong indication that: (1) the geometry optimization and TS detection should be conducted with model solvent or (2) the model is too small and the model solvent penetrates two close the reactants which result with high sensitivity of the reaction to the model solvent. Here it should be underlined that the enzymes usually try to prevent solvent from accessing the reagents to (1) decrease the entropy of binding by the release of reagents solvatation shell and expulsion of solvent molecules bound in the active site and (2) prevent solvent-mediated side reactions.
9. Although it is possible to introduce G correction, one has to bear in mind that those values may introduce an error in entropy estimation associated with the way we constructed the cluster (i.e., constraints).

10. This helps for the detection of potential errors that inevitably occur in the course of modeling and provides additional information on factors stabilizing energy of particular transition states and intermediates (like solvatation correction or higher basis set).
11. The chosen values of radii (of 20 and 15 Å) for selecting the ONIOM model and its optimization zone, as well as the origin of these zones (here Fe atom bound in the active site), can be modified according to the needs of the system studied. For example, one might choose these zones based on the distance from any atom of selected residue(s), e.g., the ones that will be included in the QM-layer and set lower values of the radii.
12. The ***vmd_hli_pdb_plus_charge_to_atq.awk*** script ascribes AMBER atom types to atoms of standard protein residues and water molecules, however, for nonstandard residues, atom types need to be provided manually, which is done in the following step.
13. The values of these charges are used in the next step to constraint the charges of H atoms replacing LAH:s in the RESP charge fitting procedure. This is necessary if we want to update the charges of the QM subsystem and at the same time preserve the correct integer total charge of the whole system.
14. For more details on the meaning of the parameters, *see* the AMBER manual.
15. We do not discuss all possible arguments for WHAM program (e.g., how to use it with periodic coordinate). Please refer to the manual of wham for more details.

Acknowledgments

M.Sz. and M.G. acknowledge the financial support from the National Center of Science, Poland under the OPUS grant number UMO-2016/21/B/ST4/03798. M.G. acknowledges the fellowship with the project no. POWR.03.02.00-00-I013/16. T.B. acknowledges the financial support from the National Science Centre, Poland under the OPUS grant number UMO-2014/15/B/NZ1/03331. The authors acknowledge the computational grants from PL-GRID (Cyfronet). The authors are grateful to Sven de Marothy for providing us with his XYZViewer program.

References

1. Rohman A, van Oosterwijk N, Thunnissen AM, Dijkstra BW (2013) Crystal structure and site-directed mutagenesis of 3-ketosteroid Delta1-dehydrogenase from Rhodococcus erythropolis SQ1 explain its catalytic mechanism. J Biol Chem 288(49):35559–35568. https://doi.org/10.1074/jbc.M113.522771
2. Wang J, Wolf RM, Caldwell JW, Kollman PA, Case DA (2004) Development and testing of a general amber force field. J Comput Chem 25

(9):1157–1174. https://doi.org/10.1002/jcc.20035

3. Seminario JM (1996) Calculation of intramolecular force fields from second-derivative tensors. Int J Quantum Chem 60(7):1271–1277. https://doi.org/10.1002/(SICI)1097-461X(1996)60:7<1271::AID-QUA8>3.0.CO;2-W
4. Nakazawa T, Igarashi T, Tsuru T, Kaji Y (2011) Density functional investigation of Fen clusters (n≤6) with Cr substitutions: UB3LYP/LanL2DZ calculation. Comput Mater Sci 50(3):982–990. https://doi.org/10.1016/j.commatsci.2010.10.037
5. Grimme S, Ehrlich S, Goerigk L (2011) Effect of the damping function in dispersion corrected density functional theory. J Comput Chem 32(7):1456–1465. https://doi.org/10.1002/jcc.21759
6. Borowski T, Szaleniec M (2019) Challenges in modelling metalloenzymes. In: Broclawik E, Borowski T, Radoń M (eds) Transition metals in coordination environments: computational chemistry and catalysis viewpoints. Springer International Publishing, Cham, pp 503–525. https://doi.org/10.1007/978-3-030-11714-6_17
7. Nilsson K, Lecerof D, Sigfridsson E, Ryde U (2003) An automatic method to generate force-field parameters for hetero-compounds. Acta Crystallogr D Biol Crystallogr 59(Pt 2):274–289. https://doi.org/10.1107/s0907444902021431
8. Wang J, Wang W, Kollman PA, Case DA (2006) Automatic atom type and bond type perception in molecular mechanical calculations. J Mol Graphics Model 25(2):247–260. https://doi.org/10.1016/j.jmgm.2005.12.005
9. Duan Y, Wu C, Chowdhury S, Lee MC, Xiong G, Zhang W, Yang R, Cieplak P, Luo R, Lee T, Caldwell J, Wang J, Kollman P (2003) A point-charge force field for molecular mechanics simulations of proteins based on condensed-phase quantum mechanical calculations. J Comput Chem 24(16):1999–2012. https://doi.org/10.1002/jcc.10349
10. Lee MC, Duan Y (2004) Distinguish protein decoys by using a scoring function based on a new AMBER force field, short molecular dynamics simulations, and the generalized born solvent model. Proteins Struct Funct Bioinf 55(3):620–634. https://doi.org/10.1002/prot.10470
11. Harvey JN (2006) On the accuracy of density functional theory in transition metal chemistry. Annu Rep Sect C Phys Chem 102:203–226. https://doi.org/10.1039/B419105F
12. Hoffken HW, Duong M, Friedrich T, Breuer M, Hauer B, Reinhardt R, Rabus R, Heider J (2006) Crystal structure and enzyme kinetics of the (S)-specific 1-phenylethanol dehydrogenase of the denitrifying bacterium strain EbN1. Biochemistry 45(1):82–93. https://doi.org/10.1024/bi051596b
13. Stawoska I, Dudzik A, Wasylewski M, Jemiola-Rzeminska M, Skoczowski A, Strzalka K, Szaleniec M (2017) DFT-based prediction of reactivity of short-chain alcohol dehydrogenase. J Comput Aided Mol Des 31(6):587–602. https://doi.org/10.1007/s10822-017-0026-5
14. Fukui K (1981) The path of chemical reactions – the IRC approach. Acc Chem Res 14(12):363–368. https://doi.org/10.1021/ar00072a001
15. Tomasi J, Mennucci B, Cammi R (2005) Quantum mechanical continuum solvation models. Chem Rev 105(8):2999–3093. https://doi.org/10.1021/cr9904009
16. Becke AD (1993) Density-functional thermochemistry. III. The role of exact exchange. J Chem Phys 98(7):5648–5652. https://doi.org/10.1063/1.464913
17. Grimme S, Antony J, Ehrlich S, Krieg H (2010) A consistent and accurate ab initio parametrization of density functional dispersion correction (DFT-D) for the 94 elements H-Pu. J Chem Phys 132(15):154104. https://doi.org/10.1063/1.3382344
18. Szaleniec M, Salwinski A, Borowski T, Heider J, Witko M (2012) Quantum chemical modeling studies of ethylbenzene dehydrogenase activity. Int J Quantum Chem 112(8):1990–1999. https://doi.org/10.1002/Qua.23143
19. Vreven T, Morokuma K (2006) Chapter 3: Hybrid methods: ONIOM(QM:MM) and QM/MM. In: Spellmeyer DC (ed) Annual reports in computational chemistry, vol 2. Elsevier, Amsterdam, pp 35–51. https://doi.org/10.1016/S1574-1400(06)02003-2
20. Gotz AW, Clark MA, Walker RC (2014) An extensible interface for QM/MM molecular dynamics simulations with AMBER. J Comput Chem 35(2):95–108. https://doi.org/10.1002/jcc.23444
21. Grossfield A. WHAM: the weighted histogram analysis method. http://membrane.urmc.rochester.edu/wordpress/?page_id=126
22. Kohen A, Limbach H-H (2006) Isotope effects in chemistry and biology. Taylor & Francis, Boca Raton
23. Nagel ZD, Klinman JP (2006) Tunneling and dynamics in enzymatic hydride transfer. Chem Rev 106(8):3095–3118. https://doi.org/10.1021/cr050301x

24. Rzepa HS (2015) KINISOT. A basic program to calculate kinetic isotope effects using normal coordinate analysis of transition state and reactants. https://doi.org/10.5281/zenodo.19272
25. Anisimov V, Paneth P (1999) ISOEFF98. A program for studies of isotope effects using Hessian modifications. J Math Chem 26 (1):75–86. https://doi.org/10.1023/A:1019173509273
26. Wigner EP (1932) Crossing of potential thresholds in chemical reactions. Z Phys Chem B19:203–216
27. Glanowski M, Wójcik P, Procner M, Borowski T, Lupa D, Mielczarek P, Oszajca M, Świderek K, Moliner V, Bojarski AJ, Szaleniec M (2021) Enzymatic δ1-Dehydrogenation of 3-Ketosteroids - reconciliation of kinetic isotope effects with the reaction mechanism. ACS Catal 11:8211–8225. https://doi.org/10.1021/acscatal.1c01479
28. Li L, Li C, Zhang Z, Alexov E (2013) On the dielectric "constant" of proteins: smooth dielectric function for macromolecular modeling and its implementation in DelPhi. J Chem Theory Comput 9(4):2126–2136. https://doi.org/10.1021/ct400065j

Chapter 11

The Role of Ligand Rebinding and Facilitated Dissociation on the Characterization of Dissociation Rates by Surface Plasmon Resonance (SPR) and Benchmarking Performance Metrics

Aykut Erbaş and Fatih Inci

Abstract

Surface plasmon resonance (SPR) is a real-time kinetic measurement principle that can probe the kinetic interactions between ligands and their binding sites, and lies at the backbone of pharmaceutical, biosensing, and biomolecular research. The extraction of dissociation rates from SPR-response signals often relies on several commonly adopted assumptions, one of which is the exponential decay of the dissociation part of the response signal. However, certain conditions, such as high density of binding sites or high concentration fluctuations near the surface as compared to the bulk, can lead to non-exponential decays via ligand rebinding or facilitated dissociation. Consequently, fitting the data with an exponential function can underestimate or overestimate the measured dissociation rates. Here, we describe a set of alternative fit functions that can take such effects into consideration along with plasmonic sensor design principles with key performance metrics, thereby suggesting methods for error-free high-precision extraction of the dissociation rates.

Key words Off-rate, Binding kinetics, Facilitated dissociation, Dissociation, Plasmonic sensors, Surface chemistry

1 Introduction

Surface plasmon resonance (SPR) is a label-free optical method that has been used heavily to characterize the kinetic properties of a wide array of biomolecular systems [1–3], including antigen–antibody interactions, host–guest interactions on molecular cavities, drug affinity, viral activity, protein–DNA kinetics, enzymatic activity, as well as biosensing applications at multiple settings, such as clinics, border security, and point-of-care [4–6]. The fundamental working principle of SPR relies on the alterations of the refractive index (RI) (combining changes in bulk and binding to the surface) of a *reaction volume* right above a surface, on which the amount of

Quentin Vanhaelen (ed.), *Computational Methods for Estimating the Kinetic Parameters of Biological Systems*, Methods in Molecular Biology, vol. 2385, https://doi.org/10.1007/978-1-0716-1767-0_11,

biomaterial varies over time [7]. This mechanism allows the detection of solution-phase ligands associating with the binding sites immobilized on the surface, or inversely, of the dissociation of the ligands off their binding sites into a ligand-free solution in the course of time.

The general methodology of the SPR protocol begins with the loading of the surface-immobilized binding sites with corresponding ligands by flowing a solution containing the ligands over the surface [8, 9]. Once the loading of the binding sites by the ligands is complete, the response signal reaches a saturation level (Fig. 1). The next step is to flow a ligand-free solution that washes off the unbound ligands in the solution. After this step, ideally the setup contains only those ligands that are bound to their binding sites. This state, where lots of bound ligands on the surface with no ligand in the solution, is not energetically favorable (ligands lose translational entropy when they are bound). Thus, the bound ligands begin to leave their binding sites to dissociate into the ligand-free solution until the (average) total free energy change per ligand becomes zero. Since the RI of the reaction volume changes as the ligands diffuse away from the surface, the dissociation of a large number of ligands from their binding sites ideally manifests themselves as a decay phase in the response signal (Fig. 1). This decay continues until fractions of bound and free ligands obey the detailed balance, that is, when chemical equilibrium is reached (Fig. 1).

The half-life of the decay phase of the signal is related to the time scale of the microscopic dissociation. The inverse of this time scale is referred to as the dissociation or off-rate, k_{off} (with the units of inverse time, 1/s). The off-rate provides key information about the frequency of unbinding events, which is of paramount importance to determine the relevant equilibrium association or dissociation (i.e., K_a or K_d) constants of the reaction. Conventionally, the dissociation rate can be extracted by fitting the decay phase of the response signal to an exponential function in the form of

$$f(t) = \exp\left(-k_{\mathrm{off}} t\right). \tag{1}$$

The use of the exponential function, Eq. 1, implicitly assumes that the dissociation process of each ligand off its binding site is independent of other ligands in the system. In other words, if one prepares a hypothetical system that contains only a single binding site with a single ligand bound to it, and then, obtain the distribution of dissociation times after ***a sufficiently high number*** of independent but completely identical experiments, the resulting distribution would be exponential. However, in typical SPR (or in similar single-molecule) setups, the binding sites are often densely placed (e.g., ~1–10 nm) in order to ensure the efficient binding of ligands to the surface during the association phase (Fig. 1). As a

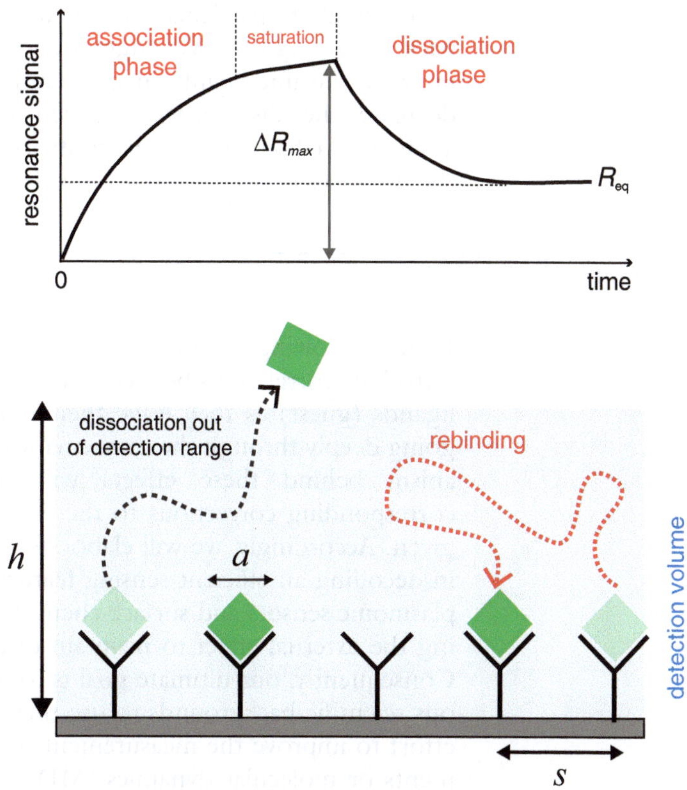

Fig. 1 Upper panel: A typical SPR signal. During the association phase, the binding sites are loaded with ligands until a saturation phase is reached. Then, a ligand-free solution is injected into the system to initiate the dissociation of the bound ligands. Lower panel: During the dissociation phase, a ligand can either diffuse out of the detection volume near the surface with a height of h or rebind to a neighboring empty binding site at a distance, s, thus, delay the dissociation time

result, once the dissociation phase starts, there would be a coexistence of free and bound ligands at any instant; some ligands would leave their binding sites, but they would be still diffusing near the surface. These free ligands can rebind to an empty binding site (including their own) or collide with stably bound nearby host–guest complexes and decrease their stability. In other words, the asynchronous dissociation of ligands could lead to a decay signal that might not be described by an exponential decay. When such data is treated by the exponential fit function (Eq. 1), the extracted value of the off-rate would be lower or higher than the intrinsic value. Most importantly, the usage of Eq. 1 could cause variations even for the same host–guest system from experiment to experiment due to several factors related to system preparation. In this chapter, we discuss two such factors that can significantly influence the entire dissociation process, and thus, argue against the usage of exponential fits, particularly for weakly binding ligands and densely packed binding sites. We suggest alternative fit functions that can improve the uncertainty in measured reaction rates and possibly decrease the discrepancy between independent measurements along with a discussion on the control of surface chemistry to adjust the distances between recognition elements for ligand binding.

2 Methods

In the following sections, we describe the two effects that can introduce correlations between the unbinding events of molecular ligands (guest) as they leave their binding sites (host). Without going deeply through the mathematical details, the physical mechanism behind these effects will be summarized, and the corresponding corrections to the exponential fit function will be given. Accordingly, we will elaborate the key performance metrics in decoding an efficient sensing feature through material types of plasmonic sensors and surface chemistry approaches and minimizing the external effect to maintain direct host–guest interactions. Consequently, our ultimate goal is to guide researchers from various scientific backgrounds to use these fit functions with minimal effort to improve the measurement quality in their kinetic experiments or molecular dynamics (MD) simulations. We also aim to guide them to choose the most appropriate options for sensor hardware, soft biological layers, and maintaining high-selectivity before using plasmonic platforms.

2.1 Rebinding of Ligands Due to High Surface Density of Binding Sites

SPR principle relies on the dissociation of a large number of ligands from their binding sites into a ligand-free reservoir. To increase the measurement precision, a large number of surface-immobilized binding sites are often needed. This inevitably results in a high grafting density of binding sites on the surface and a reduction of the average distance between the sites. We may estimate the average distance between binding sites, s, via calculating the amount ligands on the surface as follows:

$$s \approx 1.5\left(\frac{M_{\mathrm{w}}}{\Delta R_{\max}}\right)^{1/2} \mathrm{nm},$$

where M_{w} is the molecular weight of ligands in the units of Dalton, and $\Delta R_{\max}$ is the maximum signal in the units of resonance units (RU) (Fig. 1). For a typical saturation value in the SPR signal of $\Delta R \approx 100$ RU (1 RU = 1 pg/mm^2 as reported in the technical features of the Biacore System [10]), the average distance between two binding sites is $s \approx 15$ nm if we assume a molecular weight of 10 kDa (*see* **Note 1**). Notice that the grafting density of binding sites is defined as $\sigma_{\mathrm{g}} = 1/s^2$, thus the equation above can also be used to obtain the average grafting density of binding sites.

While a high grafting density can improve the statistics of the dissociation phase, a ligand, upon its unbinding, can rebind to its own or another nearby empty binding site rather than exiting outside of the detection volume (Fig. 1). Those ligands would not contribute to the dissociation signal even though they have evacuated their original binding sites since the fraction of ligands within the evanescent field (reaction volume) does not change. A quantitative estimate can be made for the time scale of the rebinding events. On average, a dissociated ligand can diffuse to neighboring binding sites located at a distance s after a diffusion time of $\tau_s = s^2/4D$, where D is the self-diffusion coefficient of the ligand [11, 12]. If the distance between the neighboring binding sites, s, is much larger than the height of detection volume (i.e., roughly the decay length of the evanescent field, $h \approx 100$ nm) [10], the dissociating ligands would have a higher probability of diffusing away from the surface and a low probability of rebinding. However, if the binding sites are placed densely (i.e., $s \ll h$), the ligands could diffuse to nearby empty binding sites before exiting the detection volume. Those ligands would exit the detection volume eventually but only after spending a certain amount of time at other binding site(s). As a result, the average residence time of the ligand within the detection volume would not reflect the actual microscopic dissociation time of the ligands off its binding site but rather provide a value about the total dwell time of the ligands near the surface (inside the detection volume).

Several MD and Monte Carlo (MC) simulation studies on model systems as well as experiments on insulin-like growth factors interacting with their binding proteins [11] systematically demonstrated that the effect of rebinding on dissociation rates is readily observable for densely placed binding sites [11–14]. To extract the intrinsic off-rates (unaffected by rebinding events) from the response signal, an analytical expression that takes the rebinding events into account has been suggested in the form of a *stretched-exponential* fit function as [11]:

$$f(t) = \exp\left(\frac{k_{\text{off}}^2 t}{k_s}\right) \text{erfc}\left(\sqrt{\frac{k_{\text{off}}^2 t}{k_s}}\right). \tag{2}$$

In Eq. 2, $k_s = 1/\tau_s = 4D/s^2 = 4D\sigma_g$ is the rate characterizing the rebinding events, and erfc(x) is the error function. The key point in Eq. 2 is that as k_s approaches to k_{off} (i.e., as the separation distance between binding sites increases), and Eq. 2 turns into the conventional exponential function given by Eq. 1. In short, Eq. 2 suggests a correction to the off-rates, which mainly increases its effect for densely packed binding sites (*see* **Note 2**).

MD simulations can directly demonstrate the effect of rebinding occurrences on off-rate measurements (Fig. 2). Since MD can track the trajectories of individual ligands to determine their

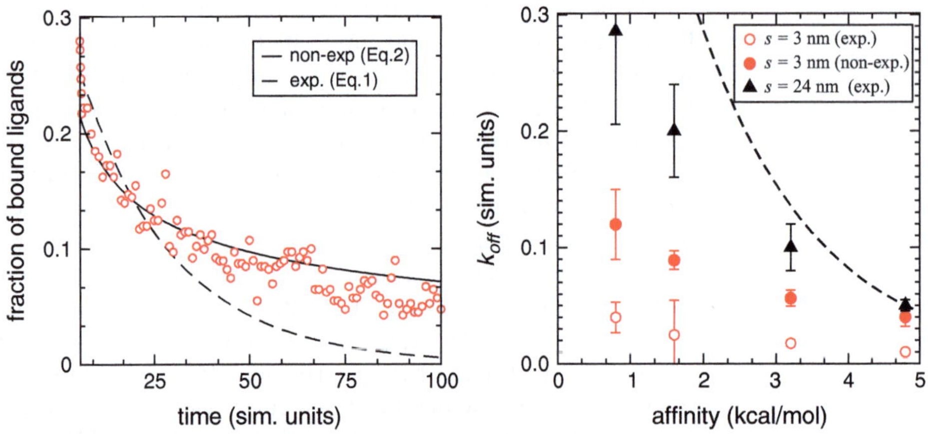

Fig. 2 MD simulations of monomeric ligands dissociating from their binding sites. Left panel: The fraction of bound ligands decreases as time progresses. The time is given in simulation units. Exponential (dashed curve) and the non-exponential (solid curve) fits to the dissociation data are plotted by releasing the k_{off} as a fit parameter. Right panel: The off-rates obtained from the fit procedure for various binding affinities by using both methods. In all cases, the off-rates decrease with increasing affinity. As the average distance, s, is increased, the off-rate of the ligand approaches the Arrhenius limit ($k_{\text{off}} \sim \exp(-\text{affinity})$). (Adapted from Ref. 12 by assuming a simulation unit length scale of $a = 1$ nm)

binding states, they can easily interrogate a range of separation distances and binding affinities. Figure 2 shows the data obtained from a set of such MD simulations [12] with spherical ligand particles of size $a \approx 1$ nm that are initially immobilized on their binding sites. Thus, the simulation in that study specifically models the beginning of the dissociation phase, at which all binding sites are already loaded by ligands (Fig. 1). The ligands interact with their binding sites weakly with affinities ranging between $E \approx 1$ kcal/mol and 5 kcal/mol, which are typical numbers for small analytes or proteins interacting with their receptors [15]. In those computational experiments, binding sites are placed on the surface with an average distance of $s = 3$ nm; at this short separation distance, we expect to see a considerable effect of rebinding events on the dissociation kinetics. Indeed, if the dissociation data is fitted both to an exponential function (Eq. 1) and to the alternative fitting equation (Eq. 2) separately, the exponential fit gives a ten times slower off-rates as compared to the non-exponential fit in Eq. 2 (Fig. 2). If the same experiments are repeated by sparsely placing the binding site with $s = 24$ nm, thus bringing the value of k_s closer to k_{off}, the extracted off-rates for exactly the same host–guest pairs are closer to the k_{off} value extracted via Eq. 2 (Fig. 2). This finding suggests that if the binding sites are sparsely placed, the effect of the rebinding events on off-rates can be reduced significantly. When simulations are repeated with higher affinities, the off-rates obtained by Eq. 2 converge better to the off-rates obtained by sparsely placed binding sites, but no improvement is observed for the conventional exponential fit (Fig. 2). These simulations readily demonstrate the critical effect of the binding-site placement on the dissociation kinetics of molecular ligands (e.g., drugs, proteins) and in the meantime validate the use of the non-exponential fit functions given in Eq. 2 in the extraction process of off-rates for any SPR setup with arbitrary surface coverage.

2.2 Facilitated Dissociation (FD) of Ligands by Solution-Phase Ligands

During the last decade, a large number of experimental and computational studies have uncovered a concentration-dependent dissociation mechanism for various biomolecular ligands [16]. Particularly, small DNA binding proteins (e.g., bacterial transcription factors) [17] as well as subunits of large protein complexes (DNA polymerase) [18] exhibit a dramatic increase in their dissociation rates at the sub-micromolar concentrations levels. Furthermore, MD simulations inspired by single-molecule experiments demonstrated that this effect can be operative in SPR-like setups [19]. Such a stoichiometry-dependent mechanism is quite surprising and unexpected since the classical textbook view assumes that the dissociation rate has no concentration dependence [20].

The underlying mechanism for the concentration-dependent dissociation kinetics is attributed to the multivalent binding nature of molecular ligands, which allows bound ligands to partially expose the binding sites for other binding competitors in solution [21]. Indeed, many molecular ligands are bound to their binding sites with such multivalent interactions. For instance, proteins bind to their receptor or to DNA via a combination of hydrogen bonds, and electrostatic and van der Walls interactions. A bound ligand such as a nucleic acid-binding protein or a drug molecule can *break* some but not all of these multivalent interactions and easily expose the binding site for the binding competitors in solution. The resulting presumably less stable complex of more than one ligand on a single binding site can promote rapid unbinding of both or one of the bound ligands. This effect is referred to as facilitated dissociation (FD) or exchange, and it is mathematically described as follows [22]:

$$k'_{\mathrm{off}} = k_{\mathrm{off}} + k_{\mathrm{exch}}[c] \quad (3)$$

The expression given by Eq. 3 suggests that as the free ligand concentration in solution, c, increases, the off-rate accelerates linearly with the concentration. The correction in Eq. 3 operates only when the ligand concentration is non-zero (i.e., when $c = 0$, $k'_{\mathrm{off}} = k_{\mathrm{off}}$). In Eq. 3, the exchange rate, k_{exch}, has the units of $M^{-1}\ s^{-1}$ and determines the strength of the facilitated dissociation effect. The typical values of $k_{\mathrm{exch}} \approx 10^4\ M^{-1}\ s^{-1}$ for mid-size (~5 nm) transcription factor proteins unbinding from dsDNA at physiological conditions [17].

In a typical SPR setup, FD can alter the intrinsic dissociation rates as follows. During the dissociation phase, free ligands (recently dissociated) and bound ligands can coexist. As a result, the free ligands can collide with the intact host–guest complexes and accelerate their dissociation via a direct FD as described by Eq. 3. FD could be more influential at the beginning of the dissociation phase due to the non-uniform distribution of the ligands inside the measurement reservoir. This is mainly because, at the beginning of the dissociation phase, more ligands are located near the surface than within the bulk due to the formation of a concentration gradient [12]. In equilibrium, this concentration gradient ideally would disappear, and the ligand concentration in the reservoir would become homogenous. However, at early and intermediate times (close to the intrinsic residence time of the ligand on the binding site), the time-dependent relaxation of the initial concentration quench could lead to FD.

Unlike rebinding, an analytical expression that can correct the side effects of FD on off-rates, as in Eq. 2, does not exist. This is mainly because such expression is expected to depend on the unbinding energy (thus, on the off-rate itself) for the reason that we describe above. However, since the spread of free ligands throughout the reservoir is mainly governed by diffusion, an expression for the time-dependent concentration difference between bulk and near-surface, $\Delta c(t)$, may be written as follows. The average distance that the ligands can diffuse in the vertical direction at time t is Z, with $Z^2 = 2Dt$, where D is the diffusion coefficient of the ligand. If the total number of ligands in the setup (on the surface at the beginning of the dissociation phase) is N, then the concentration of ligands inside a volume, whose height is determined by the diffused distance Z, may be written as $\Delta c(t) \sim N/A \times Z = \sigma_g/(2Dt)^{1/2}$, where A is the area of the reaction surface, and $\sigma_g = N/A = 1/s^2$ is grafting density. In this case, the concentration difference is simplified to

$$\Delta c(t)\frac{\tilde{1}}{s^2\sqrt{t}}. \tag{4}$$

Equation 4 suggests that, for large period of time, $\boldsymbol{\Delta c(t)} \rightarrow 0$, and consequently the concentration difference throughout the reservoir becomes uniform. For large separation distances, $s \gg h$, Eq. 4 also converges to zero. This indicates that at least the effect of FD on off-rates could be reduced by preparing sparsely immobilized binding sites. Indeed, in the fluorescence studies of transcription factors, the binding sites that are separated by roughly 1 micron apart were shown to lead to exponential decays [16]. This is mainly because FD, as in the rebinding case that we discuss above, also requires the diffusion of free ligands to neighboring binding sites. However, unlike in the rebinding effect (which requires a neighboring binding site to be empty), FD affects only those binding sites that are occupied by ligands.

2.3 Deciphering the Key Performance Metrics of Plasmonic Sensing

Common strategies on plasmonic sensors rely on (1) sensitivity within the detection volume (Fig. 1), where the resonance angle or wavelength shifts according to RI changes; (2) thin-film sensitivity, which is caused by the loaded mass, resulting in resonance shifts proportional to the changes in the film thickness; and (3) figure-of-merit (FOM) is the ratio of resonance shifts with linewidth [23]. As elaborated in the earlier subsections, the SPR sensing scheme is considered as a gold standard method to evaluate binding kinetics and estimate interactions between ligands and their binding sites reliant on the adsorption isotherm models (Scatchard, Langmuir, and Freundlich) as follows [24]:

Master equation:

$$\frac{\mathrm{d}\Delta R}{\mathrm{d}t} = kc(\Delta R_{\max} - \Delta R) - k_{\text{off}}\Delta R \tag{5}$$

Scatchard:

$$\frac{\Delta R\text{ex}}{c} = K_{\text{a}}\left(\Delta R_{\max} - \Delta R_{eq}\right) \tag{6}$$

Langmuir:

$$\Delta R = \frac{\Delta R_{\max}[c]}{K_{\text{d}} + [\text{c}]} \tag{7}$$

Freundlich:

$$\Delta R = \Delta R_{\max}[c]^{1/n} \tag{8}$$

where ΔR is the binding response; k_{c} and k_{off} are respective on and off-rates with units of 1/s. In Scatchard and Langmuir isotherms, K_{a} and K_{d} are also the forward and reverse equilibrium constants; subscripts *ex*, *max*, and *eq* indicate experimental, maximum, and equilibrium, respectively. The exponent *1/n* is known as the Freundlich exponent. Scatchard isotherm exhibits freely reversible host–guest interactions and denotes the number of binding sites once the host is at equilibrium [25, 26]. Langmuir isotherm model states a homogeneous binding occurred on the sensor surface [27, 28], whereas the Freundlich model displays a heterogeneous model, being favorable with the low concentration regime and elucidating multilayer formation on the surface [29].

While considering the detection limits, specificity, and sensitivity, there are the other metrics impacting the performance of sensor that include hardware (e.g., sensor structure/material, detector, light source, and optical path), soft biological layer (surface chemistry), as well as external factors (RI and complexity of solution). Nanoplasmonic sensors, for instance, denote tunable fashions through probing more binding sensitivity and exhibiting the reduced sensing in the bulk. Upgraded sensors that are mostly integrated with advanced materials like graphene or designed with well-confined nanostructures provide sharper resonance signals than that of conventional SPR systems, enabling easy probing the resonance changes in minute concentrations of the sample [30]. In the following subsections, we comprehensively reckon all inherent factors affecting key performance metrics, i.e., material types for sensor (hardware), surface functionalization (soft biological layer), and antifouling strategies (for minimizing external factors) that can be used to reduce the side effects in the measurements that we describe in the previous sections.

Recent advances in materials science have empowered the performance and versatility of plasmonic sensors by integrating new-fangled materials into sensor manufacturing [31]. The sensors are mainly fabricated via (1) top-down approaches to create nanostructures or produce metamaterials and hyperbolic surfaces, and alternatively and/or via (2) bottom-up approaches to self-assemble nanoparticles with plasmonic features. Before considering any chemical approaches, material type (gold, silver, aluminum, and palladium) and composition profoundly impact on sensing performance since surface property defines the choice of chemistry to properly sense target analytes.

2.3.1 Surface Functionalization

The sensor surface consists of two major areas: (1) analyte-specific unit (bioactive area), where recognition elements (binding sites) are localized to provide specific host–guest interactions; and (2) inert area that is blocked for non-specific interactions (i.e., phantom binding sites). Signals derived from the surface is usually proportional to the yield of binding events influenced by the surface density of binding sites and the specificity between this area and inert area [32]. In this regard, the specificity of target recognition, the orientation of immobilized binding sites, and the selectivity of the surface coverage (the integration of spatial ratio of bioactive and inert areas addressed by antifouling agents) are three major strategies for designing surface chemistry on the sensor systems [31].

The "specificity" parameter relies on the affinity of molecules (binding sites and ligands), being proportional to a function of the number of both physical and chemical interactions between them. The first interaction type consists of the immobilization of binding sites without any disruption or changes in the native conformation [32]. The latter interactions are the combination of weak molecular interactions between ligands and binding sites. Since these interactions play pivotal signals for the kinetic systems, the analytes are mostly captured on the surface by a bioactive unit, and also labeled with another bioactive unit in order to leverage the specificity of the sensor [33]. The recognition elements on the bioactive area are mostly chosen within the families of immunoglobulins (Ig) or antibodies, in particular, from the IgG family. They are in the glycoprotein structure and consist of two major polypeptide chains: heavy (–COOH end) and light chains (–NH_2 end). Briefly, two heavy chains form fragment crystallizable (Fc) region of an antibody (constant part), whereas two light chains exhibit fragment antigen-binding (Fab) region of an antibody for recognition, thereby acting as a crucial spot for specific recognition. Besides, aptamers, other commonly employed bioactive units, are rationally designed on nucleic acid format, but acting as antibody-type recognition. They are however produced through lengthy and expensive procedures, limiting their expansion to the field. Among these units, nucleic acids, carbohydrates, and molecularly imprinted polymers are utilized [34]. Regardless of the structure and chemical composition of binding sites, the most essential step in the surface chemistry is to orient these units in the right direction to capture the target ligands properly. The "orientation" of binding sites holds two major roles: (1) properly oriented units exhibit stronger efficiency for target analyte recognition, as well as minimize any nonspecific interactions [33, 35]; and (2) highly oriented units enable high recognition and specificity, increasing sensor sensitivity and maximizing the sensor performance. Oriented self-assembled coupling (chemisorption), covalent coupling with chemical linkers, and bioaffinity coupling reactions are usually projected for this manner.

Surfaces coated with Ag or Au allow weak interactions with antibodies and also allow forming dative bonds via disulfide bridges originated from cysteine residues of antibodies or thiol groups of aptamers [32, 36, 37]. Despite easy application of chemisorption, the orientation and the specificity cannot be controlled since this process is random, and most likely, metal-reactive groups of other biomolecules might bind to both the active and inert areas of the sensor. As multiple washing steps are performed during the experimentation, such weak interactions might be liberated from the surface. In such cases, antibodies could be spliced into two fragments with a reducing agent (2-mercaptoethylamine (2-MEA), tris (2-carboxyethyl)phosphine (TCEP), 2-mercaptoethanol (2-ME), or 1,4-dithiotreithol (DTT)) that reacts with disulfide bond of Fc portion, allowing the fragments to be bound via thiol groups to the metal surface, and also ensuring to preserve the right orientation [31, 38]. Despite being easily applicable, this strategy requires strictly controlled conditions to keep the antibody structure native and to maximize the surface coverage (i.e., grafting density); hence, such a strategy could be a better option for immobilizing aptamers that are easily engineered according to this scenario.

Covalent coupling with chemical linkers is one of the most commonly employed strategies during the activation and immobilization of binding sites on the sensors. This process not only improves the stability but also creates a well-harmonized structure by anchoring a bifunctional self-assembled monolayer (SAM) (one binds to the metal surface and the latter acts as a linker), whose linker-end complements with bioactive unit molecules. The chain size and functionality of SAM molecules are chosen according to the reactive groups of bioactive units. For this process, in-solution method depends on multiple parameters, e.g., pH, temperature, local roughness, crystal orientation, and oxidation of the metal surface [31]. As an example, a gentle washing with piranha solution or treating with oxygen plasma helps oxidization on the surface that improves the strength of the S–Au bonds while forming SAM [39]. Vapor functionalization—another strategy for covalent coupling—is carried out at ambient pressure [40] or under high-vacuum conditions in order to achieve well-ordered monolayers, yet the utility of this method is limited due to the need for complex setups and multistep procedures [41].

The latter strategy, bioaffinity coupling, secures the oriented immobilization, reliant on weak interactions between the linker and the recognition element that allow a highly specific and strong recognition mechanism. Avidin–biotin interaction is a commonly used strategy exhibiting extremely high affinity with a dissociation constant $K_d \approx 10^{-14}$ mol/L. In practice, avidin molecules are immobilized on the surface, while biotin molecules are linked to the recognition elements (mostly an antibody, nucleic acid, or aptamer). The order of these molecules might be vice versa. However, constructing the right orientation relies on the position and number of biotin molecules on the recognition elements. Another bioaffinity coupling is applied via antibody-binding proteins such as Protein A and Protein G that selectively bind to the Fc portion of antibodies, guiding them to be anchored in the correct orientation [42].

2.3.2 Antifouling Strategies

Ideally, the binding sites are immobilized on the sensing area, whereas all the surrounding needs to be inert for any kind of non-specific interactions. Forming inert areas have pivotal roles in developing high sensitivity capable of detecting a drastically low number of target of interest from complex biospecimens (e.g., whole blood, serum, urine, sweat, lavage fluid, and saliva) made of highly dense media that results in high RI changes during the measurement by a conventional SPR platform. When this is a case, constructing an antifouling layer help passivating the non-active area, minimizing or ideally avoiding any nonspecific interactions. Protein-stemmed strategies, such as applying bovine serum albumin (BSA) or casein, are usually preferred for this aim because of their high abundance and easy binding through physio-adsorption. As described above, passive adsorption of molecules could be liberated typically during the multistep washing. Chemical binding of polymeric antifouling agents is thereby preferred to ensure stable conjugation and form flexibility like observed in cell membrane systems [43]. Zwitterionic polymers and poly(ethylene glycol)

(PEG) are the most employed antifouling agents, owing to exhibiting strong surface hydration [44, 45]. While zwitterionic polymers are used, the hydration layer is created by electrostatic-induced hydrogen bonding. On the other side, this layer is formed by hydrogen bonding between water and the ether oxygen atoms while PEG is immobilized on the surface [45–47]. Theoretically, this hydration layer creates a physical and energy barrier, where water molecules cannot be replaced by any biomolecule or larger entities like cells. In some cases, the surface roughness forms superhydrophobic surfaces that aid target molecules to be captured, whereas the adsorption of other molecules in solution is eliminated [45]. Besides, many printing strategies are employed to improve selectivity in sensors by (1) conventional direct printing via an automatized inked tips to form either continuous or discrete areas on the sensor; (2) noncontact printing (NCP) reliant on lithographic stamps, mostly made of PDMS, that transfers biomolecules onto the bioactive area once they are immersed in an inking solution [48]. Nano-stencil lithography (NSL) is another strategy to bring bio-reactive layer in contact with biomolecules through a stencil mask consisting of sub-100-nm patterns [49]. The mask employed in this strategy could be reused multiple times, and it is able to pattern large superficial areas rapidly [49].

3 Final Remarks

In this chapter, we described two effects that can influence the measured dissociation rates by SPR or in general similar setup relying on kinetic mass changes. The first effect arises from the rebinding of ligands to binding sites before exiting the detection volume, thus, adds a delay on the dissociation data. This delay results in an underestimation of off-rates (Fig. 2). The second effect is due to accelerated dissociation of bound ligands by free ligands near the binding site (FD). FD in principle leads to higher off-rates. As we discussed, both effects diminish as the average distance between the binding sites increases (or at low grafting densities). When the measurements are performed for strongly bound ligands (e.g., >5 kcal/mol), the alternative fit function, Eq. 2 (e.g., the stretched exponential [11]), can correct the rebinding effects. However, for weakly binding small ligands (<3–4 kcal/mol), even the improved fit function underestimates the off-rates (Fig. 2). Thus, sparsely placed binding sites seem to be the best option to extract the dissociation rates of ligands with very weak affinity in both experiments and computer simulations of SPR systems. The large separation will also reduce the unwanted effect of FD.

Considering the fundamental and experimental interfaces of SPR-type sensors, material type holds a pivotal characteristic on the sensitivity along with the integration of optical setup and provides an opportunity to choose more surface-sensitive or bulk-sensitive measurements [50]. In our calculations, we think that enhancing the sensitivity layer at the vicinity of the sensor will help us monitoring more interactions (especially diffusion-controlled reactions) between target binding sites and ligands. Besides, the surface chemistry and antifouling strategies will enable to tune surface grafting. For instance, polymer-brush type of agents will increase surface area, potentially forming more binding sites and dynamic structure compared to the SAM-based backbone structures (particularly rigid constructions). At the same time, integrating an antifouling layer to the sensor surface potentially minimizes nonspecific binding and increases the direct host–guest interactions during the measurements, thereby enabling more secure data collection to understand the binding, dissociation, and diffusion events. Furthermore, the separation distance between the binding sites could be achieved using a specific ratio of antifouling agents and recognition elements [51].

4 Notes

1. In Biocore systems, 1 RU corresponds to 1 pg/mm^2, whereas in other systems (Autolab) 0.01 ng/mm^2 corresponds to 1 m° (millidegree). Hence, the conversion from RU to m° is 1 RU = 0.1 m°.
2. Once the raw data is obtained in (time vs. signal) format, by using excel or other available mathematical manipulation software (e.g., MATLAB, Mathematica), the data can be fit to the function, $f(t)$ in Eq. 2. For instance, in Mathematica, the signal can be read in a datafile with the command "datafile=**Import ["signal", "Table"]**". Then, this array can be fit by using the command "**FindFit[datafile, $f(t)$, k_{off}, t]**" to extract the off-rate.

References

1. Karlsson R (2004) SPR for molecular interaction analysis: a review of emerging application areas. J Mol Recognit 17(3):151–161. https://doi.org/10.1002/jmr.660
2. Inci F, Saylan Y, Kojouri AM, Ogut MG, Denizli A, Demirci U (2020) A disposable microfluidic-integrated hand-held plasmonic platform for protein detection. Appl Mater Today 18:100478. https://doi.org/10.1016/j.apmt.2019.100478
3. Saylan Y, Denizli A (2018) Molecular fingerprints of hemoglobin on a nanofilm chip. Sensors 18(9):3016. https://doi.org/10.3390/s18093016
4. Li Z, Leustean L, Inci F, Zheng M, Demirci U, Wang S (2019) Plasmonic-based platforms for diagnosis of infectious diseases at the point-of-care. Biotechnol Adv 37(8):107440. https://doi.org/10.1016/j.biotechadv.2019.107440

5. Tokel O, Inci F, Demirci U (2014) Advances in plasmonic technologies for point of care applications. Chem Rev 114(11):5728–5752. https://doi.org/10.1021/cr4000623
6. Inci F, Tokel O, Wang S, Gurkan UA, Tasoglu S, Kuritzkes DR, Demirci U (2013) Nanoplasmonic quantitative detection of intact viruses from unprocessed whole blood. ACS Nano 7(6):4733–4745. https://doi.org/10.1021/nn3036232
7. McDonnell JM (2001) Surface plasmon resonance: towards an understanding of the mechanisms of biological molecular recognition. Curr Opin Chem Biol 5(5):572–577. https://doi.org/10.1016/S1367-5931(00)00251-9
8. Saylan Y, Erdem Ö, Cihangir N, Denizli A (2019) Detecting fingerprints of waterborne bacteria on a sensor. Chemosensors 7(3):33. https://doi.org/10.3390/chemosensors7030033
9. Inci F, Celik U, Turken B, Özer HÖ, Kok FN (2015) Construction of P-glycoprotein incorporated tethered lipid bilayer membranes. Biochem Biophys Reports 2:115–122. https://doi.org/10.1016/j.bbrep.2015.05.012
10. de Mol NJ, Fischer MJE (2010) Surface Plasmon resonance: methods and protocols. Humana Press, Totowa, New Jersey. https://doi.org/10.1007/978-1-60761-670-2
11. Gopalakrishnan M, Forsten-Williams K, Cassino TR, Padro L, Ryan TE, Täuber UC (2005) Ligand rebinding: self-consistent mean-field theory and numerical simulations applied to surface plasmon resonance studies. Eur Biophys J 34:943–958. https://doi.org/10.1007/s00249-005-0471-2
12. Erbaş A, de la Cruz MO, Marko JF (2019) Receptor-ligand rebinding kinetics in confinement. Biophys J 116(9):1609–1624. https://doi.org/10.1016/j.bpj.2019.02.033
13. Gopalakrishnan M, Forsten-Williams K, Nugent MA, Täuber UC (2005) Effects of receptor clustering on ligand dissociation kinetics: theory and simulations. Biophys J 89(6):3686–3700. https://doi.org/10.1529/biophysj.105.065300
14. Carroll J, Raum M, Forsten-Williams K, Täuber UC (2016) Ligand-receptor binding kinetics in surface plasmon resonance cells: a Monte Carlo analysis. Phys Biol 13(6):066010. https://doi.org/10.1088/1478-3975/13/6/066010
15. Murray JB, Roughley SD, Matassova N, Brough PA (2014) Off-rate screening (ORS) by surface plasmon resonance. An efficient method to kinetically sample hit to lead chemical space from unpurified reaction products. J Med Chem 57(7):2845–2850. https://doi.org/10.1021/jm401848a
16. Erbaş A, Marko JF (2019) How do DNA-bound proteins leave their binding sites? The role of facilitated dissociation. Curr Opin Chem Biol 53:118–124. https://doi.org/10.1016/j.cbpa.2019.08.007
17. Kamar RI, Banigan EJ, Erbas A, Giuntoli RD, de la Cruz MO, Johnson RC, Marko JF (2017) Facilitated dissociation of transcription factors from single DNA binding sites. Proc Natl Acad Sci U S A 114(16):E3251–E3257. https://doi.org/10.1073/pnas.1701884114
18. Lewis JS, Spenkelink LM, Jergic S, Wood EA, Monachino E, Horan NP, Duderstadt KE, Cox MM, Robinson A, Dixon NE, van Oijen AM (2017) Single-molecule visualization of fast polymerase turnover in the bacterial replisome. Elife 6:e23932. https://doi.org/10.7554/eLife.23932
19. Erbaş A, De La Cruz MO, Marko JF (2018) Effects of electrostatic interactions on ligand dissociation kinetics. Phys Rev E 97(2):022405. https://doi.org/10.1103/PhysRevE.97.022405
20. Berg HC (1984) Random walks in biology – new and expanded edition. Princeton University Press, Princeton, New Jersey. ISBN: 9780691000640
21. Sing CE, De La Cruz MO, Marko JF (2014) Multiple-binding-site mechanism explains concentration-dependent unbinding rates of DNA-binding proteins. Nucleic Acids Res 42(6):3783–3791. https://doi.org/10.1093/nar/gkt1327
22. Graham JS, Johnson RC, Marko JF (2011) Concentration-dependent exchange accelerates turnover of proteins bound to double-stranded DNA. Nucleic Acids Res 39(6):2249–2259. https://doi.org/10.1093/nar/gkq1140
23. Oh SH, Altug H (2018) Performance metrics and enabling technologies for nanoplasmonic biosensors. Nat Commun 9:5263. https://doi.org/10.1038/s41467-018-06419-3
24. Li X, Husson SM (2006) Adsorption of dansylated amino acids on molecularly imprinted surfaces: a surface plasmon resonance study. Biosens Bioelectron 22(3):336–348. https://doi.org/10.1016/j.bios.2006.04.016
25. Rocha MS (2015) Extracting physical chemistry from mechanics: a new approach to investigate DNA interactions with drugs and proteins in single molecule experiments. Integr Biol 7(9):967–986. https://doi.org/10.1039/c5ib00127g

26. Dibekkaya H, Saylan Y, Yılmaz F, Derazshamshir A, Denizli A (2016) Surface plasmon resonance sensors for real-time detection of cyclic citrullinated peptide antibodies. J Macromol Sci Part A Pure Appl Chem 53 (9):585–594. https://doi.org/10.1080/10601325.2016.1201756
27. Langmuir I (1916) The constitution and fundamental properties of solids and liquids. Part I. Solids. J Am Chem Soc 38(11):2221–2295. https://doi.org/10.1021/ja02268a002
28. Fornstedt T, Forssén P, Samuelsson J (2013) Modeling of preparative liquid chromatography. Liq Chromatogr:573–592. https://doi.org/10.1016/B978-0-12-805393-5.00024-5
29. Saylan Y, Yılmaz F, Derazshamshir A, Yılmaz E, Denizli A (2017) Synthesis of hydrophobic nanoparticles for real-time lysozyme detection using surface plasmon resonance sensor. J Mol Recognit 30(9):e2631. https://doi.org/10.1002/jmr.2631
30. Dahlin AB, Wittenberg NJ, Höök F, Oh SH (2013) Promises and challenges of nanoplasmonic devices for refractometric biosensing. Nanophotonics 2(2):83–101. https://doi.org/10.1515/nanoph-2012-0026
31. Oliverio M, Perotto S, Messina GC, Lovato L, De Angelis F (2017) Chemical functionalization of plasmonic surface biosensors: a tutorial review on issues, strategies, and costs. ACS Appl Mater Interfaces 9(35):29394–29411. https://doi.org/10.1021/acsami.7b0158
32. Jung Y, Jeong JY, Chung BH (2008) Recent advances in immobilization methods of antibodies on solid supports. Analyst 133:697–701. https://doi.org/10.1039/B800014J
33. Reimhult E, Höök F (2015) Design of surface modifications for nanoscale sensor applications. Sensors 15(1):1635–1675. https://doi.org/10.3390/s150101635
34. Saylan Y, Erdem Ö, Inci F, Denizli A (2020) Advances in biomimetic systems for molecular recognition and biosensing. Biomimetics 5 (2):20. https://doi.org/10.3390/biomimetics5020020
35. Kalia J, Raines R (2010) Advances in bioconjugation. Curr Org Chem 14(2):138–147. https://doi.org/10.2174/138527210790069839
36. Ouellette D, Alessandri L, Chin A, Grinnell C, Tarcsa E, Radziejewski C, Correia I (2010) Studies in serum support rapid formation of disulfide bond between unpaired cysteine residues in the VH domain of an immunoglobulin G1 molecule. Anal Biochem 397(1):37–47. https://doi.org/10.1016/j.ab.2009.09.027
37. Algar WR (2011) The controlled display of biomolecules on nanoparticles: a challenge suited to bioorthogonal chemistry. Bioconjug Chem 22(5):825–858. https://doi.org/10.1021/bc200065z
38. Karyakin AA, Presnova GV, Rubtsova MY, Egorov AM (2000) Oriented immobilization of antibodies onto the gold surfaces via their native thiol groups. Anal Chem 72 (16):3805–3811. https://doi.org/10.1021/ac9907890
39. Baba A, Taranekar P, Ponnapati RR, Knoll W, Advincula RC (2010) Electrochemical surface plasmon resonance and waveguide-enhanced glucose biosensing with N-alkylaminated polypyrrole/glucose oxidase multilayers. ACS Appl Mater Interfaces 2(8):2347–2354. https://doi.org/10.1021/am100373v
40. Choi Y, Jeong Y, Chung H, Ito E, Hara M, Noh J (2008) Formation of ordered self-assembled monolayers by adsorption of octylthiocyanates on Au(111). Langmuir 24 (1):91–96. https://doi.org/10.1021/la701302g
41. Cossaro A, Dell'Angela M, Verdini A, Puppin M, Kladnik G, Coreno M, de Simone M, Kivimäki A, Cvetko D, Canepa D, Floreano L (2010) Amine functionalization of gold surfaces: ultra high vacuum deposition of cysteamine on Au(111). J Phys Chem C 114 (35):15011–15014. https://doi.org/10.1021/jp104824c
42. Wang S, Esfahani M, Gurkan UA, Inci F, Kuritzkes DR, Demirci U (2012) Efficient on-chip isolation of HIV subtypes. Lab Chip 12 (8):1508–1515. https://doi.org/10.1039/C2LC20706K
43. Leng C, Hung HC, Sun S, Wang D, Li Y, Jiang S, Chen Z (2015) Probing the surface hydration of nonfouling zwitterionic and peg materials in contact with proteins. ACS Appl Mater Interfaces 7(30):16881–16888. https://doi.org/10.1021/acsami.5b05627
44. Jiang S, Cao Z (2010) Ultralow-fouling, functionalizable, and hydrolyzable zwitterionic materials and their derivatives for biological applications. Adv Mater 22(9):920–932. https://doi.org/10.1002/adma.200901407
45. Lin PH, Li BR (2020) Antifouling strategies in advanced electrochemical sensors and biosensors. Analyst 145:1110–1120. https://doi.org/10.1039/C9AN02017A
46. Leung BO, Yang Z, Wu SSH, Chou KC (2012) Role of interfacial water on protein adsorption at cross-linked polyethylene oxide interfaces. Langmuir 28(13):5724–5728. https://doi.org/10.1021/la204805x

47. Galvin CJ, Dimitriou MD, Satija SK, Genzer J (2014) Swelling of polyelectrolyte and polyzwitterion brushes by humid vapors. J Am Chem Soc 136(36):12737–12745. https://doi.org/10.1021/ja5065334
48. Tran H, Killops KL, Campos LM (2013) Advancements and challenges of patterning biomolecules with sub-50 nm features. Soft Matter 9:6578–6586. https://doi.org/10.1039/C3SM00149K
49. Huang M, Galarreta BC, Artar A, Adato R, Aksu S, Altug H (2012) Reusable nanostencils for creating multiple biofunctional molecular nanopatterns on polymer substrate. Nano Lett 12(9):4817–4822. https://doi.org/10.1021/nl302266u
50. Mataji-Kojouri A, Ozen MO, Shahabadi M, Inci F, Demirci U (2020) Entangled nanoplasmonic cavities for estimating thickness of surface-adsorbed layers. ACS Nano 14(7):8518–8527. https://doi.org/10.1021/acsnano.0c02797
51. Zareie HM, Boyer C, Bulmus V, Nateghi E, Davis TP (2008) Temperature-responsive self-assembled monolayers of oligo(ethylene glycol): control of biomolecular recognition. ACS Nano 2(4):757–765. https://doi.org/10.1021/nn800076h

Chapter 12

Computational Tools for Accurate Binding Free-Energy Prediction

Maria M. Reif and Martin Zacharias

Abstract

A quantitative thermodynamic understanding of the noncovalent association of (bio)molecules is of central importance in molecular life sciences. An important quantity characterizing (bio)molecular association is the binding affinity or absolute binding free energy. In recent years, the computational prediction of absolute binding free energies has evolved considerably in terms of accuracy, computational speed, and user-friendliness. In this chapter, we first give an overview of how absolute free energies are defined and how they can be determined with computational means. We proceed with an outline of the theoretical basis of the two most reliable methods, potential of mean force, and double decoupling calculations. In particular, we describe how the sampling problem can be alleviated by application of restraints. Finally, we provide step-by-step instructions of how to set up corresponding molecular simulations with a commonly employed molecular dynamics simulation engine.

Key words Molecular dynamics simulations, Free-energy calculation, Binding free energy, Potential of mean force, Double decoupling

1 Introduction

1.1 General Considerations

The physical foundation of most processes in molecular biology is the reversible noncovalent interaction of molecules. Two binding partners, the target T and the ligand L, may associate and form a complex TL *via* the binding reaction

$$T + L \rightleftharpoons TL. \tag{1}$$

For example, T and L may be an enzyme and a substrate, a drug target and a drug molecule, or an antibody and an antigen. The reaction is reversible, i.e., the complex can repeatedly associate and dissociate. At equilibrium, the law of mass action states that the ratio of product and educt concentrations c_i of the involved species i is constant,

Quentin Vanhaelen (ed.), *Computational Methods for Estimating the Kinetic Parameters of Biological Systems*, Methods in Molecular Biology, vol. 2385, https://doi.org/10.1007/978-1-0716-1767-0_12,

$$\frac{c_{TL}}{c_T c_L} = const = K_b \quad . \tag{2}$$

This constant is referred to as the binding constant K_b, and its inverse is the dissociation constant $K_d = K_b^{-1}$.

A complete understanding of the association process involves the characterization of its thermodynamic and kinetic features, which allows to answer questions such as: How strong is the binding between T and L? How rapidly does the binding between T and L occur? Answering these questions also allows to make comparisons between multiple possible binding partners, e.g., Is ligand L or ligand L' a stronger binder to target T? Is the complex TL or the complex TL' more long-lived? Such questions are of central importance to the detailed physico-chemical understanding of biology and to applications in pharmacy, medicine, and technology.

This chapter focuses on noncovalent stoichiometric biomolecular association processes in aqueous solvent. Situations at constant pressure and temperature are considered; hence, the relevant thermodynamic quantity characterizing a binding event is the Gibbs binding free energy ΔG_{bnd}. Usually, it is referred to as an "absolute" binding free energy to distinguish it from the so-called relative binding free energies $\Delta\Delta G_{rel} = \Delta G_{bnd}(L) - \Delta G_{bnd}(L')$ that are a difference of two binding free energies $\Delta G_{bnd}(L)$ and $\Delta G_{bnd}(L')$ for two different ligands L and L' binding to the target. The standard concentration in solution is $c^\circ = 1$ mol·L^{-1} (1 M), and the corresponding quantities are indicated with a "∘" superscript, e.g., the standard Gibbs binding free energy ΔG°_{bnd}. In computational work, one usually deals with one molecule sampling a certain volume in solution, and hence, one uses the volume available to one molecule at the standard concentration, $V^\circ = 1.66$ nm^3, to convert into standard quantities.

The equilibrium condition (Eq. 2) can also be written in terms of chemical potentials μ_i of the involved species i in solution,

$$\mu_{TL} = \mu_T + \mu_L, \tag{3}$$

where

$$\mu_i = \mu_i^\circ + RT \ln\left(\frac{\gamma_i c_i}{c^\circ}\right), \tag{4}$$

R is the ideal gas constant, T the absolute temperature, μ_i° the standard chemical potential, and γ_i is the activity coefficient of species i (assumed unity in the following, which applies to the limit of vanishing c_i in the solvent). In the following, all activity coefficients and concentrations refer to the 1 M concentration scale. Defining $\Delta G^\circ_{bnd} = \mu^\circ_{TL} - \mu^\circ_T - \mu^\circ_L$ and combining Eqs. 3 and 4, we find

$$\Delta G^\circ_{bnd} = -RT \ln\left(\frac{c^\circ c_{TL}}{c_T c_L}\right), \tag{5}$$

where the c_i are equilibrium concentrations. Along with Eq. 2,

$$\Delta G^{\circ}_{bnd} = -RT \ln \left(c^{\circ} K_b \right) \quad . \tag{6}$$

From Eqs. 2 and 6, it follows that the bound complex TL is favored over the unbound components T and L if ΔG°_{bnd} is negative. One can distinguish weak binders with K_d in the range of 10^{-5} to 10^{-3} M and strong binders with K_d lower than 10^{-6} M. In computational work, weak binders pose challenges because of their usually transient and weak interactions with the target molecule. Weak binders also have dissociation rate constants k_{off} usually higher than 0.1 s^{-1}, meaning that the dissociation of the TL complex happens faster than within a second. In this context, and to draw a link to kinetics, note that since at equilibrium there is a balance of the forward and reverse reactions in Eq. 1, with reaction rates k_{on} and k_{off}, respectively, one can write $k_{on} c_T c_L = k_{off} c_{TL}$, and thus, it follows from Eq. 2 that

$$K_d = k_{off} / k_{on}, \tag{7}$$

i.e., the binding affinity is related to the ratio of association and dissociation rate constants.

To establish a link to statistical mechanics, we consider a system at equilibrium (at temperature T and pressure P°), where one molecule of species i is solvated in a large number N of solvent molecules [1]. The volume of this system is V. The partial molar volume of i is $\bar{v}_i$. We write μ_i° in terms of partition functions [1–3]

$$\mu_i^{\circ} = -RT \ln \left[\left(V c^{\circ} \right)^{-1} \frac{Q_i}{Q_o} \right] + P^{\circ} \bar{v}_i, \tag{8}$$

where Q_i is the canonical partition function for the considered system and Q_o is the canonical partition function for N solvent molecules at equilibrium at temperature T and pressure P°. One can see that μ_i° contains the work of solvating the solute i in N solvent molecules at constant temperature and pressure.

In classical statistical mechanics, the partition function of a N_{tot}-particle system can be written as a phase space integral using the classical Hamiltonian $H(\mathbf{p}^{3N_{tot}}, \mathbf{r}^{3N_{tot}}) = U(\mathbf{r}^{3N_{tot}}) + \sum_{i=1}^{N_{tot}} (2m_i)^{-1} p_i^2$, where $\mathbf{p}^{3N_{tot}}$ and $\mathbf{r}^{3N_{tot}}$ are the vectors of momenta and positions, respectively, and m_i is the mass of atom i [4],

$$Q_i = \int\int d\mathbf{p}^{3N_{tot}}\, d\mathbf{r}^{3N_{tot}} \exp\left[-\beta H(\mathbf{p}^{3N_{tot}}, \mathbf{r}^{3N_{tot}}) \right], \tag{9}$$

where $\beta = (k_B T)^{-1}$ and k_B is the Boltzmann constant. Similarly, one can write down the partition function of the pure-solvent system, Q_o. Both Q_i and Q_o can be simplified by carrying out the integration over momenta analytically. Plugging Eq. 9 for Q_i as well as its analog for Q_o into Eq. 8, one can cancel the solvent momenta contributions in the numerator and denominator, such that only the solute momenta integral for the $N_{u,i}$ solute atoms,

$\Pi_{j=1}^{N_{u,i}}(2\pi m_j k_B T)^{\frac{3}{2}}$, remains. Furthermore, considering only the $3N_{u,i}-6$ internal solute atom coordinates $\mathbf{q}^{3N_{u,i}-6}$ and carrying out the integration over the three Euler angles defining the solute orientation gives a term $8\pi^2 V$, such that [1]

$$\mu_i^\circ = -RT\ln\left[8\pi^2(c^\circ)^{-1}\Pi_{j=1}^{N_{u,i}}(2\pi m_j k_B T)^{\frac{3}{2}}\frac{Z_i}{Z_o}\right] + P^\circ \bar{v}_i, \quad (10)$$

where the argument of the logarithm now contains a ratio of configurational integrals Z_i/Z_o,

$$\frac{Z_i}{Z_o} = \frac{\int\int d\mathbf{q}^{3N_{u,i}-6}\, d\mathbf{r}^{3N_s}\exp\left[-\beta U(\mathbf{q}^{3N_{u,i}-6}, \mathbf{r}^{3N_s})\right]}{\int d\mathbf{r}^{3N_s}\exp\left[-\beta U(\mathbf{r}^{3N_s})\right]}, \quad (11)$$

where N_s is the number of solvent atoms. It follows that [1]

$$\Delta G_{bnd}^\circ = -RT\ln\left[(8\pi^2)^{-1}c^\circ\frac{Z_o Z_{TL}}{Z_T Z_L}\right] + P^\circ \Delta\bar{v}, \quad (12)$$

where $\Delta\bar{v} = \bar{v}_{TL} - \bar{v}_L - \bar{v}_T$ is the difference in partial molar volumes of the solvated species. Hence, the absolute binding free energy is related to the ratio of configurational integrals of the solvated T and L molecules and that of the solvated TL complex.

1.2 Computational Tools to Predict Binding Free Energies

Since the characterization of the thermodynamics of molecular recognition is such an important topic in many areas of life sciences, there is a plethora of computational tools available for this purpose [5–9]. Most tools rely on a classical molecular mechanics (MM) description [10] of the system, and these methods are the main focus here. Quantum-mechanical (QM) methods still come with a prohibitively large computational expense and are hence not routinely applied for biological systems [11]. Mixed quantum-mechanical molecular-mechanical (QM/MM) methods are of importance for studies involving reactions with covalent bond breaking or formation [12, 13]. They will not be considered here because the majority of molecular binding events in biology is governed by noncovalent interactions.

At the broadest level, one can distinguish molecular docking [14, 15] and molecular simulation methods [5, 6, 9, 16] to assess absolute binding free energies. A recent overview is provided in Ref. [8]. Common to all the methods is that there is always a tradeoff between accuracy and wallclock time. Moreover, certain prerequisites have to be fulfilled for a computational absolute binding free-energy prediction to be meaningful [6]: The relevant interactions in the computational system have to be reliably captured by a classical MM description, and one has to have reasonably accurate structures of the target–ligand complex with defined protonation states of relevant residues (although this point may be somewhat relaxed in certain molecular simulation methods [17]).

In addition, while advanced molecular simulation methods can deal with conformational changes in the target as well as with issues concerning non-unique ligand binding modes, for straightforward molecular docking it is essential that the ligand binding process does not involve these complications. The limitations may, however, be overcome by a combination of molecular docking and molecular simulation [18–20].

Molecular docking is a two-step process that first generates a so-called pose of the target–ligand complex and then scores the interactions in the complex to estimate the binding affinity. The method is in principle simple and fast. However, it is not only because of the above limitations that molecular docking is not a very reliable method to predict binding free energies. There are several other shortcomings. Most docking programmes neglect the effect of explicit water molecules [21] and do not give an accurate account of the changes in configurational entropy of the ligand and the target upon binding [22]. In addition, many of the commonly used scoring functions [23] essentially merely count important target–ligand interactions, which is a highly simplified quantitative account of the thermodynamic changes that occur during target–ligand binding. Nevertheless, molecular docking methods are a useful qualitative tool in that different ligands can be ranked according to their binding affinity and that important target–ligand interactions can still be identified.

Molecular simulation tools to compute absolute binding free energies are commonly separated into three classes:

1. Endpoint methods: Among the molecular simulation methods, endpoint methods are the least accurate but commonly the most speedy. They only consider the endpoints of the binding process, i.e., the isolated binding partners and the bound complex. The three most important endpoint methods are the linear interaction energy method (LIE) [24–26], the molecular mechanics Poisson–Boltzmann surface area method (MM/PBSA) [27, 28], and the related molecular mechanics Generalized Born surface area method (MM/GBSA) [26, 28, 29]. In the LIE method, the interactions between the ligand with the environment are assessed for multiple, Boltzmann-sampled conformations in the explicitly solvated endpoint states with a classical molecular mechanics force field. While a linear-response assumption [24, 30] is used to convert the changes in ligand–environment electrostatic interactions into a free-energy contribution, an empirical relation is used for the ligand–environment van der Waals interactions [24] and a possible third reorganization free-energy [30] contribution. Changes in target–solvent interactions during binding are neglected. Hence, the LIE method is rather approximate and often works only well for similar ligands and a system-specific

choice of empirical parameters. In this sense, it is somewhat reminiscent of a scoring function used in docking. However, it has the advantages of accounting for explicit ligand solvation effects and conformational flexibility in the target.

MM/PBSA and MM/GBSA methods share the features of a molecular mechanics representation of intra-target, intra-ligand, and intermolecular target–ligand interactions, of an implicit solvent representation [31] (*via* the solvent dielectric constant) and of a simple evaluation of the nonpolar solvation free-energy contribution through a parameterized dependence on the solvent-exposed surface area [32]. They differ in the calculation of the electrostatic solvation free-energy contribution, which is either done by numerically solving the Poisson–Boltzmann equation in the MM/PBSA method (e.g., *via* finite differences [33]) or analytically from a parameterized model [34] derived from the Born equation [35] in the MM/GBSA method. The latter method is generally computationally more affordable and somewhat less accurate [36]. Similar to the LIE method, all the above terms are assessed for multiple, Boltzmann-sampled conformations of the isolated (implicitly or explicitly solvated) binding partners and the (implicitly or explicitly solvated) complex. From their difference, the binding free energy can be deduced. A disadvantage of MM/PBSA and MM/GBSA methods is the neglect of explicit water molecules in the ligand binding pocket. These water molecules may be released upon binding, which can give rise to a non-negligible free-energy contribution, or they may be important in bridging ligand–target interactions and overlooking this effect may give a spurious estimate of the binding affinity [37, 38]. Another disadvantage is the handling of the entropy change during binding. It is often neglected because its converged calculation, e.g., based on a normal-mode analysis of the sampled configurations, is computationally expensive [39].

2. Pathway methods: The characteristic feature of a pathway method is that one considers the bound target–ligand complex and drives the unbinding process *via* user-defined non-physical force-field energy terms called restraints along a certain reaction coordinate, usually a one-dimensional distance between the centers of masses of a set of target and ligand atoms. In the commonly used umbrella sampling [40] method, harmonic restraining potentials are added in multiple simulations along the reaction coordinate to bias the sampling such that ligand unbinding occurs. Additional restraints can be added to ameliorate sampling problems [41] in the sense that the sampling problem is shifted to simulations where convergence is reached at lower computational cost. Besides the simulations associated with the ligand unbinding process, extensive additional post-

processing, and possibly simulation work, is required to convert the raw free energy of unbinding into a standard binding free energy [41, 42] (Subheading 1.3.2). Pathway methods have the advantage that both the target and the ligand can be biomacromolecules (i.e., encompass hundreds of solute atoms) and that the free-energy profile (related to the so-called potential of mean force (PMF) [43]) of the binding/unbinding process along the chosen reaction coordinate is obtained, which allows a mechanistic understanding of target–ligand association in terms of thermodynamic and partially also kinetic features [44]. The former point is of great relevance because knowledge about biomacromolecular association, especially protein–protein binding, is central to the understanding of cellular processes [45] and offers attractive options for the development of novel drugs [46]. A disadvantage is that special care has to be taken when the binding/unbinding path involves high free-energy barriers, e.g., due to burial of the binding pocket [47], or when structural changes occur during unbinding [48]. Furthermore, large computational boxes are required to adequately describe the noninteracting unbound state.

3. Alchemical methods: The characteristic feature of an alchemical method is that one considers the bound target–ligand complex and the ligand in solution and in both systems gradually decouples the ligand from its environment leading to a ligand molecule that is no more interacting with its environment [49]. Such a molecule is called a "ghost" or "dummy" molecule, and the method is generally referred to as "double decoupling" (DD) [1]. In the decoupling simulation of the bound complex, restraints are commonly added to restrain the ligand molecule to the binding pocket [1, 50]. Besides the two decoupling simulations, extensive additional post-processing, and possibly simulation work, is required to convert the raw free energy of unbinding into a standard binding free energy [1, 50, 51] (Subheading 1.3.3). Alchemical methods have the advantage that they can be easily carried out in the case of buried binding pockets. Disadvantages are the restriction to relatively small ligands (not larger than ca. 40 atoms) and that special care has to be taken in the case of charged ligand molecules [52–54].

The following sections will explain in detail how the potential of mean force and double decoupling methods are used to obtain absolute binding free energies in the framework of classical atomistic molecular dynamics (MD) simulations. MD simulation is a widely used computational simulation method that can yield a high-resolution picture of (bio)molecular systems and accurately quantify thermodynamic and kinetic characteristics of (bio)-molecular binding processes [8, 44, 55, 56]. There is a plethora of MD simulation packages available, e.g., AMBER [57], CHARMM [58], GROMACS [59], GROMOS [60], or NAMD [61]. The code in these packages is continuously being maintained and updated with respect to advances in simulation techniques and computer hardware. Especially computation on graphics processing units (GPUs) [62, 63] or parallel programme execution on multiple compute nodes [59, 64] makes binding affinity calculation *via* MD simulation a versatile and efficient means to gain basic understanding in molecular biology or to support practical applications such as drug design.

1.3 Theory

1.3.1 Molecular Dynamics Simulation

Classical MD simulations use a classical Hamiltonian and Newton's equations of motion to propagate atom coordinates in the system of interest for a given time, usually at constant temperature and pressure [10, 16]. The simulated system is of microscopic size and hence not genuinely representative of a typical wet-lab sample in a test tube, but algorithmic features during the simulation (periodic boundary conditions [10, 65]) and possibly analysis in post-processing can address issues that may incur [66, 67]. The Hamiltonian contains a potential-energy contribution called the "force field" [68, 69]. It captures bonded interactions in a molecular mechanics fashion through pairwise potential-energy terms, e.g., using harmonic potentials for bonds and angles and cosine series for rotations (torsions) around bonds. Van der Waals interactions are usually described with a Lennard–Jones potential [70, 71], and electrostatic interactions may be calculated using lattice summation [72]. The derivative of the potential energy with respect to particle positions yields forces that are used in a numerical integration algorithm to calculate coordinates for a finite time step (typically 2 fs) further in time, and this procedure is repeated to give a collection of thousands of simulation frames called a trajectory [10]. The trajectory can then be further analyzed to compute macroscopic properties *via* the tools of statistical mechanics [65]. Since MD simulations provide microscopic insight into the investigated system, one can then understand how the macroscopic properties emerge from the underlying microscopic interactions. Thus, MD simulations can be extraordinarily helpful to understand, complement, or predict experimental findings [73, 74].

In principle, MD simulations could deliver an absolute binding free energy from a long simulation where the two binding partners are immersed in a large solvent box, and the user observes how much time the system spends in the bound (t_b) and unbound (t_u) states [75], where the two molecules form the complex and where they are widely separated in solvent, respectively,

$$K_b = \frac{t_b}{t_u c^\circ} \quad . \tag{13}$$

However, it may take large compute times until sufficient reversible binding/unbinding events are observed to get a statistically accurate binding free energy, which is why this approach is not generally applicable but rather limited to weakly bound complexes with fast association and dissociation rate constants. Nevertheless, it was shown that application of enhanced sampling methods allows to sample a sufficient number of association and dissociation events such that Markov models can be constructed from which information about the kinetics and thermodynamics of the binding process can be extracted [76].

If, in the above brute-force counting approach, one also counts how often the two molecules sample a certain intermolecular distance r, one can obtain a r-dependent free-energy profile, from the corresponding probability $P(r)$,

$$\Delta G(r) = -RT \ln P(r) \quad . \tag{14}$$

Strictly, this free-energy profile is not identical to the PMF because it decreases at large intermolecular radial separation r. The decrease pertains to the Jacobian contribution of $-2RT \ln r$ and is absent in the PMF, which hence levels off at large intermolecular separation [43, 77]. Again, due to the large computational effort required to get a converged result, such simulations are usually only applicable to weak binders or small systems [78]. This is why computational tricks are used to calculate binding free energies more efficiently, as explained in the following.

1.3.2 Calculation of a Potential of Mean Force—A Pathway Method

An absolute binding free energy can be expressed in terms of the PMF along the center-of-mass distance r between the two binding partners. There are many ways to calculate a PMF. They differ in the way the sampling is done (with or without restraints), the way a free energy is obtained from the sampled configurations (from the probability distribution of sampled intermolecular distances or from integration of the average force along the reaction coordinate), and the way the reaction coordinate is defined [77]. In a widely employed setup, a series of N_w simulations ("windows") distributed along the reaction coordinate r is simulated. They differ in special harmonic potential-energy restraint terms that are added to the physical force field,

$$U_i(r) = \frac{1}{2} k (r - r_{0,i})^2, \tag{15}$$

where $i = 1, 2, \ldots N_w$, k is the force constant, and $r_{0,i}$ are distances distributed along the reaction coordinate such that they encompass the bound and unbound states. The harmonic restraint potential-energy terms are also referred to as umbrella bias potentials and the method as umbrella sampling [40].

The distance probability distributions sampled in each window can then be unbiased with the Zwanzig equation [79],

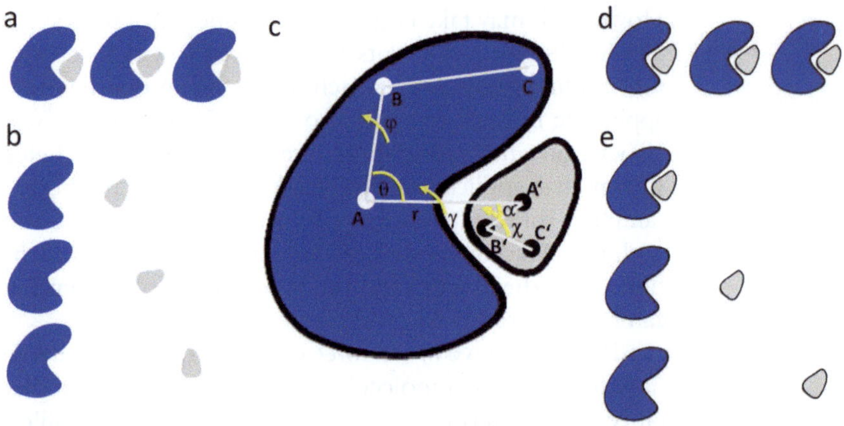

Fig. 1 Schematic depiction of a target–ligand system. The target and ligand molecules are depicted in blue and gray, respectively. (**a**) In the unrestrained state, a ligand in the binding pocket samples different positions and orientations. (**b**) In the unrestrained state, a ligand dissociating from the target molecule samples different positions and orientations. (**c**) Definition of geometric features to limit the axial and orientational degrees of freedom of the target–ligand complex. Three atoms *A*, *B*, *C* are defined in the target molecule, and three atoms *A*′, *B*′, and *C*′ are defined in the ligand molecule. The intermolecular distance is given by the distance between atoms *A* and *A*′. Angles θ and α are defined between atoms *B*, *A*, *A*′ and *A*, *A*′, and *B*′, respectively, and torsions φ, γ, and χ are defined around the *BA*, *AA*′, and *A*′*B*′ bonds, respectively. (**d**) In the restrained state, a ligand in the binding pocket is kept in a fixed position and orientation. (**e**) In the restrained state, a ligand dissociating from the target molecule moves along an axis and is kept in a fixed orientation

$$\Delta G_{A \to B} = -RT \ln \langle \exp[-\beta(H_B - H_A)] \rangle_A, \quad (16)$$

where a free-energy difference $\Delta G_{A\to B} = G_B - G_A$ between two states B and A with Hamiltonians H_B and H_A, respectively, can be calculated from an ensemble average $\langle \ldots \rangle_A$ over a simulation with Hamiltonian A. In the above case of umbrella-biased PMF calculations, H_A corresponds to the physical Hamiltonian augmented with the umbrella restraining potential and H_B corresponds to the unbiased physical Hamiltonian. This unbiasing can be done for each of the N_w windows, and the resulting free energies per window can be combined by setting free-energy constants per window such that a smooth PMF curve is obtained. For instance, the weighted-histogram analysis method (WHAM) [80] algorithm implements this procedure, but there are also several other methods to calculate a PMF from the N_w umbrella-biased simulations [77, 81, 82].

In practice, using the above setup, one may encounter difficult sampling problems. In each window, the distance probability has to be converged, and this may be tough if, for instance, conformational changes happen along the unbinding pathway. Moreover, exhaustive sampling of the translational and rotational motion of the binding partners may be very time-consuming (Fig. 1a, b). The sampling problem can be greatly reduced by further application of restraints to the system, as depicted in Fig. 1c [41]. Thus, the orientation of the ligand molecule w.r.t. the target molecule is

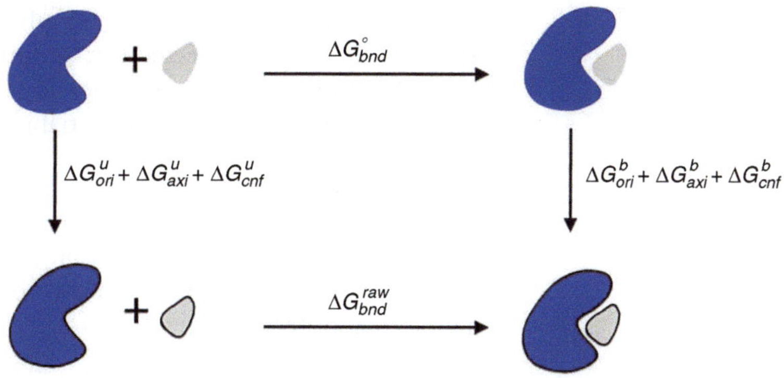

Fig. 2 Thermodynamic cycle used in the PMF method to calculate the absolute standard binding free energy ΔG°_{bnd} between two molecules. The target and ligand molecules are depicted in blue and gray, respectively. Solid bounding lines denote the presence of restraints. In the lower horizontal leg of the cycle, the raw binding free energy ΔG^{raw}_{bnd} expresses the reversible work of ligand–-target association where orientational, axial, and conformational degrees of freedom are restrained. The vertical legs of the cycle pertain to the free energy of installing these restraints in the bound and unbound states, $\Delta G^{b}_{ori} + \Delta G^{b}_{axi} + \Delta G^{b}_{cnf}$ and $\Delta G^{u}_{ori} + \Delta G^{u}_{axi} + \Delta G^{u}_{cnf}$, respectively

kept fixed, and when the ligand is pulled out of the binding pocket, it is forced to move along a fixed axis (Fig. 1d, e). The absolute standard binding free energy is then obtained from the thermodynamic cycle depicted in Fig. 2. The approach was originally suggested by Woo and Roux [41] and uses (*i*) axial restraining to restrict the motion along the reaction coordinate to an axis along the direction of separation, (*ii*) orientational restraining to fix the orientation of the binding partners with respect to each other, and (*iii*) conformational restraining to limit possible conformational changes in the binding partners. The presence of the restraints during configurational sampling in the N_w window simulations has to be ultimately corrected, meaning that the corresponding free-energy contributions have to be calculated to get the absolute binding free energy pertaining to the unrestrained, physical system. These calculations involve the sampling of the unrestrained molecules, albeit only in the endpoints of the PMF calculation. Hence, the sampling problem is shifted to endpoint calculations rather than having to be addressed along the entire unbinding pathway. This approach has recently been widely employed to calculate absolute binding free energies of bio(macro)molecules (*see*, e.g., refs. [7, 48, 83–86]).

In the following, superscripts "*b*" and "*u*" denote the bound and unbound states, respectively, and subscripts "*ori*," "*cnf*," and "*axi*" denote restraints pertaining to orientational, conformational, and axial degrees of freedom, respectively.

Orientational and axial motion of the ligand molecule with respect to the target molecule can be limited *via* restraints on the two angles θ and α and the three torsions φ, γ, and χ as depicted in Fig. 1c. The axial restraining potential is

$$U_{axi}(\theta,\varphi) = \frac{1}{2}k\left[(\theta-\theta_o)^2 + (\varphi-\varphi_o)^2\right], \quad (17)$$

where the angle θ and torsion φ along with their respective target values θ_o and φ_o are defined as depicted in Fig. 1c. Similarly, the orientational potential is

$$U_{ori}(\alpha,\chi,\gamma) = \frac{1}{2}k\left[(\alpha-\alpha_o)^2 + (\chi-\chi_o)^2 + (\gamma-\gamma_o)^2\right], \quad (18)$$

where the angle α and torsions χ and γ along with their respective target values α_o, χ_o, and γ_o are defined as depicted in Fig. 1c. With application of these restraints, along with conformational restraints that may be defined, e.g., in terms of restraints on Cα atom distances [85] or in terms of a distance root-mean-square deviation (dRMSD; *see* Subheading 3.2, **Note 4**) [86], a PMF curve is obtained from which a "raw" free energy of binding may be deduced as

$$\Delta G_{bnd}^{raw} = -RT\ln\left[\frac{\int_b dr\ e^{-\beta\Delta G(r)}}{\int_u dr\ e^{-\beta\Delta G(r)}\left(\int_u dr\right)^{-1}}\right] = -RT\ln\left[\frac{\int_b dr\ e^{-\beta\Delta G(r)}}{e^{-\beta\Delta G(r_u)}}\right], \quad (19)$$

where the simplification was done using r_u as the target–ligand distance in the unbound regime, where $\Delta G(r) = const.$ Integrals $\int_b$ and $\int_u$ refer to integration over the bound and unbound regimes of the PMF, respectively. The absolute standard binding free energy is given as

$$\Delta G_{bnd}^{\circ} = \Delta G_{bnd}^{raw} + \Delta G_{ori}^{u} + \Delta G_{axi}^{u} + \Delta G_{cnf}^{u} - \Delta G_{ori}^{b} - \Delta G_{axi}^{b} - \Delta G_{cnf}^{b} \quad . \quad (20)$$

The terms added to ΔG_{bnd}^{raw} correct for the presence of restraints and convert to standard concentration in the unbound state. The free energy associated with the axial restraining potential in the unbound state is calculated by integration over the entire possible range of angle θ and torsion φ for the ligand at a distance r_u from the target molecule,

$$\Delta G_{axi}^{u} = -RT\ln\left[(V^{\circ})^{-1} r_u^2 \int_0^{\pi} d\theta\ \sin\theta\ \int_0^{2\pi} d\varphi\ e^{-\beta U_{axi}(\theta,\varphi)}\right] \quad . \quad (21)$$

Through the presence of V°, this term also converts the concentration in the unbound state to the standard concentration 1 M. Note that certain simulation codes implicitly allow the use of an axial reaction coordinate by application of restraint forces along a certain coordinate only. In this case, the axial restraining potential in

Eq. 17 is not required. One may then correct to the standard-state volume in the unbound state through [42]

$$\Delta G^u_{axi} = -RT \ln \left[(V^\circ)^{-1} \Delta r_{x,u} \Delta r_{y,u} \Delta r_{z,u} \right], \tag{22}$$

where $\Delta r_{i,u}$ is the distance along the coordinate axis $i = x, y, z$ sampled in the unbound, restrained state. The length $\Delta r_{j,u}$ along the reaction coordinate j cancels with a corresponding length for the unbound regime in the calculation of ΔG^{raw}_{bnd} when the division by $\int_u dr$ in the denominator of Eq. 19 is omitted. Similarly, a free-energy contribution for the bound state reads [85]

$$\Delta G^b_{axi} = -RT \ln \left[\frac{\Delta r_{y,b} \Delta r_{z,b}}{4\pi r^2_{ave,b}} \right], \tag{23}$$

where the restrained axis is the x-axis and $r_{ave,b}$ is the average target–ligand distance sampled in the unrestrained, bound state. The argument of the logarithm is the fraction of sampled areas available to the ligand in the bound restrained and bound unrestrained states. An alternative to Eq. 23 is to use a fraction of sampled volumes in the logarithm,

$$\Delta G^b_{axi} = -RT \ln \left[\frac{\Delta r_{x,b} \Delta r_{y,b} \Delta r_{z,b}}{V_b} \right], \tag{24}$$

where V_b is the volume sampled by the ligand in the bound unrestrained state. As discussed in Ref. [85], Eqs. 23 and 24 imply essentially equivalent corrections.

Assuming that the solvent moiety is isotropic, orientational restraints in the unbound state can, similarly to Eq. 21, be accounted for analytically by integration over the Euler angles,

$$\Delta G^u_{ori} = -RT \ln \left[\frac{1}{8\pi^2} \int_0^{\pi} d\alpha \, \sin\alpha \int_0^{2\pi} d\chi \int_0^{2\pi} d\gamma \; e^{-\beta U_{ori}(\alpha,\chi,\gamma)} \right]. \tag{25}$$

The terms ΔG^b_{ori} and ΔG^b_{axi} are conveniently evaluated with the Zwanzig equation [79] from a simulation of the complex without axial and orientational restraints,

$$\Delta G^b_{ori} + \Delta G^b_{axi} = -RT \ln \left\langle e^{-\beta [U_{axi}(\theta,\varphi) + U_{ori}(\alpha,\chi,\gamma)]} \right\rangle_{unres}, \tag{26}$$

where $U_{ori}(\alpha, \chi, \gamma)$ is the orientational restraining potential. The subscript "*unres*" implies that the averaging is performed over an unrestrained simulation. The free energy associated with conformational restraints in the unbound and bound states can be calculated again from the Zwanzig equation [79],

$$\Delta G^i_{cnf} = -RT \ln \left\langle e^{-\beta U^i_{cnf}(\mathbf{r})} \right\rangle_{unres}, \tag{27}$$

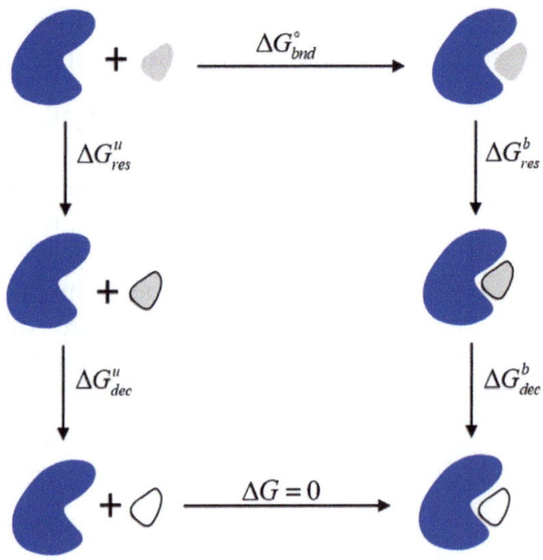

Fig. 3 Thermodynamic cycle used in the double decoupling method to calculate the absolute standard binding free energy ΔG°_{bnd} between two molecules. The target and ligand molecules are depicted in blue and gray, respectively. Solid bounding lines denote the presence of restraints, and white-colored molecules are dummy molecules. The top vertical legs of the cycle pertain to the installation of restraints to keep the ligand in the binding pocket (ΔG^{b}_{res}) and an analogous restraint in the solvated state (ΔG^{u}_{res}). The bottom vertical legs of the cycle express the reversible work of decoupling the restrained ligand from its environment in the bound and solvated states, ΔG^{b}_{dec} and ΔG^{u}_{dec}, respectively. The binding free energy of the decoupled (dummy) ligand to the target molecule vanishes

where i denotes the unbound or bound state and $U^{i}_{cnf}(\mathbf{r})$ is a special restraint term that fixes the conformation of chosen atoms in the unbound and bound states, $\mathbf{r}$ denoting the vector of all atom positions.

1.3.3 Double Decoupling Calculations—An Alchemical Method

The double decoupling method employs the thermodynamic cycle depicted in Fig. 3. Along the bottom vertical legs of the thermodynamic cycle, the interactions of the ligand molecule with its environment are gradually annihilated, and the reversible work along this annihilation process is called the decoupling free energy, ΔG^{b}_{dec} and ΔG^{u}_{dec}, where the superscripts again refer to the bound and unbound states, respectively. Often, the term "decoupling" is used to refer to deinstallation of intermolecular interactions only, whereas "annihilation" also implies deinstallation of intramolecular interactions. Decoupling and annihilation free energies can be calculated with "alchemical" methods, e.g., thermodynamic integration (TI) [87, 88] or stepwise application of the Zwanzig free-energy perturbation equation [79]. Here, the former method will be explained.

To convert the ligand into a dummy molecule, its van der Waals and electrostatic interactions with the environment, $U^{LE}_{vdW}(\mathbf{r})$ and $U^{LE}_{ele}(\mathbf{r})$, respectively, are made dependent on a parameter λ that interpolates between the fully interacting state ($\lambda=0$) and the dummy state ($\lambda=1$). The λ-dependent part of the Hamiltonian is then, assuming linear interpolation,

$$H(\lambda,\mathbf{r})=(1-\lambda)\left[U^{LE}_{vdW}(\mathbf{r})+U^{LE}_{ele}(\mathbf{r})\right] \quad . \tag{28}$$

Other parameterization options, e.g., using nonlinear functions of λ, have been investigated before [89, 90]. In the TI method [87, 88], multiple simulations at different λ-values, $\lambda=0\ldots1$, are performed, and integration over ensemble averages of Hamiltonian λ-derivatives evaluated in each of the simulations gives the free-energy difference,

$$\Delta G^{i}_{dec}=\int_{0}^{1}d\lambda' \left\langle \frac{\partial H(\lambda,\mathbf{r})}{\partial\lambda} \right\rangle_{\lambda'}, \tag{29}$$

where i denotes the bound or unbound state. The integration is usually done numerically based on the trapezoidal rule. Double decoupling simulations are commonly done for relatively small ligands, and conformational sampling problems can in general be tackled with state-of-the-art computational resources and simulation software. However, exhaustive configurational phase space sampling of the decoupled ligand molecule in the complex is highly impractical since in principle the whole computational box needs to be explored by the ligand to converge the associated Hamiltonian λ-derivative. Therefore, the ligand is kept in the binding pocket with restraint terms [1, 50, 51]. This can either be just a harmonic distance restraint between the center of mass of the ligand and the center of mass of chosen target atoms that lies as close as possible to the ligand center of mass, or it can be a set of different harmonic restraints that do not only restrain the target–ligand distance, but also the orientation of the ligand molecule. The latter situation is illustrated in Fig. 1c, where a distance restraint on the distance r between two atoms in the target and the ligand is used,

$$U_{dis}(r)=\frac{1}{2}k_r(r-r_o)^2, \tag{30}$$

as well as two angle restraints involving angles θ and α,

$$U_{ang}(\theta,\alpha)=\frac{1}{2}k_\theta(\theta-\theta_o)^2+\frac{1}{2}k_\alpha(\alpha-\alpha_o)^2, \tag{31}$$

and three torsion restraints involving torsions φ, γ, and χ,

$$U_{tor}(\varphi,\gamma,\chi)=\frac{1}{2}k_\varphi(\varphi-\varphi_o)^2+\frac{1}{2}k_\gamma(\gamma-\gamma_o)^2+\frac{1}{2}k_\chi(\chi-\chi_o)^2 \quad . \tag{32}$$

The angles and torsions are defined as depicted in Fig. 1c, k_i are the restraint force constants, and the distance, angle, and torsion parameters with subscript "o" are the target values of the restraints. These restraints may be reversibly grown during the decoupling simulation in the complex such that the free-energy contribution of growing the restraint can be directly estimated with the TI method [87, 88]. The restraints may also be fully present during the whole decoupling simulation in the complex, and the corresponding free-energy contribution may then be evaluated with the Zwanzig equation [79] based on a simulation of the unrestrained complex,

$$\Delta G^b_{res} = -RT\ln \left\langle e^{-\beta[U_{dis}(r)+U_{ang}(\theta,\alpha)+U_{tor}(\varphi,\gamma,\chi)]} \right\rangle_{unres} \quad . \tag{33}$$

This equation can analogously be used if the restraining solely involves a distance restraining potential. In the unbound state, an analytical equation gives the free-energy contribution [51]

$$\Delta G^u_{res} = RT\ln \left[\frac{8\pi^2 V^\circ (k_r k_\theta k_\alpha k_\varphi k_\gamma k_\chi)^{\frac{1}{2}}}{r_o^2 \sin\theta_o \sin\alpha_o (2\pi k_B T)^3} \right] \tag{34}$$

or, in the distance-restraint-only approach [50],

$$\Delta G^u_{res} = RT\ln \left[\frac{V^\circ k_r^{\frac{3}{2}}}{(2\pi k_B T)^{\frac{3}{2}}} \right] \quad . \tag{35}$$

The absolute binding free energy involves all contributions along the thermodynamic cycle in Fig. 3,

$$\Delta G^\circ_{bnd} = \Delta G^u_{dec} + \Delta G^u_{res} - \Delta G^b_{dec} - \Delta G^b_{res} \quad . \tag{36}$$

2 Methods

2.1 Technical Prerequisites

A MD simulation and analysis engine has to be installed on a local desktop engine and/or a compute cluster. The GROMACS package [59] includes tools to set up, perform, and analyze MD simulations. It can be downloaded from the Internet at www.gromacs.org, where also installation instructions and the programme manual are provided. As explained elsewhere, it is useful to install versions of the programme that allow parallel execution on multiple compute nodes and/or execution involving GPUs because this comes with a significant increase in compute performance. A collection of tutorials concerning simulations with the GROMACS programme can be found elsewhere [91].

For free-energy simulations with the GROMACS programme, it may be beneficial to link to the PLUMED library [92], which is a versatile tool to carry out simulations with restraints.

2.2 Calculation of a Potential of Mean Force

2.2.1 Prior Considerations

A structure of the target–ligand complex has to be solvated in a computational water box, and if necessary a neutralizing set of counterions is added. The system can be energy-minimized, slowly heated up to room temperature, and brought to equilibrium at ambient conditions. Standard protocols are available for this equilibration [91]. Analogous to the setup of the solvated target–ligand complex above, one needs to prepare solvated systems of only the target and the ligand if conformational restraints are to be applied.

Preparation of the PMF simulations involves the definition of:

- The pulling pathway. In case of non-buried binding pockets, visual inspection of the target–ligand complex allows to identify an axis along which the ligand may be pulled away from the target such that minimal steric clashes occur. Movement along the axis can be enforced by an axial restraint (Fig. 1c) or by rotating the target–ligand complex such that the pulling axis coincides with a given coordinate axis (e.g., the *x*-axis) and pulling along that axis. Note that along the chosen pathway, it may still be possible that target–ligand interactions are installed or broken, which complicates the sampling problem.
- The employed restraints. The atoms defining the restraints in Fig. 1c can be chosen as real atoms (e.g., protein Cα atoms) or virtual atoms (e.g., centers of masses of a set of Cα atoms), lying sufficiently far apart from each other in stable regions of the molecules. Stable secondary structure elements are useful regions to define the restraints, whereas flexible solvent-exposed loops are not. For each restraining potential, one needs to define a force constant and a reference value. For the latter, visual inspection of the initial target–ligand complex can be used. Alternatively, one can perform a short simulation (e.g., 10 ns) of the target–ligand complex and determine the equilibrium values of the distances, angles, and torsions defined in Fig. 1c. This is highly recommended because: (*i*) analysis of the mean-square fluctuation of a geometric variable ξ, $\langle(\xi - \langle\xi\rangle)^2\rangle$, can also help to deduce a force constant k_ξ according to

$$k_\xi = \frac{RT}{\langle(\xi - \langle\xi\rangle)^2\rangle} \tag{37}$$

and (*ii*) one can examine whether the chosen geometric variables are really stably fluctuating around a given equilibrium value, i.e., display a unimodal distribution. Typical values for force constants are around 1500–4000 kJ$\cdot$mol$^{-1}\cdot$nm^{-2} for a distance restraint and 40–200 kJ$\cdot$mol$^{-1}\cdot$rad^{-2} for angle or torsion restraints.

Note that the definition of restraints in the input of a MD simulation programme usually relies on special input sections in the molecular topology file or other restraint-specification files. Prior to specifying a restraint, the user has to make sure that he/she knows the functional form of the restraining potential and the units of the involved quantities.

- The data to be written to file. For simulation times in the range of hundreds of nanoseconds to microseconds, coordinates and energies may be written to file, e.g., every 5–10 ps. For the further analysis of the PMF simulations, it is recommended to write out more frequently (e.g., every 2–3 ps) the sampled reaction coordinate value and the associated restraint force. In some MD simulation programmes, it may not be possible to write out the latter quantities so that they have to be calculated *a posteriori* from the coordinate trajectory.

2.2.2 Simulations

Two types of simulations have to be performed to calculate the binding free energy according to Eq. 20, the N_w restrained (umbrella sampling) simulations pertaining to the bottom horizontal leg in Fig. 2 (ΔG_{bnd}^{raw}) and unrestrained simulations of the target–ligand complex and (in case conformational restraints were present) also of the free, solvated target and ligand molecules. Note that the latter simulations have to be rather long (e.g., on the order of 100 ns) because calculation of a PMF with the described restraining procedure shifts the bulk of the sampling problem to the end states. In particular if the free energy pertinent to conformational restraints is to be evaluated, one has to sufficiently sample the phase space of the unrestrained molecules which may be very time-consuming.

Note, in addition, that the solvated target–ligand system prepared as described above can serve as initial coordinates for only the first few PMF windows, for which the reference target–ligand distance is close to the one in the bound state. For windows with a larger reference distance, very high forces would result in case of a bound initial structure. Therefore, the corresponding initial coordinates can be generated, e.g., in a prior simulation where the reference distance of the harmonic distance restraining potential is continuously varied, or (better) by an initial set of N_w short simulations, each window starting with the end structure of the previous window. Note, finally, that the very first window may be chosen to have a reference distance slightly lower than the target–ligand distance in the bound complex to thoroughly depict the bound minimum in the final PMF.

2.2.3 Analysis

Here, the two required main analyses are explained. First, one has to analyze the umbrella sampling simulations to calculate ΔG_{bnd}^{raw}. There are several ways to analyze the umbrella sampling done in the N_w PMF windows to extract a free-energy profile along the desired reaction coordinate. Two commonly employed methods are WHAM [80, 81] and umbrella integration [82, 93, 94] (*see* Subheading 3.2, **Note 1**). In the former method, the biased probability distributions of sampled reaction coordinate values in each window are converted into unbiased free energies and concatenated. This is done by solving a set of nonlinear coupled equations in a self-consistent manner. In the latter method, the free-energy profile is obtained from integration of its derivative, minus the average physical force along the reaction coordinate (i.e., essentially, the average restraint force), where the averaging is over the remaining degrees of freedom (hence the term "potential of mean force"). Regardless of the employed analysis method, i.e., the free-energy estimator, a prerequisite for obtaining a proper free-energy profile is of course that the underlying simulations are converged, which has to be thoroughly checked by monitoring how the free-energy profile changes upon stepwise prolongation of the simulations (e.g., is the free-energy profile obtained from a 50% longer simulation time the same within chemical accuracy?). Convergence issues can also be seen in individual windows by examining the corresponding probability distributions of the reaction coordinate. Besides the overall issue of convergence, there are other free-energy estimator-specific issues that have to be verified, e.g., for the WHAM method, the probability distributions have to overlap to concatenate the free-energy profile. As a rough guideline, the distribution of successive windows should overlap by 20–50%. Usually, it is more efficient to sample more windows for a shorter amount of time than to sample less windows for a long time because the former approach is more parallelizable. Finally, statistical error estimates on the free-energy profile can be obtained from bootstrapping [95]. The raw binding free energy ΔG_{bnd}^{raw} is obtained according to Eq. 19 and a statistical error on it from addition of errors in the unbound and bound regimes. Note that the PMF should level off in the unbound regime because the ligand is no more interacting with the target and thus essentially alone in solvent. However, strictly, this plateau is only observed if the reaction coordinate is really axial (as is the case in the present description). For a pure internal-distance reaction coordinate, the free-energy profile decreases at larger distances due to the entropic contribution of a larger volume. This can be corrected with a corresponding Jacobian term $+2RT \ln r$ capturing the conversion of a reaction coordinate defined in terms of an internal coordinate into a reaction coordinate defined in terms of Cartesian coordinates [43, 77].

The second main analysis task involves processing the unrestrained simulations to calculate the free energy associated with the restraints (vertical legs in Fig. 2). While ΔG_{axi}^{u} and ΔG_{ori}^{u} can be calculated analytically (Eqs. 21 and 25, respectively), the four other quantities ΔG_{cnf}^{b}, ΔG_{axi}^{b}, ΔG_{ori}^{b}, and ΔG_{cnf}^{u} have to be evaluated from the free simulations of the bound (target–ligand complex) and unbound states (free target and free ligand in solution). This can be done with TI [87, 88], multi-step, or single-step application of the

Zwanzig equation [79]. The latter is the computationally most efficient way, which relies, however, on the assumption that the free simulations also sample sufficiently the relevant regions of the restrained system phase space. If this is granted, one has to calculate the restraint energies from the configurations sampled in the free simulations. The ensemble-averaged exponential in Eqs. 26 and 27 then delivers the corresponding free-energy contributions. Finally, all calculated free energies can be combined to give ΔG°_{bnd} *via* Eq. 20. This quantity may be compared to experimental values pertaining to a concentration of 1 M for the ligand in solution because it is normalized in Eq. 21 with respect to the standard volume V°.

2.2.4 Summary

The general steps to calculate ΔG°_{bnd} from PMF calculations with (*i*) umbrella sampling [40] of a distance reaction coordinate, (*ii*) the WHAM method [80, 81] for analysis, and (*iii*) the use of axial, orientational, and conformational restraints during the umbrella sampling [41] are summarized in Fig. 4. These are general steps that pertain to any MD simulation package. In addition, standard protocols for the system preparation typically follow the same procedure independent of the employed MD simulation package. They can, e.g., be found in online tutorials [91]. Note that the packages may differ in the specification of restraints and the implementation of analysis tools. In Subheading 2.3.4, an example is provided and files and programmes to be used in the context of the `GROMACS` package [59] and the `PLUMED` library [92, 96] are mentioned.

2.2.5 Example

As an example, we consider the calculation of the binding free energy of a peptide to a mutant of the voltage-dependent anion channel (VDAC) [48]. The channel protein corresponds to PDB ID 3EMN [97], albeit with deletion of the N-terminal segment that normally protrudes into the channel. Instead, a 14-residue helical peptide derived from the N-terminal segment is noncovalently bound to the channel interior [48]. The channel protein and the short peptide are thus the target molecule and the ligand, respectively. The system can be set up according to standard protocols as described in Ref. [48]. It consists of the channel protein inserted into a solvated phospholipid membrane (oriented in the x, y-plane) and the noncovalently bound peptide residing in the channel interior. The reaction coordinate is defined as the z-component of the center-of-mass (com) distance between the channel and the peptide. Initial coordinates for the PMF calculations can be extracted from a short pulling simulation employing the `pull` code implemented in the GROMACS package [98]. The `pull` code is activated with the corresponding `pull` input flags in the `mdp` file [98]. In the resulting pulling simulation, the peptide is slowly

1. Set up a solvated, neutral system of the target-ligand complex, as well as the target and ligand alone.
2. Equilibrate these systems.
3. Determine atoms defining the restraint terms depicted in Fig. 1c. This may be done by visual inspection and/or a short simulation of the target-ligand complex.
4. Specify the restraints to be used in the adequate MD input file and the specific file section therein.
5. Generate input coordinates for PMF windows differing from the bound configuration.
6. Perform the umbrella sampling simulations.
7. Verify the action of the restraints by inspection of time series of the involved geometric quantities.
8. Determine the time series and probability distributions of the sampled reaction-coordinate (distance) values.
9. Analyze whether addition of windows is required to improve the overlap of probability distributions.
10. Possibly simulate more PMF windows.
11. Apply the WHAM algorithm to obtain a free-energy profile and perform an error analysis of the free-energy profile.
12. Monitor simulation convergence by inspection of distance probability distributions and change of the free-energy profile upon simulation prolongation. The simulation of each PMF window has to be converged.
13. Numerically integrate over the bound and unbound state of the free-energy profile and thus obtain the corresponding free-energy change. This is the "raw" binding free energy ΔG_{bnd}^{raw} (Eq. 19).
14. Perform free (unrestrained) simulations of the target-ligand complex, as well as the target and ligand alone.
15. From the obtained trajectories, determine the time series of the geometric quantities involved in restraints during the PMF simulations. Calculate the corresponding restraining potential energy values.
16. Apply the Zwanzig equation to calculate the axial and orientational restraint free-energy contributions pertaining to the bound state (Eq. 26) and a possible conformational restraint free-energy contribution pertaining to the bound and unbound states (Eq. 27).
17. Calculate analytically the free-energy contribution of axial and orientational restraints in the unbound state (Eqs. 21 and 25).
18. Combine all quantities to calculate ΔG_{bnd}° (Eq. 20).

Fig. 4 Steps to calculate ΔG_{bnd}° (Eq. 20) from PMF calculations with umbrella sampling [40] of a distance reaction coordinate, the WHAM method [80, 81] for analysis, and the use of axial, orientational, and conformational restraints during the umbrella sampling [41]

pulled out of the channel into the solvent. A complication arises because the peptide tends to unfold in an aqueous environment. Therefore, it is beneficial to apply a dRMSD restraint during the following PMF calculations to keep the peptide in the folded, helical structure. This restraint may be defined in a `plumed.dat` file that can be linked to the `GROMACS` MD simulation engine `mdrun` as `-plumedplumed.dat` [96]. This requires that `GROMACS` is installed with linking to the `PLUMED` library [92, 96]. The `PLUMED` library [92] can also be used to define other restraints involved in the PMF calculation. For the present system, a restraint force is applied along the *z*-axis, i.e., the membrane normal. This can be specified in the `plumed.dat` file with the `RESTRAINT` keyword [96]. The employed force constant was 1500 kJ·mol^{-1}·nm^{-2}. This is also specified in the `plumed.dat` file [96]. Note that for each PMF window, a separate `plumed.dat` file needs to be prepared (e.g., `plumed0.dat`, `plumed1.dat`, etc.), differing in the reaction coordinate value, here, values of − 0.20, − 0.12, − 0.05, ...3.66, 3.76, and 3.86 nm. In the `plumed.dat` files, one can specify the name of output files containing time series of the reaction coordinate values sampled in the different PMF windows, e.g., `DIS0`, `DIS1`, etc. For construction of a free-energy profile, the WHAM programme package of Grossfield [95] can be used. It contains the `wham` programme to run the weighted-histogram analysis. The gist of the input to the `wham` programme is a metadata file containing information about the reaction coordinate time series (contained in the above `DIS` files) and the parameters of the harmonic restraining potential (specified in the `plumed.dat` files). The `wham` programme outputs the free-energy profile along with error estimates. From this raw free-energy profile, the raw binding free energy ΔG_{bnd}^{raw} is obtained according to Eq. 19. Free-energy contributions ΔG_{cnf}^{b} and ΔG_{cnf}^{u} pertaining to the dRMSD restraint in the bound and unbound states, respectively, have to be calculated in a next step. This can be done by calculating PMFs along the dRMSD coordinate as described in Ref. [48]. The Zwanzig equation [79] was not applicable here because in the unbound, solvated state, the peptide unfolds and hence its sampled configurations are substantially different from those of the restrained peptide. Finally, since during the unbinding PMF calculations there was no orientational or axial restraining potential, one merely needs to normalize the unbound-state volume w.r.t. the standard concentration to obtain a standard binding free energy

$$\Delta G_{bnd}^{\circ} = \Delta G_{bnd}^{raw} - \Delta G_{cnf}^{b} + \Delta G_{cnf}^{u} - RT \ln \frac{V_u}{V_o}, \tag{38}$$

where $V_u = L_x L_y \Delta r_z$, $L_x = L_y$ are the average box-edge lengths along the *x*- and *y*-dimensions, respectively, and Δr_z is the distance range along the *z*-axis sampled by the peptide in the unbound regime [48]. In this example, the following results are obtained (from simulations of 40–150 ns length per PMF window):

- $\Delta G_{bnd}^{raw} = -72.8$ kJ·mol^{-1}
- $\Delta G_{cnf}^{b} = 4.1$ kJ·mol^{-1}
- $\Delta G_{cnf}^{u} = 9.4$ kJ·mol^{-1}
- $-RT \ln \frac{V_u}{V_o} = -6.8$ kJ·mol^{-1}

It follows that $\Delta G_{bnd}^{\circ} = -74.3$ kJ·mol^{-1} (Eq. 38).

2.3 Double Decoupling Calculations

2.3.1 Prior Considerations

Analogous to the setup of PMF calculations, here, too, the first steps involve solvating the structure of the target–ligand complex in a computational water box. If necessary, a neutralizing set of counterions is added. The system can be energy-minimized, slowly heated up to room temperature, and brought to equilibrium at ambient conditions. Standard protocols are available for this equilibration [91]. Analogous to the setup of the solvated target–ligand complex above, one needs to prepare a solvated system of only the ligand. The computational box can be significantly smaller than the box used for the target–ligand complex. Then, the DD simulations have to be prepared. This involves the definition of:

- The decoupling procedure. A common way is to first eliminate electrostatic interactions with the environment because the pertinent TI curve is approximately linear and hence a low number of simulations are sufficient to obtain a precise integral. In a second step, van der Waals interactions with the environment are eliminated. In this step, a soft-core van der Waals potential [99, 100] is used to avoid issues at the endpoint of vanishing ligand size attributable to the singularity in the repulsive hard-core interaction in the Lennard–Jones potential. Without soft-core potentials, this singularity leads to a divergent Hamiltonian λ-derivative at the endpoint that renders the TI curve difficult to integrate and, moreover, leads to high forces, which complicates numerical integration of the equation of motion. Note that for the decoupling in the bound state, during both steps restraints are applied to keep the ligand properly in the binding pocket [1, 50, 51]. An alternative is to grow the restraint prior to the decoupling TI simulations in the bound state in a separate free-energy simulation (*see* Subheading 3.3, **Note 4**). Note, in addition, that annihilating solely the interactions with the environment is only reasonable in case of rigid ligand molecules (*see* Subheading 3.3, **Note 2**) and that instead of annihilating interactions, the reverse direction, i.e., growing interactions, may also be considered. Both ways have advantages and disadvantages (*see* Subheading 3.3, **Note 1**). Note, finally, that the definition of the decoupling procedure in the input of a MD simulation programme usually relies on special input sections in the input parameter file, molecular topology file, or other files.

- The employed restraints. The considerations in the analogous bullet of Subheading 2.3 also apply to the restraints employed here. For a discussion of the choice of force constants for the restraints on the ligand during decoupling in the bound state see Ref. [101].
- The thermostatting algorithm. A non-trivial issue that may be encountered in simulations featuring a decoupled molecule is a possibly spurious distribution of the kinetic energy because the almost or fully decoupled molecule does barely or no more exchange kinetic energy with its environment. Due to this violation of equipartition of kinetic energy, the molecule may sample unphysical configurations, which entails spurious results for the sampled Hamiltonian λ-derivatives. In general, spurious distribution of kinetic energy [102, 103] and improper sampling of a constant-temperature ensemble [104] can occur with simple velocity scaling thermostats. Therefore, in decoupling simulations, it is recommended [101] to use a stochastic thermostat to ensure that also the weakly or noninteracting ligand molecule has a proper temperature for all molecular degrees of freedom by virtue of stochastic collisions and balancing friction.
- The data to be written to file. For simulation times in the range of hundreds of nanoseconds to microseconds, coordinates and energies may be written to file, e.g., every 5–10 ps. For the further analysis of the DD simulations, it is recommended to write out more frequently (e.g., every 2–3 ps) the value of the Hamiltonian λ-derivative.

2.3.2 Simulations

Two types of simulations have to be performed to calculate the binding free energy according to Eq. 36, TI simulations and simulations to calculate the free energy of the restraint contributions. Two TI simulations have to be carried out. They pertain to the bottom vertical legs in Fig. 3 (ΔG^b_{dec} and ΔG^u_{dec}), i.e., consider the bound and unbound ligand molecules. Each TI simulation consists of several individual simulations that differ in the value of λ. For instance, one can, as described above for both the bound and unbound ligands, do a TI that eliminates electrostatic interactions with the environment and involves five λ-states ($\lambda = 0.0$, 0.25,. . .1.0) and then a TI that eliminates van der Waals interactions with the environment and involves 21 λ-states ($\lambda = 0.0$, 0.05, 0.1,. . .1.0). In the latter simulations, soft-core van der Waals interactions [99, 100] must be used. Alternatively, rather than doing two TI calculations, one may also simultaneously decouple electrostatic and van der Waals interactions in a single TI involving, e.g., 21 λ-states ($\lambda = 0.0$, 0.05, 0.1,. . .1.0). In this case, both soft-core electrostatic and van der Waals interactions [99, 100] must be used. One-step and two-step approaches in TI calculations were compared in Ref. [105]. As long as appropriate soft-core parameters are used, the one-step approach can be equally precise as the two-step approach and is more convenient for the user [105]. In the subsequent analysis of the TI curve (Eq. 29), one verifies whether the number of employed λ-states was sufficient to yield a precise value of the integral. Note that it makes sense to not start the simulations of all λ-states from the same initial coordinates but to first generate initial coordinates for each λ-state, e.g., in a prior simulation where the value of λ is continuously varied, or (better) by an initial set of short simulations for each λ-state, where each simulation starts with the end structure of the previous window. This may avoid instabilities in the simulations because it improves equilibration to the new Hamiltonian at a different λ-state.

Since the TI simulation involving the ligand in the bound state uses restraints to keep the ligand in the binding pocket, one has to perform an unrestrained simulation of the target–ligand complex to evaluate ΔG^{b}_{res} (Eq. 33).

2.3.3 Analysis

Here, the two required main analyses are explained. First, one has to analyze the TI simulations to calculate ΔG^{b}_{dec} and ΔG^{u}_{dec}. For each λ-state in each TI simulation, the ensemble average of the Hamiltonian λ-derivative is obtained. Convergence of the ensemble average can be assessed by inspecting the time series and calculating a statistical error estimate, e.g., *via* block averaging [65]. One may also prolong the simulations and monitor possible changes. When the simulations have converged, the TI curve has to be examined with respect to its numerical integrability. If necessary, additional simulations at other λ-states have to be carried out. This may be the case in high-curvature regions or at local extrema. Finally, the integration can be done with, e.g., the trapezoidal rule, and an error estimate on the free energy can be obtained from integration of the individual errors on the ensemble averages of λ-derivatives.

Note that the way the λ-parameterization of the Hamiltonian is done and the choice of soft-core function can significantly influence the shape of the TI curve and hence alter the data point requirements for a reasonable numerical integration. Novel functional forms of the soft-core potentials deviating from the original formalism [99, 100] have been suggested to make free-energy calculations more efficient and robust [106, 107]. Different scaling schemes and soft-core functions are compared in Refs. [89, 90, 107, 108]. Note also that ΔG^{b}_{dec} and ΔG^{u}_{dec} can be calculated with other methods than TI, e.g., multi-step free-energy perturbation or Multistate Bennett Acceptance Ratio (MBAR) [109] (*see* Subheading 3.3, **Note 5**). The second main task involves analysis of the unrestrained target–ligand complex simulation to calculate the free energy of installing the restraints ΔG^{b}_{res}. This can be done with TI [87, 88], multi-step or single-step application of the Zwanzig equation [79]. The latter is the computationally most efficient way, which relies however, on the assumption that the free simulation samples sufficiently the relevant regions of the restrained system phase space. This should be no problem if the restraints were set up in a meaningful way (Subheading 2.5). Based on the configurations sampled in the free simulation of the target–ligand complex, one has to calculate the restraint energies. The ensemble-averaged exponential in Eq. 33 then delivers ΔG^{b}_{res}. The restraint free-energy contribution pertaining to the unbound ligand, ΔG^{u}_{res}, can be calculated analytically (Eq. 34 or 35).

Finally, all calculated free energies can be combined to give ΔG°_{bnd} *via* Eq. 36. This quantity may be compared to experimental values pertaining to a concentration of 1 M for the ligand in solution because it is normalized in Eqs. 34 and 35 with respect to the standard volume V°.

2.3.4 Summary

The general steps to calculate ΔG°_{bnd} from DD calculations with the TI method [87, 88] and the use of restraints [1, 50, 51] during the decoupling in the bound state are summarized in Fig. 5. These are general steps that pertain to any MD simulation package. In addition, standard protocols for the system preparation typically follow the same procedure independent of the employed MD simulation package. They can, e.g., be found in online tutorials [91]. Note that the packages may differ in the specification of restraints and the implementation of analysis tools. In Subheading 2.5.4, an example is provided and files and programmes to be used in the context of the `GROMACS` package [59] are mentioned.

2.3.5 Example

As an example, we consider the calculation of the binding free energy of *n*-phenylglycinonitrile to the L99A, M102Q mutant of T4 lysozyme (PDB ID 2RBN) [110]. This protein–ligand system also serves as an example for the DD method in Refs. [91, 101], and the system can be set up as described in Ref. [91]. In addition, one needs to prepare a system containing solely the *n*-phenylglycinonitrile molecule immersed in water. This can be done according to standard protocols as described in Ref. [91]. In a next step, one can perform a short simulation of the protein–ligand complex to determine values for restraints. According to Fig. 1c, three atoms in the protein and the ligand have to be chosen, e.g., as in Ref. [91], the N, Cα, and C atoms of Y77 in the protein and the aromatic carbon atom attached to the functional group as well as the two neighboring aromatic carbon atoms in the ligand. The restraint specifications are written into the `intermolecular_interactions` section of the `GROMACS` topology file for the protein–ligand complex. The force constants can be set to 4000 kJ·mol^{-1}·nm^{-2}

1. Set up a solvated, neutral system of the target-ligand complex and the ligand alone.
2. Equilibrate these systems.
3. Determine atoms defining the restraint terms depicted in Fig. 1c. This may be done by visual inspection and/or a short simulation of the target-ligand complex.
4. Specify the restraints to be used in the adequate MD input file and the specific file section therein.
5. Determine the details of the decoupling procedure and edit corresponding files.
6. Perform the TI simulations for the bound and unbound states.
7. Verify the action of the restraints in the bound state by inspection of time series of the involved geometric quantities.
8. Determine the time series and averages of the sampled Hamiltonian λ-derivative values.
9. Plot the TI curve and analyze whether addition of λ-states is required to improve the quality of the numerical integration. Perform an error analysis of the individual λ-states and the resulting free energy.
10. Possibly simulate more λ-states.
11. Monitor simulation convergence by inspection of Hamiltonian λ-derivatives and change of the TI curve upon simulation prolongation. The simulation of each λ-state has to be converged.
12. Numerically integrate the TI curve and thus obtain the corresponding free-energy change. This is a decoupling free energy ΔG^{i}_{dec} (Eq. 29), where i refers to the bound or unbound state.
13. Perform a free (unrestrained) simulation of the target-ligand complex.
14. From the obtained trajectory, determine the time series of the geometric quantities involved in restraints during the TI simulation. Calculate the corresponding restraining potential energy values.
15. Apply the Zwanzig equation to calculate the restraint free-energy contribution pertaining to the bound state (Eq. 33).
16. Calculate analytically the free-energy contribution of restraints in the unbound state (Eq. 34).
17. Combine all quantities to calculate ΔG°_{bnd} (Eq. 36).

Fig. 5 Steps to calculate ΔG°_{bnd} (Eq. 36) from DD calculations based on TI [87, 88] and the use of restraints during the TI simulation of the bound state [1, 50, 51]

and 40 kJ·mol^{-1}·rad^{-2} for the distance and angle or torsion restraints, respectively. Since the ligand molecule is rather rigid, it is sufficient to decouple interactions with the environment only while leaving intramolecular interactions installed. The decoupling simulations can be done by simultaneously decoupling ligand–environment intermolecular electrostatic and van der Waals interactions in a series of 21 λ-states ($\lambda = 0.0, 0.05, 0.1, \ldots, 1.0$). Soft-core potentials [99, 100] need to be used, and in the bound state simulation, the restraints are present throughout. The `GROMACS` programme writes out `dhdl.xvg` files containing time series of the Hamiltonian λ-derivatives. The data can be averaged, error estimates calculated from block averaging [65] and the TI curves can be integrated with the trapezoidal rule to get decoupling free energies ΔG^b_{dec} and ΔG^u_{dec} for the bound and unbound states, respectively (Eq. 29). The restraint free-energy contribution for the unbound state, ΔG^u_{res}, is obtained from Eq. 34. An unrestrained simulation of the bound state needs to be performed, time series of the geometric quantities involved in the restraints need to be calculated (e.g., with the `gmx distance` and `gmx gangle` programmes) along with corresponding energies and Eq. 33 delivers ΔG^b_{res}. Finally, Eq. 36 gives the binding free energy ΔG°_{bnd}. In this example, the following results are obtained (from relatively short production simulations of 6 ns length):

- $\Delta G^b_{dec} = 86.4$ kJ·mol^{-1}
- $\Delta G^u_{dec} = 32.5$ kJ·mol^{-1}
- $\Delta G^u_{res} = 28.6$ kJ·mol^{-1}
- $\Delta G^b_{res} = 6.2$ kJ·mol^{-1}

It follows that $\Delta G^\circ_{bnd} = -31.5$ kJ·mol^{-1} (Eq. 36). Adding a symmetry correction [101] of $-RT\ln 2$ to account for the two equivalent configurations of the ligand, a binding free energy of − 33.2 kJ·mol^{-1} ensues.

3 Notes

3.1 General Notes

1. Weakly binding ligand molecules ($K_d \approx 10^{-5} \ldots 10^{-3}$ M) interact transiently with the target molecule, spend less time in the binding pocket than strong binders, and may not sample a unique bound configuration. This renders the definition of the bound state difficult, and hence, also the definition of restraints as demonstrated in Fig. 1c may not be unambiguous.
2. To check the convergence of simulations involved in the calculation of an absolute binding free energy, it is important to examine, besides the quantities that directly enter the analysis, also other degrees of freedom that might change slowly during the binding/unbinding process or the decoupling procedure

and that may thus not be sampled adequately. For example, this may include amino acid side-chain dihedral angle rotations, large-scale domain motions of the target molecule, or changes in the ion binding or hydration of the ligand binding pocket. Especially, the latter issue is frequently encountered and can give rise to severe convergence issues in case of occluded binding pockets [111, 112]. Methods have been suggested to ameliorate the sampling problem [111–113]. Conformational sampling issues in the ligand or the target may be addressed with conformational restraints [41] as described above. To adequately sample the conformational phase space in the free end-state simulations, or to address other sampling issues that may occur, one can resort to a plethora of enhanced sampling methods [114], e.g., the Hamiltonian replica-exchange approach [115, 116]. In general, convergence of a simulation can be assessed by analyzing possible changes in the results or other properties upon simulation prolongation or by comparison of simulations that were started with different initial conditions. For free-energy calculations involving multiple windows or λ-states, one can also generate initial structures in the reverse direction.

3. The methods presented here are based on a classical molecular mechanics force field and hence not trivially applicable to covalently or metal-bound ligand molecules. In such cases, one has to resort to QM or QM/MM methods. For example, QM/MM PMFs were calculated with umbrella sampling and WHAM to evaluate the affinity for a series of reversible covalent inhibitors of rhodesain [117]. Other QM-based approaches in the realm of binding free-energy calculation are, e.g., to use quantum-mechanical bespoke force fields for a given system [118] or to apply QM-derived free-energy corrections to molecular mechanics results or to QM results pertaining to a lower level of theory [119].
4. The Zwanzig equation [79] is a computationally efficient means to estimate the free-energy contribution of restraints. It is applicable when the restraints are chosen in a "reasonable" way and no substantial structural rearrangements happen, meaning that the restrained configuration has to be accessed sufficiently often in an unrestrained simulation. Nevertheless, it is recommended to always verify that application of the Zwanzig equation is appropriate and the result statistically meaningful. This can be done by assessing the number of configurations that contribute significantly to the ensemble average and by performing an error analysis on the calculated free energy [120]. Especially in case of conformational restraints, the phase space overlap requirement of the Zwanzig equation [79], may not always be fulfilled because an unrestrained

molecule may sample different regions of conformational phase space than the restrained one. In such situations, one has to resort to staged approaches, e.g., multi-step free-energy perturbation or PMF calculations (also *see* Subheading 3.2, **Note 4**).

3.2 Notes Concerning Potential of Mean Force Calculations

1. The WHAM algorithm [80] is only one possible method to create a free-energy profile from umbrella sampling simulations. Another frequently used approach is umbrella integration [82, 93, 94]. It avoids the iterative processing of the sampled probability distributions as done in WHAM and instead relies on integration of the average sampled restraint forces. An important limitation is that the sampled probability distributions are assumed to be normal distributions that may not always be the case. In general, it was found that force-integration approaches, in comparison to the alternative of using histograms along the reaction coordinate, are advantageous because of better convergence [77].
2. Instead of restraining the system along the reaction coordinate, one may also use constraints, i.e., entirely eliminate the degree of freedom along the reaction coordinate [77]. Integration of the constraint force or application of the Zwanzig equation [79] to calculate the free-energy difference to nearby reaction coordinate values delivers a free-energy profile. The constraint-based approach has the advantage that one does not need to define the parameters for the umbrella restraining potentials, but the disadvantage that so-called metric-tensor effects [121] attributable to the elimination of the momentum conjugate to the reaction coordinate degree of freedom emerge in the PMF [122]. These metric-tensor effects can be investigated as described elsewhere [77, 122–125].
3. The choice of reaction coordinate in a PMF calculation may not be clear a priori. The extent of a potential sampling problem encountered in PMF calculations can crucially depend on the choice of the reaction coordinate. For example, for an unfortunate choice of reaction coordinate, large-scale target structural rearrangement may be necessary to allow ligand dissociation, and this may be tough to sample. Moreover, the choice of atoms involved in the definition of the reaction coordinate may be ambiguous. It has been shown that, in the absence of orientational restraints, an unfortunate choice of atoms can give rise to sampling issues in the orientation of the ligand, which in turn leads to a spurious free-energy profile [126].
4. Conformational restraints may be defined, e.g., in terms of restraints on Cα atom distances [85] or in terms of a distance root-mean-square deviation (dRMSD) [86]. The latter is particularly useful when large conformational changes occur, for

instance (partial), unfolding of a peptide ligand in the solvent moiety [48]. In such a case, application of the Zwanzig equation [79] to evaluate the free energy associated with installation of the dRMSD restraint is not feasible because the unrestrained ligand samples a huge unfolded conformational ensemble that has no phase space overlap with the folded structure. However, separate PMF calculations along a dRMSD reaction coordinate between the folded and unfolded states can be employed to evaluate the free energy associated with the dRMSD restraint [48].

3.3 Notes Concerning Double Decoupling Calculations

1. Instead of annihilating interactions of the ligand molecule with the environment, one can also start with a dummy ligand molecule and grow the interactions with the environment. Such a growth simulation also needs to use soft-core interaction potentials [99, 100]. However, despite this trick, sampling may still be tougher than in a decoupling simulation because of steric ligand–environment clashes that need to be overcome during the growth. This may be particularly pronounced for large ligand molecules. On the other hand, in the realm of free-energy perturbation (FEP) calculations, growth of a molecule (especially in solvent) is considered more advantageous than decoupling because of better convergence. In physical terms, convergence is better if one perturbs from a high-entropy state to a low-entropy state [127, 128], and in technical terms, this can be understood by the fact that the state that is sampled has to also sample configurations representative for the target state for FEP to be statistically meaningful. Whereas a small solute in solvent can sample adequate solvent configurations for a larger (perturbed) solute molecule, a large solute does not sample adequate solvent configurations for a smaller (perturbed) solute molecule, meaning that the preferred direction for FEP is growth rather than decoupling [129]. Optimal FEP growth protocols were investigated in Ref. [129].
2. For a ligand that samples similar conformations in the bound and unbound states, i.e., that is relatively rigid, the decoupling simulations only involve annihilation of ligand–environment interactions, while intra-ligand interactions are unaltered. In contrast, for ligand molecules that change conformation upon change of environment, the free energy of annihilating covalent and noncovalent intra-ligand interactions is different in the bound and unbound states and needs to be considered [130–132].
3. Usually, one uses the same decoupling procedures in the bound and unbound states. In terms of number and nature of interactions with the environment, the unbound state based on water as a solvent is relatively similar to the ligand captured in

a biomolecular binding pocket such that the qualitative characteristics of the TI curve do not differ too much in the two environments [132].

4. Figure 3 depicts the decoupling in the bound state in the presence of restraints to keep the ligand in the binding pocket, i.e., the restraint is fully present during the decoupling TI simulation. An alternative is to grow the restraint in the course of the TI or prior to the decoupling TI in another TI simulation. The free energy associated with growing the restraint is then obtained from integration over the λ-derivative of the restraint potential, analogous to Eq. 29.

5. The TI approach [87, 88] is only one possible method to address alchemical free-energy changes. Other free-energy estimators like (multi-step) FEP [79], Bennett Acceptance Ratio (BAR) [133], and MBAR [109] may also be used. A comparison of the methods is provided in Ref. [134]. Multi-step FEP is a staged application of the Zwanzig equation [79] that relies, similar to TI [87, 88] on a coupling parameter formalism. BAR [133], unlike FEP, relies on sampling and energy evaluation in both endpoints and exploits, in a statistically optimal way, the information from both states to assess the free-energy difference. It is superior to FEP in that less phase space overlap is needed between the two states. MBAR [109] augments the BAR method by simultaneously considering not only two but multiple states and derives weighting functions to minimize the error in the free-energy differences between the states.

   In the `GROMACS` package, the `gmx bar` programme can be conveniently used to analyze the `dhdl.xvg` files written out in a free-energy simulation and calculate a free-energy difference.

6. Ligand molecules presenting a net charge pose a problem in alchemical free-energy calculations where the charge is altered, such as in a TI decoupling simulation. This is because of artifacts introduced by the approximate electrostatic interaction scheme used during the simulations and the microscopic size of the simulated system. The concomitant errors in decoupling free energies are non-negligible, do generally not cancel in the DD thermodynamic cycle (Fig. 3) but are well understood and can be corrected for [52–54].

References

1. Gilson MK, Given JA, Bush BL, McCammon JA (1997) The statistical thermodynamic basis for computation of binding-affinities. A critical review. Biophys J 72:1047–1069
2. Hill TL (1985) Cooperativity in biochemistry edition 1. Springer Series in Mol. Biol. Springer, New York
3. Hill TL (1986) An introduction to statistical thermodynamics, 11st edn. Dover, New York

4. McQuarrie DA (2000) Statistical mechanics. University Science Books, Sausalito
5. Maximova T, Moffatt R, Ma B, Nussinov R, Shehu A (2016) Principles and overview of sampling methods for modeling macromolecular structure and dynamics. PLOS Comput Biol 12:e1004619/1-e1004619/70
6. Mobley DL, Gilson MK (2017) Predicting binding free energies: Frontiers and benchmarks. Annu Rev Biophys 46:531–558
7. Siebenmorgen T, Zacharias M (2019) Computational prediction of protein-protein binding affinities. WIREs Comput Mol Sci 10:e1448/1–e1448/18
8. Limongelli V (2020) Ligand binding free energy and kinetics calculation in 2020. WIREs Comput Mol Sci 10:e1455/1–e1455/32
9. de Ruiter A, Oostenbrink C (2020) Advances in the calculation of binding free energies. Curr Opin Struct Biol 61:207–212
10. Leach A (2001) Molecular modelling: principles and applications, 2nd edn. Prentice Hall, New York
11. Schmitz S, Seibert J, Ostermeir K, Hansen A, Göller AH, Grimme S (2020) Quantum chemical calculation of molecular and periodic peptide and protein structures. J Phys Chem B 124:363–3646
12. Senn HM, Thiel W (2007) QM/MM methods for biological systems. Topics Curr Chem 268:173–290
13. Sousa SF, Ribeiro AJM, Neves RPP, Brás NF, Cerqueira NMFSA, Fernandes PA, Ramos MJ (2017) Application of quantum mechanics/molecular mechanics methods in the study of enzymatic reaction mechanisms. WIREs Comput Mol Sci 7:e1281/1–e1281/29
14. Ferreira LG, Dos Santos RN, Oliva G, Andricopulo AD (2015) Molecular docking and structure-based drug design strategies. Molecules 20:13384–13421
15. Brooijmans N, Kuntz ID (2003) Molecular recognition and docking algorithms. Annu Rev Biophys Biomol Struct 32:335–373
16. Berendsen HJC (2007) Simulating the physical world. Cambridge University Press, Cambridge
17. Chen W, Morrow BH, Shi C, Shen JK (2014) Recent development and application of constant pH molecular dynamics. Mol Simul 40:830–838
18. Stjernenschantz E, Oostenbrink C (2010) Improved ligand-protein binding affinity predictions using multiple binding modes. Biophys J 98:2682–2691
19. Luitz MP, Zacharias M (2014) Protein-ligand docking using Hamiltonian replica exchange simulations with soft core potentials. J Chem Inf Model 54:1669–1675
20. Salmaso V, Moro S (2018) Bridging molecular docking to molecular dynamics in exploring ligand-protein recognition process: an overview. Front Pharmacol 9:923/1–923/16
21. de Beer SBA, Vermeulen NPE, Oostenbrink C (2010) The role of water molecules in computational drug design. Curr Topics Med Chem 10:55–66
22. Limongelli V, Marinelli L, Cosconati S, La Motta C, Sartini S, Mugnaini L, Da Settimo F, Novellino E, Parrinello M (2012) Sampling protein motion and solvent effect during ligand binding. Proc Natl Acad Sci USA 109:1467–1472
23. Li J, Fu A, Zhang L (2019) An overview of scoring functions used for protein-ligand interactions in molecular docking. Interdiscpl Sci Comput Life Sci 11:320–328
24. Åqvist J, Medina C, Samuelsson J-E (1994) A new method for predicting binding affinity in computer-aided drug design. Protein Eng 7:385–391
25. Hansson T, Marelius J, Åqvist J (1998) Ligand binding affinity prediction by linear interaction energy methods. J Comput-Aided Mol Des 12:27–35
26. Genhenden S, Ryde U (2011) Comparison of the efficiency of the LIE and MM/GBSA methods to calculate ligand-binding energies. J Chem Theory Comput 7:3768–3778
27. Kollman PA, Massova I, Reyes C, Kuhn B, Huo S, Chong L, Lee M, Lee T, Duan Y, Wang W, Donini O, Cieplak P, Srinivasan J, Case DA, Cheatham III TE (2000) Calculating structures and free energies of complex molecules: combining molecular mechanics and continuum models. Acc Chem Res 33:889–897
28. Wang E, Sun H, Wang J, Wang Z, Liu H, Zhang JZH, Hou T (2019) End-point binding free energy calculation with MM/PBSA and MM/GBSA: strategies and applications in drug design. Chem Rev 119:9478–9508
29. Onufriev AV, Case DA (2019) Generalized Born implicit solvent models for biomolecules. Annu Rev Biophys 48:275–296
30. Lee FS, Chu ZT, Bolger MB, Warshel A (1992) Calculations of antibody-antigen interactions: microscopic and semi-microscopic evaluation of the free energies of binding of phosphorylcholine analogs to McPC603. Protein Eng 5:215–228

31. Kleinjung J, Fraternali F (2014) Design and application of implicit solvent models in biomolecular simulations. Curr Opin Struct Biol 25:126–134
32. Eisenberg D, McLachlan AD (1986) Solvation energy in protein folding and binding. Nature 319:199–203
33. Davis ME, McCammon JA (1988) Solving the finite difference linearized Poisson-Boltzmann equation: a comparison of relaxation and conjugate gradient methods. J Comput Chem 10:386–391
34. Izadi S, Harris RC, Fenley MO, Onufriev AV (2018) Accuracy comparison of generalized Born models in the calculation of electrostatic binding free energies. J Chem Theory Comput 14:1656–1670
35. Born M (1920) Volumen und Hydrationswärme der Ionen. Z Phys 1:45–48
36. Feig M, Onufriev A, Lee MS, Im W, Case DA, Brooks III CL (2004) Performance comparison of generalized Born and Poisson methods in the calculation of solvation energies for protein structures. J Comput Chem 25:265–284
37. Li Z, Laziridis T (2007) Water at biomolecular binding interfaces. Phys Chem Chem Phys 9:573–581
38. Baron R, Setny P, McCammon JA (2010) Water in cavity-ligand recognition. J Am Chem Soc 132:12091–12097
39. Sun H, Duan L, Chen F, Liu H, Wang Z, Pan P, Zhu F, Zhang JZH, Hou T (2018) Assessing the performance of MM/PBSA and MM/GBSA methods. 7. Entropy effects on the performance of end-point binding free energy calculation approaches. Phys Chem Chem Phys 20:14450/1–14450/11
40. Torrie GM, Valleau JP (1977) Nonphysical sampling distributions in Monte Carlo free-energy estimation: umbrella sampling. J Comput Phys 23:187–199
41. Woo H-J, Roux B (2005) Calculation of absolute protein-ligand binding free energy from computer simulations. Proc Natl Acad Sci USA 102:6825–6830
42. Doudou S, Burton NA, Henchman RH (2009) Standard free energy of binding from a one-dimensional potential of mean force. J Chem Theory Comput 5:909–918
43. den Otter WK (2013) Revisiting the exact relation between potential of mean force and free-energy profile. J Chem Theory Comput 9:3861–3865
44. Ferruz N, De Fabritiis G (2016) Binding kinetics in drug discovery. Mol Inf 35:216–226
45. Marsh JA, Teichmann SA (2015) Structure, dynamics, assembly, and evolution of protein complexes. Annu Rev Biochem 84:551–575
46. Fry C (2015) Targeting protein-protein interactions for drug discovery. In: Meyerkord C, Fu H (eds) Protein-protein interactions, methods Mol. Biol. Humana Press (Springer), New York, p 93–106
47. Villarreal OD, Yu L, Rodriguez RA, Chen LY (2017) Computing the binding affinity of a ligand buried deep inside a protein with the hybrid steered molecular dynamics. Biochem Biophys Res Commun 483:203–208
48. Reif MM, Fischer M, Fredriksson K, Hagn F, Zacharias M (2019) The N-Terminal segment of the voltage-dependent anion channel: a possible membrane-bound intermediate in pore unbinding. J Mol Biol 431:223–243
49. Jorgensen WL, Buckner JK, Boudon S, Tirado-Rives J (1988) Efficient computation of absolute free energies of binding by computer simulations. Application to the methane dimer in water. J Chem Phys 89:3742–3746
50. Roux B, Nina M, Pomès R, Smith JC (1996) Thermodynamic stability of water molecules in the Bacteriorhodopsin proton channel: a molecular dynamics free energy perturbation study. Biophys J 71:670–681
51. Boresch S, Tettinger F, Leitgeb M, Karplus M (2003) Absolute binding free energies: a quantitative approach for their calculation. J Phys Chem B 107:9535–9551
52. Rocklin GJ, Mobley DL, Dill KA, Hünenberger PH (2013) Calculating the binding free energies of charged species based on explicit-solvent simulations employing lattice-sum methods: an accurate correction scheme for electrostatic finite-size effects. J Chem Phys 139:184103/1–184103/32
53. Reif MM, Oostenbrink C (2014) Net charge changes in the calculation of relative ligand-binding free energies *via* classical atomistic molecular dynamics simulations. J Comput Chem 35:227–243
54. Öhlknecht C, Lier B, Petrov D, Fuchs J, Oostenbrink C (2019) Correcting electrostatic artifacts due to net-charge changes in the calculation of ligand binding free energies. J Comput Chem 41:986–999
55. Wereszczynski J, McCammon JA (2012) Statistical mechanics and molecular dynamics in evaluating thermodynamic properties of biomolecular recognition. Q Rev Biophys 45:1–25
56. Baron R, McCammon JA (2013) Molecular recognition and ligand association. Annu Rev Phys Chem 64:151–175

57. Salomon-Ferrer R, Case DA, Walker RC (2013) An overview of the Amber biomolecular simulation package. WIREs Comput Mol Sci 3:198–210
58. Brooks BR, Brooks III CL, Nilsson L, Petrella RJ, Roux B, Won Y, Archontis G, Bartels C, Boresch S, Caflisch A, Caves L, Cui Q, Dinner AR, Feig M, Fischer S, Gao J, Hodoscek M, Im W, Kuczera K, Lazaridis T, Ma J, Ovchinnikov V, Paci E, Pastor RW, Post CB, Pu JZ, Schaefer M, Tidor B, Venable RM, Woodcock HL, Wu X, Yang W, York DM, Karplus M (2009) CHARMM: the biomolecular simulation program. J Comput Chem 30:1545–1614
59. Abraham MJ, Murtola T, Schulz R, Pálla S, Smith JC, Hess B, Lindahl E (2015) GROMACS: high performance molecular simulations through multi-level parallelism from laptops to supercomputers. SoftwareX 1–2:19–25
60. Schmid N, Christ CD, Christen M, Eichenberger AP, van Gunsteren WF (2012) Architecture, implementation and parallelisation of the GROMOS software for biomolecular simulation. Comput Phys Commun 183:890–903
61. Phillips JC, Braun R, Wang W, Gumbart J, Tajkhorshid E, Villa E, Chipot C, Skeel RD, Kalé L, Schulten K (2005) Scalable molecular dynamics with NAMD. J Comput Chem 26:1781–1802
62. Taufer M, Ganesan N, Patel S (2013) GPU-enabled macromolecular simulation: challenges and opportunities. Comput Sci Eng 15:56–65
63. Glaser J, Nguyenc TD, Anderson JA, Lui P, Spiga P, Millan JA, Morsef DC, Glotzer SC (2015) Strong scaling of general-purpose molecular dynamics simulations on GPUs. Comput Phys Commun 192:97–107
64. Rapaport DC (1988) Large-scale molecular dynamics simulation using vector and parallel computers. Comput Phys Rep 9:1–53
65. Allen MP, Tildesley DJ (1987) Computer simulation of liquids. Oxford University Press, New York
66. Hummer G, Pratt LR, Garcia AE (1996) Free energy of ionic hydration. J Phys Chem 100:1206–1215
67. Hünenberger PH, McCammon JA (1999) Effect of artificial periodicity in simulations of biomolecules under Ewald boundary conditions: a continuum electrostatics study. Biophys Chem 78:69–88
68. Riniker S (2018) Fixed-charge atomistic force fields for molecular dynamics simulations in the condensed phase: an overview. J Chem Inf Model 58:565–578
69. Lemkul JA (2020) Pairwise-additive and polarizable atomistic force fields for molecular dynamics simulations of proteins. In: Strodel B, Barz B (eds) Computational approaches for understanding dynamical systems: protein folding and assembly. Academ. Press–Elsevier Sci., London, p 1–71
70. Jones JE (1924) On the determination of molecular fields. II. From the equation of state of a gas. Proc R Soc Lond Ser A 106:463–477
71. Jones JE (1924) On the determination of molecular fields. I. From the variation of the viscosity of a gas with temperature. Proc R Soc Lond Ser A 106:441–462
72. Hünenberger PH (1999) Lattice-sum methods for computing electrostatic interactions in molecular simulations. In: Hummer G, Pratt LR (eds) Simulation and theory of electrostatic interactions in solution: computational chemistry, biophysics, and aqueous solution, vol 492. American Institute of Physics, New York, p 17–83
73. van Gunsteren WF, Dolenc J, Mark AE (2008) Molecular simulation as an aid to experimentalists. Curr Opin Struct Biol 18:149–153
74. Kumari I, Sandhu P, Ahmed Mushtaq, Akhter Y (2017) Molecular dynamics simulations, challenges and opportunities: a biologists prospective. Curr Prot Pept Sci 18:1163–1179
75. de Jong DH, Schäfer L, de Vries AH, Marrink SJ, Berendsen HJC, Grubmüller H (2011) Determining equilibrium constants for dimerization reactions from molecular dynamics simulations. J Comput Chem 32:1919–1928
76. Plattner N, Doerr S, De Fabritiis G, Noé F (2017) Complete protein-protein association kinetics in atomic detail revealed by molecular dynamics simulations and Markov modelling. Nature Chem 9:1005–1011
77. Trzesniak D, Kunz A-PE, van Gunsteren WF (2007) A comparison of methods to compute the potential of mean force. Chem Phys Chem 8:162–169
78. Zeller F, Zacharias M (2017) Thermodynamics and kinetics of nucleobase stacking oligomerization revealed by molecular dynamics simulations. J Chem Theory Comput 13:3005–3011
79. Zwanzig RW (1954) High-temperature equation of state by a perturbation method. I. Nonpolar gases. J Chem Phys 22:1420–1426

80. Kumar S, Bouzida D, Swendsen RH, Kollman PA, Rosenberg JM (1992) The weighted histogram analysis method for free-energy calculations on biomolecules. I. The method. J Comput Chem 13:1011–1021
81. Roux B (1995) The calculation of the potential of mean force using computer simulation. Comput Phys Commun 91:275–282
82. Kästner J, Thiel W (2005) Bridging the gap between thermodynamic integration and umbrella sampling provides a novel analysis method: umbrella integration. J Chem Phys 123:144104/1–144104/5
83. Gumbart JC, Roux B, Chipot C (2013) Efficient determination of protein-protein standard binding free energies from first principles. J Chem Theory Comput 9:3789–3798
84. Gumbart JC, Roux B, Chipot C (2013) Standard binding free energies from computer simulations: what is the best strategy. J Chem Theory Comput 9:794–802
85. Perthold JW, Oostenbrink C (2017) Simulation of reversible protein-protein binding and calculation of binding free energies using perturbed distance restraints. J Chem Theory Comput 13:5697–5708
86. Siebenmorgen T, Zacharias M (2019) Evaluation of predicted protein-protein complexes by binding free energy simulations. J Chem Theory Comput 15:2071–2086
87. Kirkwood JG (1933) Quantum statistics of almost classical assemblies. Phys Rev 44:31–37
88. Kirkwood JG (1934) Quantum statistics of almost classical assemblies. Phys Rev 45:116–117
89. Pitera JW, van Gunsteren WF (2002) A Comparison of non-bonded scaling approaches for free energy calculations. Mol Simul 28:45–65
90. Steinbrecher T, Mobley DL, Case DA (2007) Nonlinear scaling schemes for Lennard-Jones interactions in free energy calculations. J Chem Phys 127:214108/1–214108/13
91. Lemkul J (2019) From proteins to perturbed Hamiltonian: a suite of tutorials for the GROMACS-2018 molecular simulation package. Living J Comput Mol Sci 1:5068/1–5068/53
92. Tribello GA, Bonomi M, Branduardi D, Camilloni C, Bussi G (2014) PLUMED 2: new feathers for an old bird. Comput Phys Commun 185:604–613
93. Billeter SR, van Gunsteren WF (2000) Computer simulation of proton transfers of small acids in water. J Phys Chem A:3276–3286
94. Kästner J, Thiel W (2006) Analysis of the statistical error in umbrella sampling simulations by umbrella integration. J Chem Phys 124:234106/1–234106/7
95. Grossfield A (2013), The weighted histogram analysis method. Version 2.0.9. http://membrane.urmc.rochester.edu/content/wham
96. PLUMED version 2.6.2 (2020). https://www.plumed.org/doc-v2.6/user-doc/html/index.html
97. Ujwal R, Cascio D, Colletier JP, Faham S, Zhang J, Toro L, Ping P, Abramson J (2008) The crystal structure of mouse VDAC1 at 2.3 Å resolution reveals mechanistic insights into metabolite gating. Proc Natl Acad Sci USA 105:17742–17747
98. GROMACS documentation (2020). Release 2020.4. https://zenodo.org/record/4054996
99. Beutler TC, Mark AE, van Schaik R, Gerber PR, van Gunsteren WF (1994) Avoiding singularities and numerical instabilities in free energy calculations based on molecular simulations. Chem Phys Lett 222:529–539
100. Zacharias M, Straatsma TP, McCammon JA (1994) Separation-shifted scaling, a new scaling method for Lennard-Jones interactions in thermodynamic integration. J Chem Phys 100:9025–9031
101. Aldeghi M, Bluck JP, Biggin PC (2018) Absolute alchemical free energy calculations for ligand binding: a beginner's guide. In: Gore M, Jagtap UB (eds) Computational drug discovery and design, Methods Mol. Biol. Humana Press (Springer), New York, p 199–232
102. Harvey SC, Tan RK-Z, Cheatham III TE (1998) The flying ice cube: Velocity rescaling in molecular dynamics leads to violation of energy equipartition. J Comput Chem 19:726–740
103. Braun E, Moosavi SM, Smit B (2018) Anomalous effects of velocity rescaling algorithms: the flying ice cube effect revisited. J Chem Theory Comput 14:5262–5272
104. Shirts MR (2013) Simple quantitative tests to validate sampling from thermodynamic ensembles. J Chem Theory Comput 9:909–926
105. Steinbrecher T, Joung I, Case DA (2011) Soft-core potentials in thermodynamic integration: comparing one- and two-step transformations. J Comput Chem 32:3253–3263
106. Buelens FP, Grubmüller H (2012) Linear-scaling soft-core scheme for alchemical free energy calculations. J Comput Chem 33:25–33

107. Gapsys V, Seeliger D, de Groot BL (2012) New soft-core potential function for molecular dynamics based alchemical free energy calculations. J Chem Theory Comput 8:2373–2382
108. de Ruiter A, Boresch S, Oostenbrink C (2013) Comparison of thermodynamic integration and Bennett's acceptance ratio for calculating relative protein-ligand binding free energies. J Comput Chem 34:1024–1034
109. Shirts MR, Chodera JD (2008) Statistically optimal analysis of samples from multiple equilibrium states. J Chem Phys 129:124105/1–124105/10
110. Graves AP, Shivakumar DM, Boyce SE, Jacobson MP, Case DA, Shoichet BK (2008) Rescoring docking hit lists for model cavity sites: predictions and experimental testing. J Mol Biol 377:914–934
111. Deng Y, Roux B (2008) Computation of binding free energy with molecular dynamics and grand canonical MC simulations. J Chem Phys 128:115103/1–115103/8
112. Wahl J, Smieško M (2019) Assessing the predictive power of relative binding free energy calculations for test cases involving displacement of binding site water molecules. J Chem Inf Model 59:754–765
113. Maurer M, Hansen N, Oostenbrink C (2018) Comparison of free-energy methods using a tripeptide-water model system. J Comput Chem 39:2226–2242
114. Luitz M, Bomblies R, Ostermeir K, Zacharias M (2015) Exploring biomolecular dynamics and interactions using advanced sampling methods. J Phys Condens Matter 27:323101/1–323101/13
115. Sugita Y, Kitao A, Okamoto Y (2000) Multidimensional replica-exchange method for free-energy calculations. J Chem Phys 113:6042–6051
116. Fukunishi H, Watanabe O, Takada S (2002) On the Hamiltonian replica exchange method for efficient sampling of biomolecular systems: application to protein structure prediction. J Chem Phys 116:9058–9067
117. da Costa CHS, Bonatto V, dos Santos AM, Lameira J, Leitão A, Montanari CA (2020) Evaluating QM/MM free energy surfaces for ranking cysteine protease covalent inhibitors. J Chem Inf Model 60:888–89
118. Cole DJ, de Vaca IC, Jorgensen WL (2019) Computation of protein-ligand binding free energies using quantum mechanical bespoke force fields. Med Chem Commun 10:1116–1120
119. Wang M, Mei Y, Ryde U (2019) Host-guest relative binding affinities at density-functional theory level from semiempirical Molecular Dynamics simulation. J Chem Theory Comput 15:2659–2671
120. Oostenbrink C (2012) Free energy calculations from one-step perturbations. In: Baron R (ed) Computational drug discovery and design, Methods Mol. Biol. Humana Press (Springer), New York, p 487–499
121. Fixman M (1974) Classical statistical mechanics of constraints: a theorem and application to polymers. Proc Natl Acad Sci USA 71:3050–3053
122. den Otter WK, Briels WJ (1998) The calculation of free-energy differences by constrained molecular-dynamics simulations. J Chem Phys 109:4139–4146
123. Sprik M, Ciccotti G (1998) Free energy from constrained molecular dynamics. J Chem Phys 109:7737–7744
124. den Otter WK, Briels WJ (2000) Free energy from molecular dynamics with multiple constraints. Mol Phys 98:773–781
125. den Otter WK (2000) Thermodynamic integration of the free energy along a reaction coordinate in Cartesian coordinates. J Chem Phys 112:7283–7292
126. Markthaler D, Jakobtorweihen S, (2019) Lessons learned from the calculation of one-dimensional potentials of mean force. Living J Comput Mol Sci 1:11073/1–11073/25
127. Kofke DA, Cummings PT (1998) Precision and accuracy of staged free-energy perturbation methods for computing the chemical potential by molecular simulation. Fluid Phase Equilib 150–151:41–49
128. Lu N, Kofke DA (1999) Optimal intermediates in staged free energy calculations. J Chem Phys 111:4414–4423
129. Cabeza de Vaca I, Zarzuela R, Tirado-Rives J, Jorgensen WL (2019) Robust free energy perturbation protocols for creating molecules in solution. J Chem Theory Comput 15:3941–3948
130. Boresch S, Karplus M (1999) The role of bonded terms in free energy simulations:

1. Theoretical analysis. J Phys Chem A 103:103–118
131. Boresch S, Karplus M (1999) The role of bonded terms in free energy simulations. 2. Calculation of their influence on free energy differences of solvation. J Phys Chem A 103:119–136
132. Shirts MR (2012) Best practices in free energy calculations for drug design. Methods Mol Biol 819:425–467
133. Bennett CH (1976) Efficient estimation of free energy differences from Monte Carlo data. J Comput Phys 22:245–268
134. Paliwal H, Shirts MR (2011) A benchmark test set for alchemical free energy transformations and its use to quantify error in common free energy methods. J Chem Theory Comput 7:4115–4134

Chapter 13

Computational Alanine Scanning Reveals Common Features of TCR/pMHC Recognition in HLA-DQ8-Associated Celiac Disease

Linqiong Qiu, Jianing Song, and John Z. H. Zhang

Abstract

In HLA-DQ8-associated celiac disease, Gliadin-γ1 or Gliadin-α1 peptide is presented to the cell surface and recognized by several types of T-cell receptor (TCR), but it is still unclear how the TCR, peptide, and the major histocompatibility complex (MHC) act together to trigger celiac disease. For now, most of the analysis is based on static crystal structures. And the detailed information about these structures based on energetic interaction is still lacking. Here, we took four types of celiac disease-related MHC-peptide-TCR structures from three patients to perform computational alanine scanning calculations using the molecular mechanics generalized born surface area (MM/GBSA) approach combined with a recently developed interaction entropy (IE) method to identify the key residues on TCR, peptide, and MHC. Our study aims to shed some light on the interaction mechanism of this complex protein interaction system. Based on detailed computational analysis and mutational calculations, important binding interactions in these triple-interaction complexes are analyzed, and critical residues responsible for TCR/pMHC recognition pattern in HLA-DQ8-associated celiac disease are presented. These detailed analysis and computational result should help shed light on our understanding of the celiac disease and the development of the medical treatment.

Key words MHC, TCR, Peptide, Protein–protein, Binding affinity, Interaction entropy, Celiac disease, HLA-DQ8, GBSA

1 Introduction

The binding of human leukocyte antigen (HLA) molecules of the major histocompatibility complex (MHC) to antigen (Ag)-specific T-cell receptor (TCR) is an important immune-related pathway in human bodies [1–9]. Some of the HLA class II (MHCII) molecules are associated with diseases like celiac disease (CD) [10–14]. CD is a chronic inflammatory disease which destroys the lining of the small intestine [10] when wheat gluten or related proteins [10, 15] are ingested. It is an autoimmune-like disorder triggered by dietary-derived gluten. It has been reported that CD is

Quentin Vanhaelen (ed.), *Computational Methods for Estimating the Kinetic Parameters of Biological Systems*,
Methods in Molecular Biology, vol. 2385, https://doi.org/10.1007/978-1-0716-1767-0_13,

predominantly limited to individuals who can express HLA-DQ2 (DQ2)and/or HLA-DQ8 (DQ8) [10, 16–19].

In DQ2-associated celiac disease, DQ2 prefers to bind peptides derived from wheat gluten and homologous sequences of barley and rye. These peptides prefer glutamine residue in P4 (position 4) or P6 or P7 positions [20, 21], for the deamidation of glutamine residue in these positions is important in TCR recognitions [22–26]. DQ2-restricted peptides are usually found to have proline residue at P1 position to resist protease degradation in intestine [15]. In DQ8-associated celiac disease, deamidation of glutamine residue is usually required at P1 and/or P9 position [11]. Although this structural information is known for now, the energetics of individual residues are not well studied, and the existence of any common mode for TCR recognition remains unclear. Here we perform molecular dynamics study to identify the key residues responsible for protein binding interaction and hope to find common features or recognition patterns among these celiac disease-related structures, which could provide insight to help design drugs for CD patients.

To understand how the TCRs interact with two important gliadin peptides, we chose four systems with four different types of TCR and two types of main gliadin peptides, as listed in Table 1. In the HLA-DQ8.5-Gliadin-γ1 systems, HLA-DQ8.5 is made of HLA-DQ2.5 α chain and HLA-DQ8 β chain [27], as HLA-DQ2 prefers proline residue at P1 position and HLA-DQ8 prefers glutamic acid at P1 and/or P9 position, probably because the core area of gliadin-γ1 sequence (PQQSFPEQE) starts with hydrophobic proline residue and ends with deamidated residue, the glutamic acid. In HLA-DQ8 systems, the core area of gliadin-α1 sequence (EGSFQPSQE) is deamidated both at P1 and P9, thus forming negatively charged residues at both terminals to facilitate electrostatic interactions.

2 Methods

2.1 System Setup and Molecular Dynamics Simulations

HLA-DQ8.5-glia-γ1-Bel602 (PDB ID: 5KSA), HLA-DQ8.5-glia-γ1-T15 (PDB ID: 5KSB), HLA-DQ8-gliadin-α1-Bel502 (PDB ID: 5KS9), and HLA-DQ8-gliadin-α1-SP3.4 (PDB ID: 4GG6) were downloaded from the PDB databank [30]. All systems were performed with AMBER16 under AMBER ff14SB force field. Each protein molecule was immersed into a truncated octahedron box of TIP3P [31] water molecules, then the water box was extended by 12 Å from the nearest atom of the protein. Na^+ and Cl^- ions were added to the solution to neutralize charges. The particle mesh Ewald (PME) [32] was used to calculate the long-range electrostatic interactions. The non-bounded interaction cut-off was set to 10 Å. And the periodic boundary condition (PBC) was used for all

Table 1
The detail information of the four systems. The structures were taken from different persons [28, 29]. 5KSA and 5KS9 were taken from patient Bel; the TCR Bel602 and Bel502 differ only in CDR3 sequence. System 5KSB was taken from patient T. And 4GG6 was taken from patient SP[1]. The core areas are shaded.

PDBID	MHCII	PEPTIDE	PEP_SEQ	TCR
5KSA	HLA-DQ8.5	Gliadin-γ1	GPQQSFPEQEA	Bel602
5KSB	HLA-DQ8.5	Gliadin-γ1	GPQQSFPEQEA	T15
5KS9	HLA-DQ8	Gliadin-α1	APSGEGSFQPSQENPQ	Bel502
4GG6	HLA-DQ8	Gliadin-α1	SGEGSFQPSQENP	SP3.4

the systems during molecular dynamics to eliminate the boundary effect. The SHAKE [33] algorithm was used to constrain bonds involving hydrogen atoms. The crystal structure was optimized by a three-step process. First, 5000 steps were used to minimize the solvent with the protein molecule being constrained. Then, 5000 steps were for the minimization of hydrogen in the protein and solvent. Last, the energy minimization is performed for the system without any constraints.

After minimization, the system was heated from 0 K to 300 K in 200 ps, and the protein backbone atoms were constrained. A Langevin thermostat with a collision frequency of 5.0 ps^{-1} was used for the temperature regulation while the pressure was maintained at 1 atm by Berendsen barostat. The system was then equilibrated in both NVT and NPT ensembles, with the temperature at 300 K. Then MD simulations were performed using CUDA procedure in AMBER16. The systems are quite stable during 20 ns MD simulations, and the last 2 ns trajectories were extracted for use in binding energy calculations.

2.2 GBSA_ASIE Method

In protein–protein interactions [34–41] (PPIs), although there are many residues on the interface of proteins, typically a few of them provide dominant energetic contributions in the binding. These key residues are called "hotspots" [42, 43]. Alanine scanning is an effective method for finding these hot residues. In this method, residues are mutated into alanine one by one, then the difference of binding free energy between the wild and mutated residue is calculated, to determine the importance of the residue. Binding free energy can be efficiently calculated by using the MM/GBSA-based methods [44–46]. In the MM/GBSA approach, the binding free energy is divided into four components as shown in Eq. (1).

$$\Delta G_{bind} = \Delta E^{mm}_{\text{gas}} + \Delta G^{p}_{solv} + \Delta G^{np}_{solv} - T\Delta S \quad (1)$$

The first term $\Delta E_{\text{gas}}^{mm}$ is the gas phase interaction energy which includes the van der Waals (VDW) and the Coulomb electrostatic interactions (ELE) between the proteins. ΔG_{solv}^{p} and ΔG_{solv}^{np} are the polar and nonpolar contributions to the de-solvation free energies [47], ΔG_{solv}^{p} represents the electrostatic interaction between the solute and the continuum solvent, calculated by using the generalized Born (GB) model. ΔG_{solv}^{np} is estimated from a linear relation to the solvent accessible surface area (SASA), as shown in Eq. (2). The γ and β values used here are the standard values of 0.00542 kcal/(mol Å^2) and 0.92 kcal/mol.

$$\Delta G_{solv}^{np} = \gamma SASA + \beta \tag{2}$$

The last term in Eq. (1) is the absolute temperature T multiplied by the entropy S. Typically, it is estimated by a normal mode (Nmode) analysis of the vibrational frequencies; however, it has usually been ignored because the calculation is time-consuming.

Recently, a more efficient method to calculate the interaction entropy (IE) has been proposed [48–54]. The IE method is defined as

$$-T\Delta S = \mathrm{k}T \ln \langle \exp(\beta \Delta E_{\text{int}}) \rangle \tag{3}$$

where $\Delta E_{\text{int}} = E_{\text{int}} - \langle E_{\text{int}} \rangle$, where E_{int} is the interaction energy between the two interacting partners and $\langle E_{\text{int}} \rangle$ is its statistical average. In the MD process, $\langle E_{\text{int}} \rangle$ can be calculated as

$$\langle E_{\text{int}} \rangle = \frac{1}{T} \int_0^T E_{\text{int}}(t)dt = \frac{1}{N} \sum_{i=1}^{N} E_{\text{int}}(t_i) \tag{4}$$

and Eq. (3) is rearranged as

$$\langle \exp(\beta \Delta E_{\text{int}}) \rangle = \frac{1}{N} \sum_{i=1}^{N} \exp(\beta \Delta E_{\text{int}}(t_i)) \tag{5}$$

Therefore, the entropy contribution can be calculated by using the formula

$$-T\Delta S = kT \frac{1}{N} \sum\nolimits_{i=1}^{N} \exp(\beta \Delta E_{\text{int}}(t_i)) \tag{6}$$

For alanine scanning, the difference of binding free energy is expressed as follows:

$$\begin{aligned} \Delta\Delta G^{(X \to A)} &= \Delta G_{bind}^{(A)} - \Delta G_{bind}^{(X)} \\ &= \left(\Delta \mathrm{G}_{gas}^{mm(A)} + \Delta \mathrm{G}_{solv}^{p(A)} + \Delta \mathrm{G}_{solv}^{np(A)} \text{-} \mathrm{T}\Delta \mathrm{S}^{(\mathrm{A})}\right) \text{-} \\ &\left(\Delta \mathrm{G}_{gas}^{mm(X)} + \Delta \mathrm{G}_{solv}^{p(X)} + \Delta \mathrm{G}_{solv}^{np(X)} \text{-} \mathrm{T}\Delta \mathrm{S}^{\mathrm{X}}\right) \end{aligned} \tag{7}$$

We combined the MM/GBSA and the IE methods in the above alanine scanning (MM/GBSA-ASIE) calculation, which is a more accurate and efficient way to calculate the binding free energy. By using alanine scanning, we could locate the key residues on the four celiac disease-related systems. The mutation of proline might have a large influence on the structure because Cα is on the proline ring, so its mutation is not performed. Glycine has not been mutated due to the accuracy.

3 Results and Discussion

3.1 Dynamic Stabilities of Interacting Residues

To investigate the stability of individual residues, we show the root mean square fluctuation (RMSF) to determine the stability of each residue in the systems. As shown in Fig. 1, the hotspots (to be discussed below) on MHCII (green stars), peptide (olive stars), and TCR (red stars) are all very stable during the dynamic simulation process. The peptide residues are represented by solid olive lines, and they are slightly flexible at both terminals but stable in the core regions. The hotspots on peptide are labeled olive stars, and as we can see from Fig. 1, they are all located in the stable area. Violet solid lines from left to right represent complementarity determining region 1α (CDR1α), CDR2α, CDR3α, CDR1β, CDR2β, CDR3β on TCR, respectively. It is reported that, in unbounded states, CDR3 loops are highly flexible and mobile compared to CDR1 and CDR2 loops [55], while in bounded states as shown

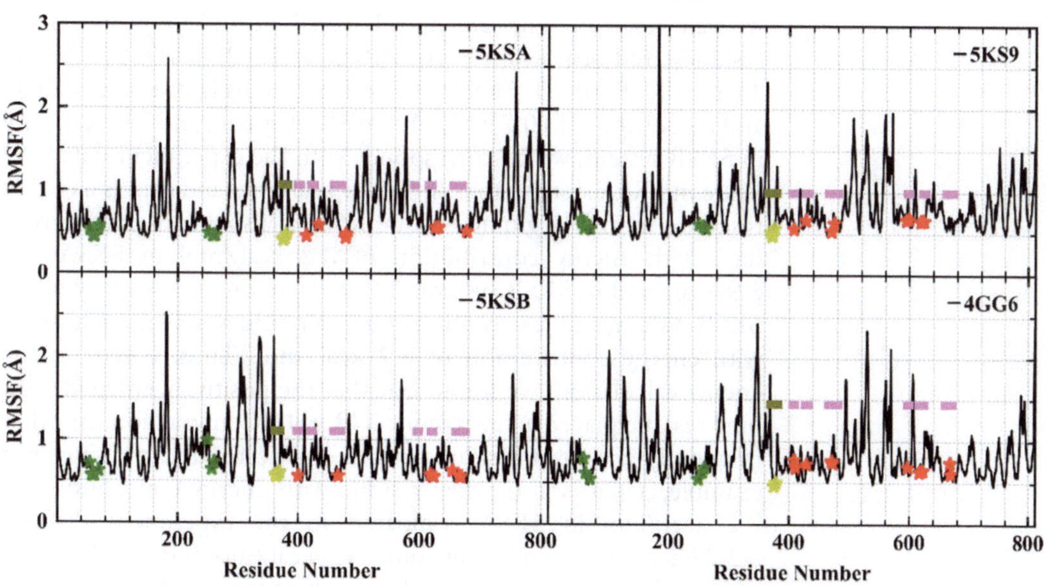

Fig. 1 RMSF of all heavy atoms of each system. The stars from left to right in each graph represent hotspots on MHC (green), peptide (olive), and TCR (red) computed during the last 2 ns of the MD. The horizontal lines (olive and pink) represent, respectively, the peptide and different TCR domains

in our data, CDR3 residues are as stable as the CDR1 and CDR2. Therefore, in its bonded state, the peptide is likely to stabilize the CDR3 loops by forming interactions with them.

3.2 MM/GBSA_ASIE Calculation

MD simulation is carried out, and residue-specific binding free energies are computed using the MM/GBSA_ASIE method described in Subheading 2. These residue-specific binding free energies provide quantitative information on the energetic contribution of individual residues to protein–protein bind. Based on these calculations, we can identify hotspots and warm spots on TCRs, MHCs, and peptides, respectively. By comparing these key residues on each system, we could find the conservative residues located individually on TCRs, MHCs, and peptides. This information is extremely useful to help elucidate important binding mechanism and recognition patterns in the complex protein interaction among TCR, MHC, and peptides, which might provide a mechanism for HLA-DQ8-related celiac disease. The detailed results and their analyses are presented in the following sections.

3.3 Key Residues on TCRs

Each system has a different TCR; however, some of them identified the same pMHC complex, and therefore must have something in common. Such common features should be highlighted.

Bel602 TCR(In 5KSA) and T15 TCR(In 5KSB) identified the same pMHC(HLA-DQ8.5-glia-γ1)structure. Even though there is no experiment data about the mutation in system 5KSB, we still can predict the result by calculating and comparing with the data in system 5KSA. When TCRs combined to HLA-DQ8.5-glia-γ1, as we can see from Fig. 2a, b, R37α, Y57α, Y113α, R66β, Y57β residues are hotspots on both Bel602 TCR (PDB ID: 5KSA) and T15 TCR (PDB ID: 5KSB). R37α is a positively charged residue, and it contributed 10.7 kcal/mol in 5KSA system and 7.4 kcal/mol in 5KSB system, which corresponds with the experiment data [29] made in 5KSA system, that is, >2.22 kcal/mol and acts as a critical residue. Another arginine residue R66β is also a conservative residue, with energy contribution of 3.2 kcal/mol in 5KSA and 5.9 kcal/mol in 5KSB system. R109β is unique that exists only in 5KSB system, and just as other arginine residues, it is also a hotspot with energy contribution of 8.2 kcal/mol. Thus, based on the energy calculation, we found out that the positively charged arginine residue played a key role in TCR pMHC interaction. Another frequently appeared residue is an aromatic residue tyrosine. For example, Y57α is also a key residue with energy contribution of 3.3 kcal/mol in 5KSA compared to the experiment data of >2.22 kcal/mol, and it contributes 1.5 kcal/mol and 5KSB system. Y57β is also a conservative residue on these two TCRs, with energy contribution of 2.7 kcal/mol in 5KSA system compared to the experiment data [29] of >2.22 kcal/mol. And it contributes 3.1 kcal/mol in 5KSB system. In 5KSA system, the calculation of

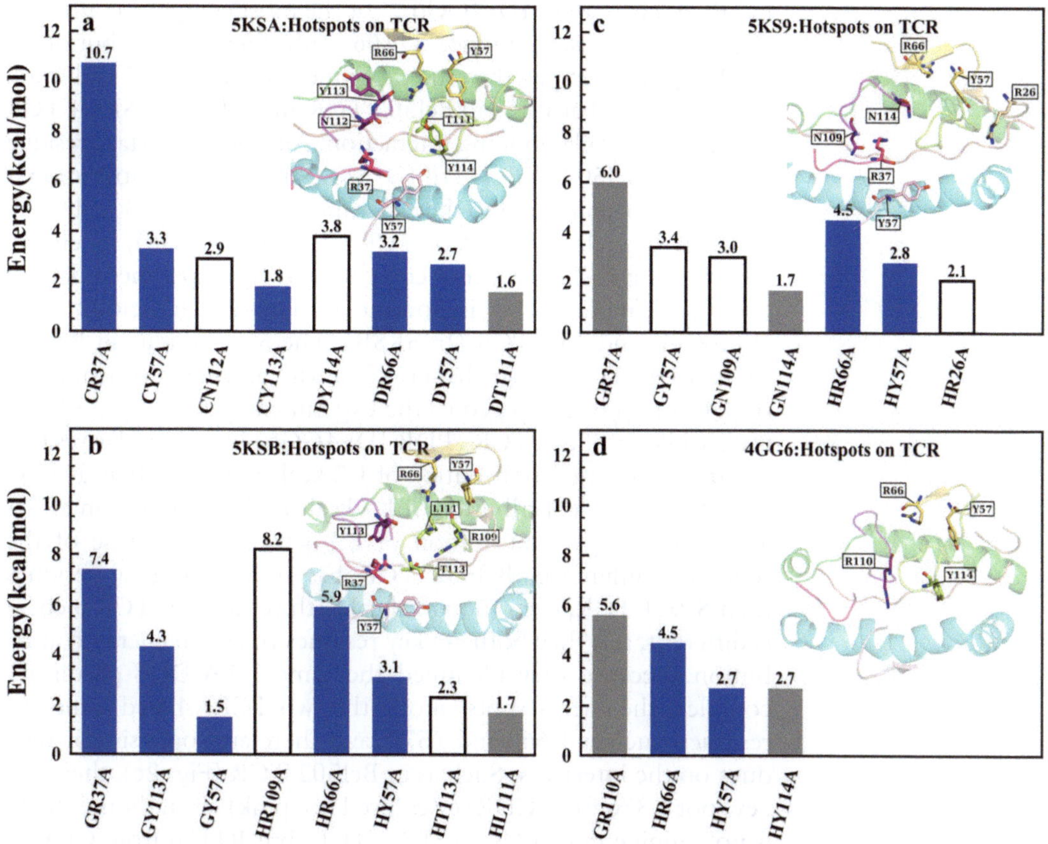

Fig. 2 Free energy contributions of main residues on TCR in TCR/pMHC bindings. (**a** and **b**) The blue solid bars represent the same hotspots in Bel602 TCR and T15 TCR, gray solid bars represent similar hotspots, and the hollow bars represent unique hotspots of each system. The *y*-axes represent free energy contributions of each residue in unit of kcal/mol. The α and β chains of MHC are colored green and cyan in the inserts, respectively. Most hotspot residues of the TCR are located in CDR1α (hot pink), CDR2α (light pink), CDR3α (purple), CDR1β (wheat), CDR2β (yellow), and CDR3β (lemon) domains. The sequence ranges of various TCR domains are as follows: CDR1α (33–38), CDR2α (56–66), CDR3α (104–117), CDR1β (26–37), CDR2β (57–68), and CDR3β (104–119). (**c** and **d**) The same as in **a** and **b** but for 5KS9 and 4GG6 systems.

energy contribution of Y113α residue is 1.8 kcal/mol compared to the experiment value of 1.93 kcal/mol, which is very close. Another key residue in 5KSA is N112α, which lies on CDR3α, and it helps to contact with MHC α chain (Fig. 4a). They are key residues with great energy contribution on TCRs.

Bel502 TCR (PDB ID: 5KS9) and SP3.4 TCR (PDB ID: 4GG6) identified the HLA-DQ8-gliadin-α1 complex. As shown in Fig. 2c, d, residues R66β and Y57β are hotspots on both Bel502 (PDB ID: 5KS9) TCR and SP3.4 TCR (PDB ID: 4GG6) with energy contribution over 2 kcal/mol. R37α is the most energetic contributing residue in Bel502 TCR (PDB ID: 5KS9) with energy contribution of 6.0 kcal/mol compared to the experiment data >2.53 kcal/mol. It is interesting to find that there is no R37α in SP3.4 TCR (PDB ID: 4GG6), but the most energetic contributing residue is also arginine, R110α with energy contribution of 5.6 kcal/mol compared to the experiment data of >1.7 kcal/mol. This indicates that R37α in Bel502 TCR and R110α in SP3.4 TCR should have similar binding interaction. Another important residue is Y57β, in Bel502 TCR (PDB ID: 5KS9) its energy contribution is 2.8 kcal/mol compared to the experiment data of >2.53 kcal/mol, which is very close. In SP3.4 TCR (PDB ID: 4GG6), its energy contribution is 2.7 kcal/mol compared to the experiment data of >1.7 kcal/mol, just as important as they are in Bel602 TCR (In 5KSA) and T15 TCR (In 5KSB). The 57th residue in Bel502 TCR (PDB ID: 5KS9) α chain is Y57α with an energy contribution of 3.4 kcal/mol compared to the experiment data of >2.53 kcal/mol, while in SP3.4 TCR (PDB ID: 4GG6) is a nonpolar residue L57α, with energy contribution of 1.2 kcal/mol. N114β in Bel502 TCR has energy contribution of 1.7 kcal/mol, and Y114β in SP3.4 TCR has energy contribution of 2.7 kcal/mol. Because of the sequence difference, Bel502 TCR has more energetic residues than SP3.4 TCR, even though the 114th residue on TCR β chain is different, but they both are key residues based on energy contribution. Because they identified the same HLA-DQ8-gliadin-α1 complex, the energy spots among the two TCRs shared some key residues, such as R66β and Y57β, and there are some similar residues on the interfaces. Such as in Bel502 TCR (Fig. 2c), there is a key spot R37α from CDR1α (colored hot pink), even though there is no arginine in CDR1α in SP3.4 TCR, but R110α from CDR3α (colored purple) has a very similar position and function as R37α in Bel502 TCR.

As shown in Fig. 2, these key residues are mainly located on flexible loops, CDR1α, CDR2α, CDR3α, CDR1β, CDR2β, and CDR3β of the TCRs. CDR1α (colored hot pink) and CDR1β (colored wheat) are located above both sides of the peptides. CDR2α (colored light pink) is mainly situated on the helix of MHC β chain (colored cyan), and CDR2β (colored yellow) is mainly situated on the helix of MHC α chain (colored green). CDR3α (colored purple) is situated on the middle of the peptides and slightly inclined to MHC α chain. CDR3β (colored lemon) is also situated on the peptides.

3.4 Key Residues on MHCs

For HLA-DQ8.5 systems (PDBID: 5KSA and 5KSB), as shown in Fig. 3a, b, the conserved hotspots on MHC β chain (colored cyan) are R70β, T77β, and H81β. The conserved key residues on MHC α chain (colored green) are F58α, R53α, H68α, and T61α; there is a slight difference because HLA-DQ8.5 contacts with different TCRs.

The main difference between HLA-DQ8 and HLA-DQ8.5 lies on the α chain (colored green), so for HLA-DQ8 systems (PDBID: 5KS9 and 4GG6), as shown in Fig. 3c, d, the conserved hotspots

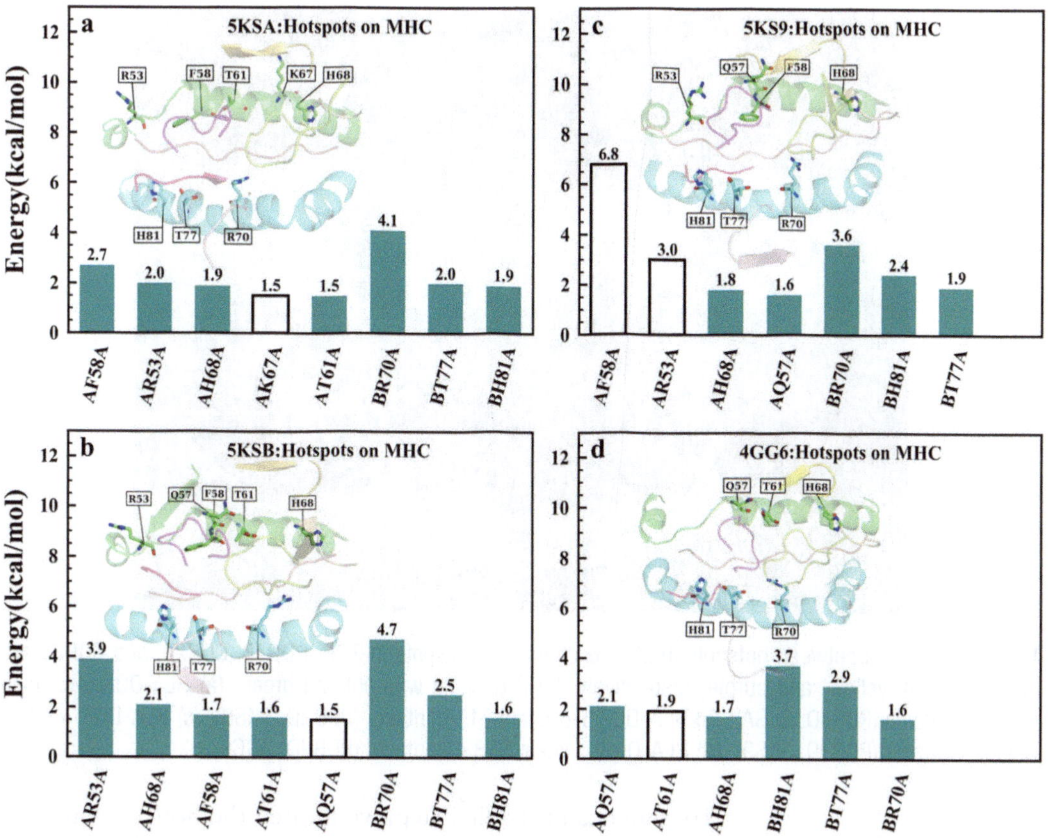

Fig. 3 Hotspots and warm spots on MHCs. The solid bars in **a** and **b** are the common key residues on HLA-DQ8.5. The same meaning in **c** and **d** for HLA-DQ8. The α and β chains of MHC are colored green and cyan, respectively. The y-axes represent free energies in unit of kcal/mol.

on MHC β chain (colored cyan) are the same as they are in HLA-DQ8.5 systems, namely residues R70β, T77β, and H81β. The conserved hotspots on MHC α chain (colored green) are H68α and Q57α with similar energy contribution in both systems. In 5KS9 system, F58α is the most energetic residue with energy contribution of 6.8 kcal/mol, and residue R53α is also an energetic residue with energy contribution of 3.0 kcal/mol. They played important role in energy contribution when contacting with Bel502 TCR. While when contact with SP3.4 TCR, they have negligible contribution in 4GG6 system.

As for the recognition mechanism of TCR, Lawren C. Wu [56] proposed in 2002 that TCR recognition can be divided into two steps. First, TCR–MHC interactions, aided by a few TCR–peptide interactions, can guide the TCR to the peptide. Second, the two CDR3 loops will rearrange the conformation to form stable contacts with the peptide. Only then, would T-cell activation be triggered.

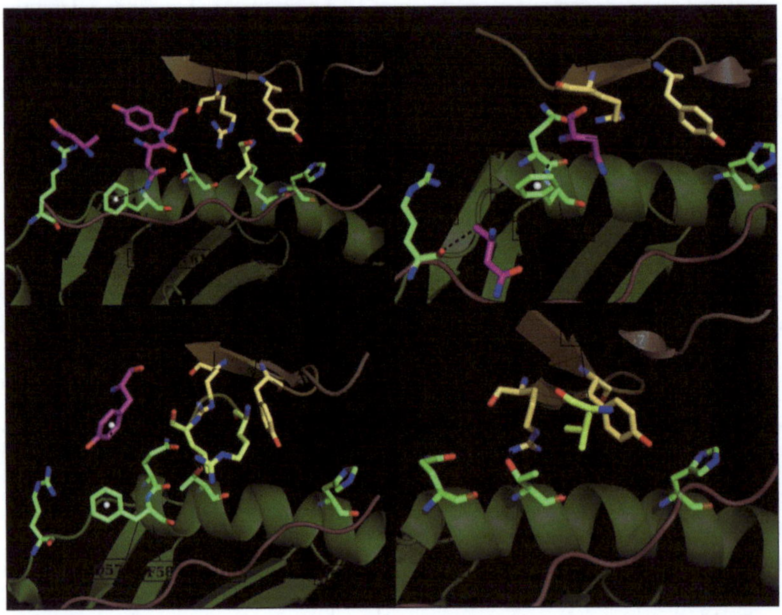

Fig. 4 Interactions between hotspots in MHC α chain and hotspots in TCR. CDR1β, CDR2β, and CDR3α are colored wheat, yellow, and purple, respectively. MHC α chain was colored green. (**a**) HLA-DQ8.5-α chain Bel602 contacts (PDB ID: 5KSA); (**b**) HLA-DQ8.5-α chain T15 contacts (PDB ID: 5KSB); (**c**) HLA-DQ8-α chain Bel502 contacts (PDB ID: 5KS9); (**d**) HLA-DQ8-α chain SP3.4 contacts (PDB ID: 4GG6)

To understand the first step, we located the hotspots on TCR and MHC based on the free energy calculation. Unveiling detailed information on these hotspots is key to understanding the first step in the TCR–MHC interaction, especially the information about the frequently appeared arginine residue and tyrosine residue on TCRs, and the highly conserved residues on MHCs needs to be analyzed.

3.5 Interaction Between MHCs and TCRs

In the first step of TCR recognition, TCR–MHC interactions play a leading role. As we can see from Fig. 2, MHC α chain (colored green) is mainly located under CDR2β (colored yellow), and it also has interactions with residues from CDR3α (purple) and CDR3β (lemon). MHC β chain (colored cyan) mainly contacts with residues from CDR2α (light pink), as well as CDR1α (hot pink) and CDR3β (lemon).

3.5.1 Interaction Between MHC α Chain and TCR

The MHC in 5KSA (Fig. 4a) and 5KSB (Fig. 4b) is HLA-DQ8.5, when HLA-DQ8.5 α chain (colored green) contacts with Bel602 TCR (Fig. 4a) and T15 TCR (Fig. 4b), R66β from CDR2β (yellow) and Y113α from CDR3α (purple) are common key residues. R66β in Bel602 TCR (Fig. 4a) formed interaction with T61, while in T15 TCR (Fig. 4b), it generates hydrogen bond with Q57 (*see* Fig. 4b and Table 3). Compared to T15 TCR (Fig. 4b), N112α is a unique hotspot of Bel602 TCR (Fig. 4a), and it contacts with F58 and T61, and occupies a large space, which leads to the change of orientation of residue Y113α. Y113α in Bel602 TCR (Fig. 4a) is comparatively parallel to the α helix, and it contacts with R53. While in T15 TCR (Fig. 4b), Y113α tilts down to the HLA-DQ8.5 α chain (colored green), located on F58α and R53, and forms interactions with them. That's probably why Y113α has a higher energy contribution in 5KSB (Fig. 4b) system. Y113α from CDR3α (purple) and Y57β from CDR2β (yellow) are hotspots on both systems, not only because they contact with HLA-DQ8.5 α chain (colored green) but also with peptide (colored salmon). In T15 TCR, positively charged R109β from CDR 3β (lemon) is a unique hotspot of T15 TCR (Fig. 4b), and it contacts with T61 and occupies huge space between TCR and peptide.

The MHC in 5KS9 (Fig. 4c) and 4GG6 (Fig. 4d) is HLA-DQ8, when HLA-DQ8 α chain (colored green) contacts with Bel502 TCR (Fig. 4c) and SP3.4 TCR (Fig. 4d), the common key residues from TCRs are Y57β and R66β from CDR2β (yellow). Y57β formed conserved interaction with H68. While the more important function for Y57β is to generate hydrogen bond with glutamine when acting with peptide. R66β contacts with Q57 in Bel502 TCR (Fig. 4c) system, and contacts with Q57 and T61 in SP3.4 TCR (Fig. 4d) system. In the Bel502 TCR (Fig. 4c) system, N109α from CDR3α (purple) forms interaction with R53, and N114α contacts with F58 from the top.

3.5.2 Interaction Between MHC β Chain and TCR

CDR2α (light pink) is mainly situated on MHC β chain, and small partial of residues from CDR1α (hot pink), CDR3β (lemon), and CDR3α (purple) are contacting with MHC β chain. The hotspots and warm spots on MHC β chain are R70β, T77β, and H81β.

When HLA-DQ8.5 β chain (colored cyan) contacts with Bel602 TCR (Fig. 5a) and T15 TCR (Fig. 5b), R37α from CDR1α (hot pink) and Y57α from CDR2α (light pink) are conserved vital residues. R37α occupies huge space, forms stable salt bridge with negatively charged residue E74 in both systems (*see* Fig. 5a, b, Tables 2 and 3), and contacts with T77β. It is the most energy-contributing residue (Fig. 2a, b) in the TCR pMHC recognition process, and it also forms strong interaction with peptide (Fig. 7a, b). In Bel602 TCR (Fig. 5a), Y114β from CDR3β (lemon) forms interaction with R70. M109α from CDR3α (purple) contacts with H81β. In T15 TCR (Fig. 5b), T113β and D114β from CDR3β (lemon) forms interaction with R70. The S109α from CDR3α (purple) contacts with H81β.

When HLA-DQ8 β chain (colored cyan) contacts with Bel502 TCR (Fig. 5c) and SP3.4 TCR (Fig. 5d), there are fewer common residues, but similar interactions were showed. In Bel502 TCR (Fig. 5c), H81 contacts with K82α. Residue S113β from CDR3β (lemon) forms interaction with R70. And R37α generates interaction with E74 and T77. While in SP3.4 TCR (Fig. 5d) system, R70 contacts with R110α and Y114, and E74 and T77 forms interaction with R110α and L57α. H81 contacts with T36α from CDR1α (hot

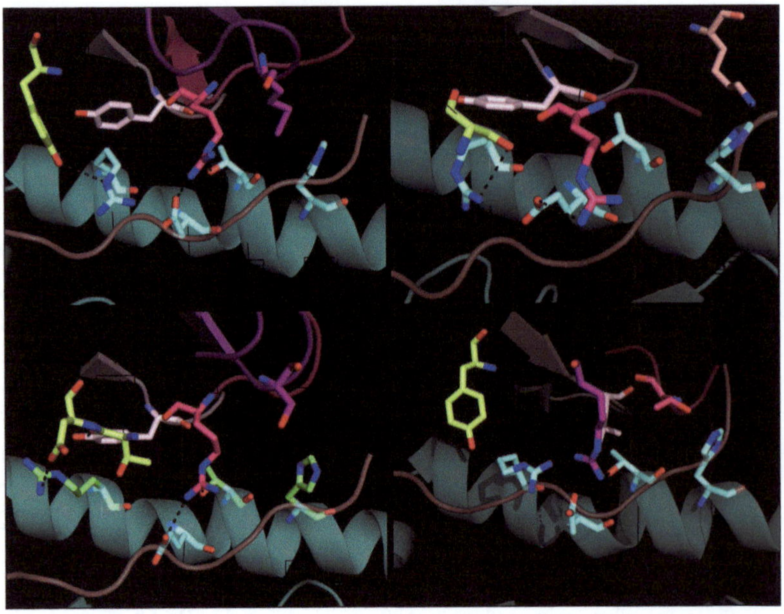

Fig. 5 The interactions between MHC β chain and TCR. CDR1α, CDR2α, CDR3α, and CDR3β are colored hot pink, light pink, purple, and lemon, respectively. MHC β chain was colored cyan. (**a**) HLA-DQ8.5-β chain Bel602 contacts (PDB ID: 5KSA); (**b**) HLA-DQ8.5-β chain T15 contacts (PDB ID: 5KSB); (**c**) HLA-DQ8-β chain Bel502 contacts (PDB ID:5KS9); (**d**) HLA-DQ8-β chain SP3.4 contacts (PDB ID:4GG6)

Table 2
Hydrogen bond information for 5KSA

Type	Accepter	Donor	Occupancy
TCR-MHC	AD55@OD1	CY113@OH	0.8862
	BE74@OE1	CR37@NH2	0.8136
	CD110@OD1	AR53@NH1	0.8118
TCR-Peptide	JE7@OE2	DY114@OH	0.9684
	DV108@O	JQ8@NE2	0.8071
	JQ8@OE1	DY57@OH	0.7397
	JQ2@OE1	CN112@ND2	0.7147
	JQ2@OE1	CR37@NH1	0.7021
	JQ3@O	CR37@NH1	0.6924
	CM109@O	JQ2@NE2	0.6254
	JQ3@O	CR37@NH2	0.5784
	JQ3@OE1	CN112@ND2	0.5732

Table 3
Hydrogen bond information for 5KSB

Type	Accepter	Donor	Occupancy
TCR-MHC	BE74@OE2	HT113@OG1	0.9294
	BE74@OE2	GR37@NH2	0.8887
	HR66@O	AQ57@NE2	0.8529
	BE74@OE2	GR37@NE	0.5932
	BE74@OE1	GR37@NH2	0.5489
	AQ57@OE1	HR66@NH2	0.5079
TCR-Peptide	JQ3@O	GR37@NH2	0.9744
	JQ8@OE1	HY57@OH	0.8712
	JQ2@OE1	GR37@NH1	0.7791
	JP6@O	HR109@NH2	0.7566
	HN108@O	JQ8@NE2	0.6231
	JP1@O	GS109@OG	0.5038
	JQ3@O	GR37@NH2	0.9744

pink). Although there is no arginine in CDR1α, R110α acts the same role as R37α in other systems that contacting with MHC β chain.

In the overview of TCR–MHC interactions, aromatic residue tyrosine and positively charged residues arginine and lysine are frequently appeared in the critical loop regions and have interactions with some specific residues.

3.6 Key Residues on Peptides

For Gliadin-γ1 systems (PDBID: 5KSA and 5KSB), as shown in Fig. 6a, b, the key residues when contacting with Bel602 TCR (PDBID: 5KSA) are Q2, Q3, F5, and Q8, their side chain are facing Bel602 TCR. When contacting with T15 TCR, the key residues on Gliadin-γ1 peptide are Q2, F5, and Q8, which lie on the peak of peptide. The side chain of Q3 tilts down and has negligible contact with T15 TCR.

For Gliadin-α1 systems (PDBID: 5KS9 and 4GG6, *see* Tables 4 and 5), the key residues in the core area of Gliadin-α1 peptide when contacting with Bel502 TCR and SP3.4 TCR are residues Q5 and Q8, which also lie on the peak of peptide.

3.6.1 Interaction Between Peptide and TCRs

In the second step of TCR p-MHC recognition, TCR–peptide interactions were thought to be dominant. The core area of peptide Gliadin-γ1 (GPQQSFPEQEA in 5KSA and 5KSB) and Gliadin-α1 (APSGEGSFQPSQENPQ in 5KS9 and SGEGSFQPSQENP in 4GG6) are differed in some spots, but they all put glutamine residue in

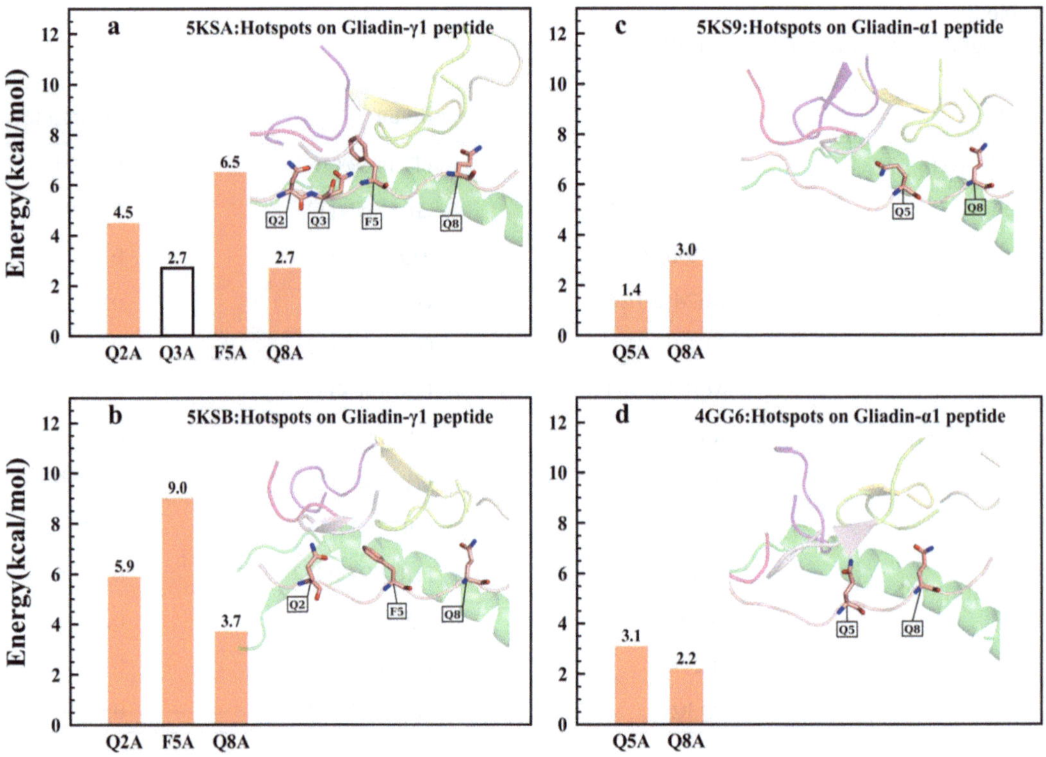

Fig. 6 Hotspots on peptides. (**a**) Hotspots of Gliadin-γ1 peptide in 5KSA system. (**b**) Hotspots of Gliadin-γ1 peptide in 5KSB system. (**c**) Hotspots of Gliadin-α1 peptide in 5KS9 system. (**d**) Hotspots of Gliadin-α1 peptide in 4GG6 system. The y-axes represent free energies in unit of kcal/mol.

Table 4
Hydrogen bond information for 5KS9

Type	Accepter	Donor	Occupancy
TCR-MHC	BD76@OD2	GS58@OG	0.6368
	HD114@OD1	BR70@NE	0.6312
TCR-Peptide	JQ8@OE1	HY57@OH	0.8996
	JQ5@OE1	GN114@ND2	0.7736
	HV108@O	JQ8@NE2	0.6637
	JS3@O	GR37@NH2	0.6218
	JS3@O	GR37@NH1	0.6036

the key spots, as a key feature of gliadin peptides, glutamine (Q) plays an important role in the interaction with TCRs, as showed in Fig. 6. By analyzing the detail interactions between the peptides and the corresponding TCRs, we may be able to unveil the mechanism in the second step.

Table 5
Hydrogen bond information for 4GG6

Type	Accepter	Donor	Occupancy
TCR-MHC	BE74@OE2	GR110@NH2	0.8025
	HR66@O	AQ57@NE2	0.5044
TCR-Peptide	JQ8@OE1	HY57@OH	0.8267
	HV108@O	JQ8@NE2	0.7480
	JS3@O	GR110@NE	0.7281

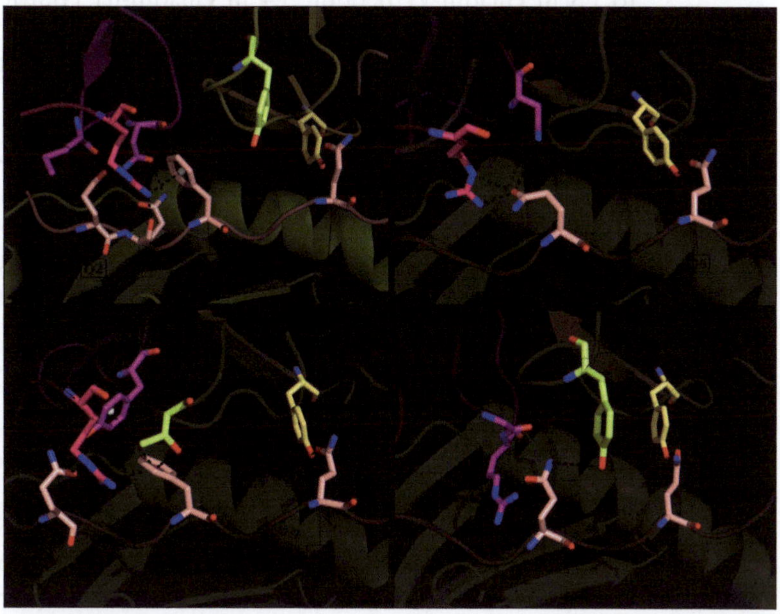

Fig. 7 The interactions between hotspots on TCR and hotspots in the core area of peptide Gliadin-γ1 (PQQS FPEQE) and Gliadin-α1(EGSFQPSQE). (**a**) Bel602 TCR Gliadin-γ1 contacts (PDB ID: 5KSA); (**b**) T15 TCR Gliadin-γ1 contacts (PDB ID: 5KSB); (**c**) Bel502 TCR Gliadin-α1 contacts (PDB ID:5KS9); (**d**) SP3.4 TCR Gliadin-α1 contacts (PDB ID: 4GG6)

3.6.2 Interaction Between Peptide Gliadin-γ1(GPQQSFPEQEA) and TCRs

When contacting with Bel602 TCR (Fig. 7a), the side chain of Q2 formed hydrogen bonds with M109α, N112α from CDR3α (purple), and R37α from CDR1α (hot pink) (*see* Table 2). Compared to Q2, residue Q3 tilts down slightly, resulting in a slightly weaker energy contribution and weaker connection with Bel602 TCR, as the side chain formed a weaker hydrogen bond with N112α, and the backbone oxygen formed weaker hydrogen bonds with R37α. The side chain of F5, an aromatic ring located between TCR α chain and TCR β chain, stands upright and occupies a lot of space and has built connections with R37α. Q8 also stands upright and generates a hydrogen bond with Y57β.

When contacting with T15 TCR (Fig. 7b), the side chain of Q2 generated a hydrogen bond with R37α as it did in Bel602 TCR. In addition, it built connections with Y113α. The side chain of Q3 tilts down and lies horizontally, thus resulting in negligible interactions with T15 TCR. F5 points upwards and lies in the middle of TCR α chain and TCR β chain as it did in Bel602 TCR. It has also formed strong interactions with R37α, Y113α from TCR α chain and T113β from TCR β chain. Residue Q8 formed interactions with Y57β.

3.6.3 Interaction Between Peptide Gliadin-α1(EGSFQPSQE) and TCRs

Gliadin-α1 (EGSFQPSQE) has one less glutamine residue in the core area compared to Gliadin-γ1 (PQQSFPEQE) peptide, and the side chain of phenylamine (F4) is facing downwards, so compared to F5 in Gliadin-γ1 peptide, F4 in Gliadin-α1 contributed much less energy when contact with TCR. As showed in Fig. 7c, d, when Gliadin-α1 (EGSFQPSQE) contacts with Bel502 TCR and SP3.4 TCR, it is characterized by glutamine residues Q5 and Q8.

When binding with Bel502 TCR (Fig. 7c), residue Q5 formed a hydrogen bond with N114α located in CDR3α (purple), and it also generated strong interactions with R37α from CDR1α (hot pink). Its side chain bends down, that is probably the reason why it has less energy contribution here than it has in 4GG6 system. The R37α residue is a vital residue as it is in Bel602 TCR (PDBID:5KSA) and T15 TCR (PDBID:5KSB), and it acts as a critical role in TCR–peptide and TCR–MHC interactions (*see* Figs. 5 and 7). Q8 forms a hydrogen bond with Y57β from CDR2β (yellow).

When binding with SP3.4 TCR (Fig. 7d), Q5 and Q8 are vital residues. Compared to the other systems, there is no arginine residue in CDR1α, but a unique residue R110α from CDR3α (purple) acts similar as R37α in other TCRs, and it forms strong interaction with Q5. Q5 also contacts with Y114β. Residue Q8 builds a hydrogen bond with Y57β, the same as Q8 acts in other systems.

In peptide Gliadin-γ1 (GPQQSFPEQEA) systems, glutamine residues are willing to face upward and contact with TCRs, Q2 contacts with positively charged residue R37α and residue from CDR3α. The fifth residue in Gliadin-γ1 is F5, and its side chain lies in the middle of TCR α chain and TCR β chain and contacts with the charged R37α. The fifth residue in Gliadin-α1 is glutamine, contacting with the positively charged arginine, too. Q8 has a conserved interaction with Y57β with similar energy contribution in all four systems.

4 Conclusion

The four different complex structures of TCR/pMHCII from three different patients were used for computational study to investigate their binding mechanism. Although every individual structure has its own unique features in some specific binding interactions and key residues, there are common features in the recognition pattern. For example, the 4GG6 system has fewer hotspot residues on its TCR–MHC and TCR–peptide interfaces relative to 5KSA, 5KSB, and 5KS9 systems. Since all four systems are pathogenic and lead to celiac disease, the conserved hotspots on these interfaces are especially important and are probably the most critical part of pathogenic causes. Thus, common features of HLA-DQ8 related celiac disease can be summarized as follows. First, TCR interacts with MHC through its CDR2 regions (Y57β, R66β, Y57α/L57α) and some of the residues from CDR1 and CDR3 regions. Second, the Q2/F5 residue from the gliadin-γ1 peptide and the Q5 residue from the gliadin-α1 peptide form vital interactions with R37α/R110α, as well as the conserved interaction between Q8 and Y57β. These residues form critical interactions in TCR–peptide recognition. As a feature of gliadin peptide, glutamine plays an indispensable role in the interaction with TCR. If there is any peptide that could maintain the binding affinity with MHCs, but cannot form interactions with these key residues on the TCR interface, the peptide will not activate the TCR even though it has been presented to the cell surface by these MHCs.

Acknowledgments

We thank Zhendong Li for help with the format of references. This work was supported by National Key R&D Program of China (grant no. 2016YFA0501700), National Natural Science Foundation of China (grant nos. 91753103, 21933010), Shanghai Sailing Program (grant no. 2016YF1408400), Shanghai Putuo District (grant 2014-A-02), Innovation Program of Shanghai Municipal Education Commission (201701070005E00020), and NYU Global Seed Grant. We thank the Supercomputer Center of East China Normal University for providing us computer time.

References

1. Broughton SE, Petersen J, Theodossis A, Scally SW, Loh KL, Thompson A, van Bergen J, Kooy-Winkelaar Y, Henderson KN, Beddoe T, Tye-Din JA, Mannering SI, Purcell AW, McCluskey J, Anderson RP, Koning F, Reid HH, Rossjohn J (2012) Biased T cell receptor usage directed against human leukocyte antigen DQ8-restricted gliadin peptides is associated with celiac disease. Immunity 37 (4):611–621. https://doi.org/10.1016/j.immuni.2012.07.013
2. Hennecke J, Wiley DC (2001) T cell receptor-MHC interactions up close. Cell 104(1):1–4. https://doi.org/10.1016/S0092-8674(01)00185-4
3. Hogquist KA, Jameson SC, Heath WR, Howard JL, Bevan MJ, Carbone FR (2012) T cell receptor antagonist peptides induce positive selection. J Immunol 188(5):2046–2056. https://doi.org/10.1016/0092-8674(94)90169-4
4. Knapp B, van der Merwe PA, Dushek O, Deane CM (2019) MHC binding affects the dynamics of different T-cell receptors in different ways. PLoS Comput Biol 15(9). https://doi.org/10.1371/journal.pcbi.1007338

5. Franckaert D, Collin R, Dooley J, Walli RH, Poussier P, Liston A, Hillhouse EE, Lesage S (2017) An orthologous non-MHC locus in rats and mice is linked to CD4(+) and CD8 (+) T-cell proportion. Genes Immun 18 (3):118–126. https://doi.org/10.1038/gene.2017.9
6. Josephs TM, Grant EJ, Gras S (2017) Molecular challenges imposed by MHC-I restricted long epitopes on T cell immunity. Biol Chem 398(9):1027–1036. https://doi.org/10.1515/hsz-2016-0305
7. Schneider-Hohendorf T, Gorlich D, Savola P, Kelkka T, Mustjoki S, Gross CC, Owens GC, Klotz L, Dornmair K, Wiendl H, Schwab N (2018) Sex bias in MHC I-associated shaping of the adaptive immune system. Proc Natl Acad Sci U S A 115(9):2168–2173. https://doi.org/10.1073/pnas.1716146115
8. Lo WL, Allen PM (2013) Self-awareness: how self-peptide/MHC complexes are essential in the development of T cells. Mol Immunol 55 (2):186–189. https://doi.org/10.1016/j.molimm.2012.10.028
9. Weissler KA, Caton AJ (2014) The role of T-cell receptor recognition of peptide:MHC complexes in the formation and activity of Foxp3(+) regulatory T cells. Immunol Rev 259(1):11–22. https://doi.org/10.1111/imr.12177
10. Abadie V, Sollid LM, Barreiro LB, Jabri B (2011) Integration of genetic and immunological insights into a model of celiac disease pathogenesis. Annu Rev Immunol 29:493–525. https://doi.org/10.1146/annurev-immunol-040210-092915
11. Henderson KN, Tye-Din JA, Reid HH, Chen Z, Borg NA, Beissbarth T, Tatham A, Mannering SI, Purcell AW, Dudek NL, van Heel DA, McCluskey J, Rossjohn J, Anderson RP (2007) A structural and immunological basis for the role of human leukocyte antigen DQ8 in celiac disease. Immunity 27(1):23–34. https://doi.org/10.1016/j.immuni.2007.05.015
12. Hovhannisyan Z, Weiss A, Martin A, Wiesner M, Tollefsen S, Yoshida K, Ciszewski C, Curran SA, Murray JA, David CS, Sollid LM, Koning F, Teyton L, Jabri B (2008) The role of HLA-DQ8 beta 57 polymorphism in the anti-gluten T-cell response in coeliac disease. Nature 456(7221):534–538. https://doi.org/10.1038/nature07524
13. Yoshida K, Corper AL, Herro R, Jabri B, Wilson IA, Teyton L (2010) The diabetogenic mouse MHC class II molecule I-A(g7) is endowed with a switch that modulates TCR affinity. J Clin Invest 120(5):1578–1590. https://doi.org/10.1172/Jci41502
14. Caio G, Volta U, Sapone A, Leffler DA, De Giorgio R, Catassi C, Fasano A (2019) Celiac disease: a comprehensive current review. BMC Med 17:142. https://doi.org/10.1186/s12916-019-1380-z
15. Jabri B, Sollid LM (2009) Tissue-specific immune responses tissue-mediated control of immunopathology in coeliac disease. Nat Rev Immunol 9(12):858–870. https://doi.org/10.1038/nri2670
16. Karell K, Louka AS, Moodie SJ, Ascher H, Clot F, Greco L, Ciclitira PJ, Sollid LM, Partanen J, D, E. G. C. C. (2003) HLA types in celiac disease patients not carrying the DQA1*05-DQB1*02 (DQ2) heterodimer: results from the European genetics cluster on celiac disease. Hum Immunol 64(4):469–477. https://doi.org/10.1016/S0198-8859(03)00027-2
17. Cecilio LA, Bonatto MW (2015) The prevalence of Hla Dq2 and Dq8 in patients with celiac disease, in family and in general population. Arq Bras Cir Dig 28(3):183–185. https://doi.org/10.1590/S0102-67202015000300009
18. Vader W, Stepniak D, Kooy Y, Mearin L, Thompson A, van Rood JJ, Spaenij L, Koning F (2003) The HLA-DQ2 gene dose effect in celiac disease is directly related to the magnitude and breadth of gluten-specific T cell responses. Proc Natl Acad Sci U S A 100 (21):12390–12395. https://doi.org/10.1073/pnas.2135229100
19. Kim CY, Quarsten H, Bergseng E, Khosla C, Sollid LM (2004) Structural basis for HLA-DQ2-mediated presentation of gluten epitopes in celiac disease. Proc Natl Acad Sci U S A 101(12):4175–4179. https://doi.org/10.1073/pnas.0306885101
20. Arentz-Hansen H, Korner R, Molberg O, Quarsten H, Vader W, Kooy YMC, Lundin KEA, Koning F, Roepstorff P, Sollid LM, McAdam SN (2000) The intestinal T cell response to alpha-gliadin in adult celiac disease is focused on a single deamidated glutamine targeted by tissue transglutaminase. J Exp Med 191(4):603–612. https://doi.org/10.1084/jem.191.4.603
21. Arentz-Hansen H, McAdam SN, Molberg O, Fleckenstein B, Lundin KEA, Jorgensen TJD, Jung G, Roepstorff P, Sollid LM (2002) Celiac lesion T cells recognize epitopes that cluster in regions of gliadins rich in proline residues. Gastroenterology 123(3):803–809. https://doi.org/10.1053/gast.2002.35381

22. Petersen J, Montserrat V, Mujico JR, Loh KL, Beringer DX, van Lummel M, Thompson A, Mearin ML, Schweizer J, Kooy-Winkelaar Y, van Bergen J, Drijfhout JW, Kan WT, La Gruta NL, Anderson RP, Reid HH, Koning F, Rossjohn J (2014) T-cell receptor recognition of HLA-DQ2-gliadin complexes associated with celiac disease. Nat Struct Mol Biol 21(5):480–488. https://doi.org/10.1038/nsmb.2817
23. Colf LA, Bankovich AJ, Hanick NA, Bowerman NA, Jones LL, Kranz DM, Garcia KC (2007) How a single T cell receptor recognizes both self and foreign MHC. Cell 129 (1):135–146. https://doi.org/10.1016/j.cell.2007.01.048
24. Rossjohn J, Gras S, Miles JJ, Turner SJ, Godfrey DI, McCluskey J (2015) T cell antigen receptor recognition of antigen-presenting molecules. Annual Rev Immunol 33 (33):169–200. https://doi.org/10.1146/annurev-immunol-032414-112334
25. Gras S, Van Rhijn I, Shahine A, Cheng TY, Bhati M, Tan LL, Halim H, Tuttle KD, Gapin L, Le Nours J, Moody DB, Rossjohn J (2016) T cell receptor recognition of CD1b presenting a mycobacterial glycolipid. Nat Commun 7:13257. https://doi.org/10.1038/ncomms13257
26. Zamoyska R (1998) CD4 and CD8: modulators of T-cell receptor recognition of antigen and of immune responses? Curr Opin Immunol 10(1):82–87. https://doi.org/10.1016/S0952-7915(98)80036-8
27. Kooy-Winkelaar Y, van Lummel M, Moustakas AK, Schweizer J, Mearin ML, Mulder CJ, Roep BO, Drijfhout JW, Papadopoulos GK, van Bergen J, Koning F (2011) Gluten-specific T cells cross-react between HLA-DQ8 and the HLA-DQ2 alpha/DQ8 beta Transdimer. J Immunol 187(10):5123–5129. https://doi.org/10.4049/jimmunol.1101179
28. Petersen J, van Bergen J, Loh KL, Kooy-Winkelaar Y, Beringer DX, Thompson A, Bakker SF, Mulder CJJ, Ladell K, McLaren JE, Price DA, Rossjohn J, Reid HH, Koning F (2015) Determinants of gliadin-specific T cell selection in celiac disease. J Immunol 194 (12):6112–6122. https://doi.org/10.4049/jimmunol.1500161
29. Petersen J, Kooy-Winkelaar Y, Loh KL, Tran M, Van Bergen J, Koning F, Rossjohn J, Reid HH (2016) Diverse T cell receptor gene usage in HLA-DQ8-associated celiac disease converges into a consensus binding solution. Structure 24(10):1643–1657. https://doi.org/10.1016/j.str.2016.07.010
30. Goodsell DS, Zardecki C, Di Costanzo L, Duarte JM, Hudson BP, Persikova I, Segura J, Shao CH, Voigt M, Westbrook JD, Young JY, Burley SK (2020) RCSB protein data Bank: enabling biomedical research and drug discovery. Protein Sci 29(1):52–65. https://doi.org/10.1002/pro.3730
31. Jorgensen WL, Chandrasekhar J, Madura JD, Impey RW, Klein ML (1983) Comparison of simple potential functions for simulating liquid water. J Chem Phys 79(2):926–935. https://doi.org/10.1063/1.445869
32. Darden T, York D, Pedersen L (1993) Particle mesh Ewald: An N· log (N) method for Ewald sums in large systems. J Chem Phys 98 (12):10089–10092. https://doi.org/10.1063/1.464397
33. Jean-Paul R, Giovanni C, Berendsen Herman JC (1977) Numerical integration of the cartesian equations of motion of a system with constraints: molecular dynamics of n-alkanes. J Comput Phys 23(3):327–341. https://doi.org/10.1016/0021-9991(77)90098-5
34. Teppa E, Zea DJ, Marino-Buslje C (2017) Protein-protein interactions leave evolutionary footprints: high molecular coevolution at the core of interfaces. Protein Sci 26 (12):2438–2444. https://doi.org/10.1002/pro.3318
35. Brender JR, Zhang Y (2015) Predicting the effect of mutations on protein-protein binding interactions through structure-based Interface profiles. PLoS Comput Biol 11(10):e1004494. https://doi.org/10.1371/journal.pcbi.1004494
36. De Moura DC, Bryksa BC, Yada RY (2014) In silico insights into protein-protein interactions and folding dynamics of the Saposin-like domain of Solanum tuberosum aspartic protease. PLoS One 9(9):e104315. https://doi.org/10.1371/journal.pone.0104315
37. Jackson RM (1999) Comparison of protein-protein interactions in serine protease-inhibitor and antibody-antigen complexes: implications for the protein docking problem. Protein Sci 8 (3):603–613. https://doi.org/10.1110/ps.8.3.603
38. Kuhlmann UC, Pommer AJ, Moore GR, James R, Kleanthous C (2000) Specificity in protein-protein interactions: the structural basis for dual recognition in endonuclease colicin-immunity protein complexes. J Mol Biol 301(5):1163–1178. https://doi.org/10.1006/jmbi.2000.3945
39. Sheinerman FB, Honig B (2002) On the role of electrostatic interactions in the design of protein-protein interfaces. J Mol Biol 318

(1):161–177. https://doi.org/10.1016/S0022-2836(02)00030-X
40. Ma BY, Elkayam T, Wolfson H, Nussinov R (2003) Protein-protein interactions: structurally conserved residues distinguish between binding sites and exposed protein surfaces. Proc Natl Acad Sci U S A 100 (10):5772–5777. https://doi.org/10.1073/pnas.1030237100
41. Li YL, Huang YP, Swaminathan CP, Smith-Gill SJ, Mariuzza RA (2005) Magnitude of the hydrophobic effect at central versus peripheral sites in protein-protein interfaces. Structure 13 (2):297–307. https://doi.org/10.1016/j.str.2004.12.012
42. Bogan AA, Thorn KS (1998) Anatomy of hot spots in protein interfaces. J Mol Biol 280 (1):1–9. https://doi.org/10.1006/jmbi.1998.1843
43. Clackson T, Wells JAJS (1995) A hot spot of binding energy in a hormone-receptor interface. Science 267(5196):383–386. https://doi.org/10.1126/science.7529940
44. Ylilauri M, Pentikainen OT (2013) MMGBSA as a tool to understand the binding affinities of Filamin-peptide interactions. J Chem Inf Model 53(10):2626–2633. https://doi.org/10.1021/ci4002475
45. Zuo ZL, Gandhi NS, Arndt KM, Mancera RL (2012) Free energy calculations of the interactions of c-Jun-based synthetic peptides with the c-Fos protein. Biopolymers 97(11):899–909. https://doi.org/10.1002/bip.22099
46. Gohlke H, Case DA (2004) Converging free energy estimates: MM-PB(GB)SA studies on the protein-protein complex Ras-Raf. J Comput Chem 25(2):238–250. https://doi.org/10.1002/jcc.10379
47. Genheden S, Ryde U (2015) The MM/PBSA and MM/GBSA methods to estimate ligand-binding affinities. Expert Opin Drug Dis 10 (5):449–461. https://doi.org/10.1517/17460441.2015.1032936
48. Huang DD, Qi YF, Song JN, Zhang JZH (2019) Calculation of hot spots for protein-protein interaction in p53/PMI-MDM2/MDMX complexes. J Comput Chem 40 (9):1045–1056. https://doi.org/10.1002/jcc.25592
49. Huang DD, Wen W, Liu X, Li Y, Zhang JZH (2019) Computational analysis of hot spots and binding mechanism in the PD-1/PD-L1 interaction. RSC Adv 9(26):14944–14956. https://doi.org/10.1039/c9ra01369e
50. Liu X, Peng L, Zhang JZH (2019) Accurate and efficient calculation of protein-protein binding free energy-interaction entropy with residue type-specific dielectric constants. J Chem Inf Model 59(1):272–281. https://doi.org/10.1021/acs.jcim.8b00248
51. Qiu LQ, Yan YN, Sun ZX, Song JN, Zhang JZH (2018) Interaction entropy for computational alanine scanning in protein-protein binding. Wiley Interdiscip Rev Comput Mol Sci 8(2):e1342. https://doi.org/10.1002/wcms.1342
52. Yang YP, He LP, Bao JX, Qi YF, Zhang JZH (2019) Computational analysis for residue-specific CDK2-inhibitor bindings. Chinese J Chem Phys 32(1):134–142. https://doi.org/10.1063/1674-0068/cjcp1901012
53. Obiol-Pardo C, Granadino-Roldan JM, Rubio-Martinez J (2008) Protein-protein recognition as a first step towards the inhibition of XIAP and Survivin anti-apoptotic 190 proteins. J Mol Recognit 21(3):190–204. https://doi.org/10.1002/jmr.887
54. Duan LL, Liu X, Zhang JZH (2016) Interaction entropy: a new paradigm for highly efficient and reliable computation of protein-ligand binding free energy. J Am Chem Soc 138(17):5722–5728. https://doi.org/10.1021/jacs.6b02682
55. Hare BJ, Wyss DF, Osburne MS, Kern PS, Reinherz EL, Wagner G (1999) Structure, specificity and CDR mobility of a class II restricted single-chain T-cell receptor. Nat Struct Biol 6(6):574–581. https://doi.org/10.1038/9359
56. Wu LC, Tuot DS, Lyons DS, Garcia KC, Davis MM (2002) Two-step binding mechanism for T-cell receptor recognition of peptide-MHC. Nature 418(6897):552–556. https://doi.org/10.1038/nature00920

Chapter 14

Umbrella Sampling-Based Method to Compute Ligand-Binding Affinity

Son Tung Ngo and Minh Quan Pham

Abstract

Many proteins have a solvent-exposed binding cleft, which permits their inhibitors to bind and unbind without significant protein conformation transforms. The binding/unbinding pathways of these protein–inhibitor complexes can be rather straightforwardly sampled by using umbrella sampling (US) simulation methods. During a US simulation, the C_α atoms of the protein are restrained via a harmonic force. The potential of mean force (PMF) along the binding pathway can be estimated by using the weighted histogram analysis method (WHAM). The binding affinity is then computed as the difference in PMF between the binding and unbinding states.

Key words Ligand-binding free energy, Biased sampling, Steered MD, Umbrella sampling, Potential of mean force, WHAM calculation

1 Introduction

The binding free energy ΔG quantifies the strength of non-covalent association between small-molecule ligands and macromolecules, which is of paramount importance for biology and medicine [1, 2]. The accurate estimation of the ligand-binding affinity can help to resolve some of the commonly encountered drug design issues such as time and cost [3, 4]. Therefore, numerous methods have been developed to probe the binding mechanism and affinity of a ligand to a protein [5]. Normally, computational methods require different amounts of computer resources, with more accurate and precise methods usually being more computationally demanding [6, 7]. From this perspective, molecular docking is among the cheapest computational methods currently available and, therefore, can be used to quickly obtain binding affinities and structures for a large set of ligands [8]. The approximate approaches are then engaged to validate the docking observation. The common schemes clarified in this group are known as methods

Quentin Vanhaelen (ed.), *Computational Methods for Estimating the Kinetic Parameters of Biological Systems*, Methods in Molecular Biology, vol. 2385, https://doi.org/10.1007/978-1-0716-1767-0_14,

based on the linear interaction energy (LIE) [9], molecular mechanics-Poisson–Boltzmann surface area (MM-PBSA) [10], and fast pulling of ligand (FPL) [11] schemes. Finally, the obtained results can then be refined using more theoretically rigorous and accurate methods such as thermodynamics integration (TI) [12], free energy perturbation (FEP) [13], and non-equilibrium molecular dynamics (NEMD) [14] simulations.

In practice, the ligand-binding free energy is most of the time calculated by using atomistic simulations [5] instead of quantum chemical calculations [15]. This is essentially due to the high computational demand of quantum chemical calculations. The real-time dynamics of biological complexes is thus sampled in phase space in order to reach the equilibrated Boltzmann distribution. Although the computing power has been rapidly increasing over the last decades allowing us to simulate a solvated system normally up to microseconds of MD simulations, it is still difficult to sample the entire phase space of large biomolecular systems, especially when ligand binding induces large conformational changes within the protein. Furthermore, protein–ligand binding and unbinding processes typically involve a wide range of timescales from milliseconds to seconds and more. With the computational resources currently available, it is impossible to reach this timescale even when relying on technology such as graphical processing unit (GPU) to optimize the performance. Therefore, the enhanced sampling methods appear as the most appropriate solutions to help the protein–ligand complexes to explore the whole phase space along the binding pathway. Several computational schemes were developed for this purpose. This includes steered MD (SMD), adiabatic bias MD (ABMD) [16], scaled MD [17], Metadynamics (MetaD) [18], and replica exchange MD (REMD) [19] methods. The ligand-binding free energy is thus determined via a scheme combining enhanced sampling and free energy calculation methods such as temperature REMD implementation of the FEP method results in the significant improvement of the accuracy and precision of ligand-binding free energy to type II dehydroquinase [20].

In this chapter, we present a study where steered MD method was used to generate the input of US simulations, and the US simulations were then applied to simulate the protein–ligand binding process [21]. The ligand–binding free energy value was then computed using the potential mean force (PMF) calculation with the help of the weighted histogram analysis method (WHAM) analysis [22].

2 Computational Ligand-Binding Affinity

In this section, we give a brief description of the methodology, and we detail the procedure to determine the ligand-binding affinity using the biased sampling simulations. The computational scheme was summarized in Fig. 1.

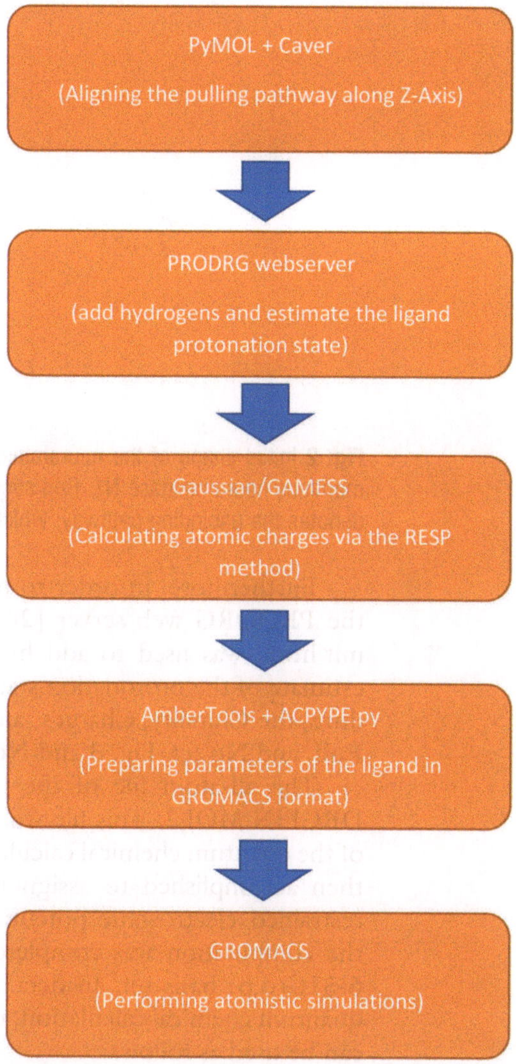

Fig. 1 Computational tactic to estimate the ligand-binding affinity via umbrella sampling simulations

2.1 Preparing the Protein–Ligand Complex

Normally, in computer-aided drug design, the molecular docking simulation is used to obtain the binding pose between a protein and a ligand, and the MD simulation is then applied to compute the obtained affinity [23]. However, in this section, for simplicity, the three-dimensional structure of the protein–ligand complex was directly obtained from the protein data bank (PDB) (PDB Identifier: 4B7Q) [24]. Moreover, it should be noted that the pulling pathway, the optimized path is predicted to mobilize the ligand from bound to unbound states, is normally predicted using the Caver 2.1 package [25] and the complex is then rotated until the pulling path is aligned along the *Z*-axis to reduce CPU time consumption (Fig. 2).

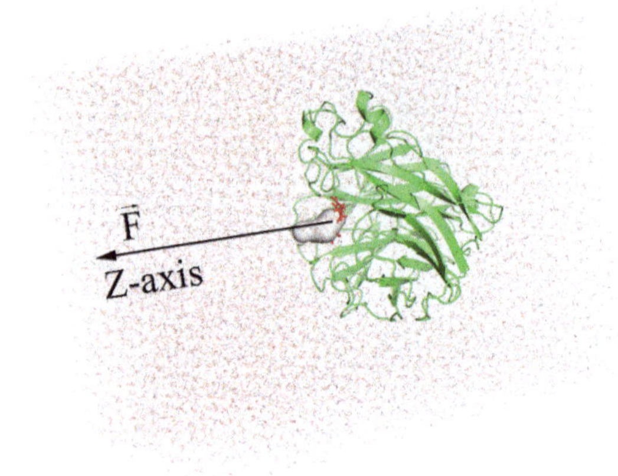

Fig. 2 Initial shape of the neuraminidase N1 versus Zanamivir. Green cartoon mentions neuraminidase N1. Red stick describes the ligand and the white tube denotes the unbinding pathway, which was generated by Caver 2.1.2 [25]

Furthermore, in order to parameterize the solvated complex, the PRODRG web server [26], http://prodrg1.dyndns.org/submit.html, was used to add hydrogen atoms and to compute an estimate of the protonation state of the ligand. During this procedure, the Chirality, Charges, and EM options were chosen as **Yes**, **Full**, and **No** (cf. Fig. 3 and **Notes 1–3**), respectively.

The all-atom file of the ligand was completed and named DRGFIN.MOL2. This file will be used later to prepare the input of the quantum chemical calculation. The electronic simulation was then accomplished to assign the ligand atomic charges via the restrained electrostatic potential (RESP) scheme [27], in which the computation was completed using the B3LYP functional at 6-31G(d,p) basis set. In detail, the example of the input file for quantum chemical calculation, used for the Gaussian package [28], can be used as follows:

```
%nprocshared=16
%mem=10GB
%chk=lig.chk
# b3lyp/6-31g(d,p) nosymm pop=mk iop(6/33=2,6/41=10,6/42=17)
scf=xqc

LIG

0 1
C 19.11300000 21.25400000 -42.59700000
H 18.94700000 21.84800000 -41.69800000
...
H 22.70800000 24.92300000 -44.03500000
```

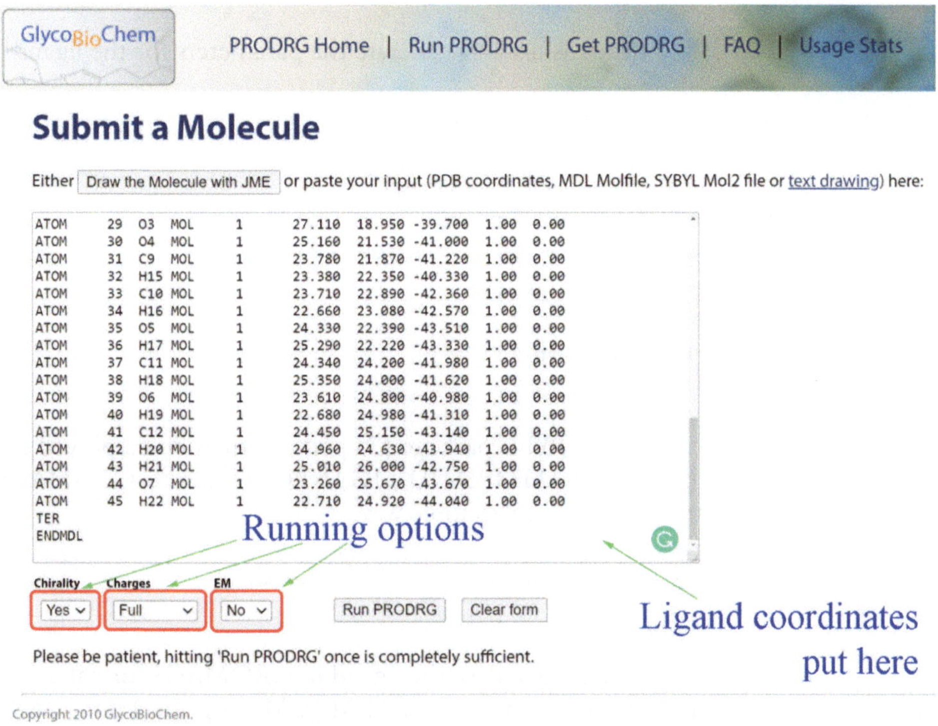

Fig. 3 The example of PRODRG webserver input and the selected parameters

In this context, we will introduce how to use the GROMACS 5.1.5 [29] to simulate the solvated complex, and the ligand parameters would be thus prepared using GROMACS format. In particular, the ligand parameters were finally generated using a combination of AmberTools18 and ACPYPE protocols [30, 31]. Particularly, the inhibitor was characterized by the general Amber force field (GAFF) [32]. The "antechamber" of AmberTools18 was utilized to produce MOL2 and PREP files of the ligand, which are required input files for "tleap" protocol. Examples of Amber commands are shown:

```
antechamber -i lig.out -fi gout -o lig.mol2 -fo mol2 -pf yes -c resp
antechamber -i lig.out -fi gout -o lig.prep -fo prepi -pf yes -c resp
```

Parmchk, a protocol of AmberTools, is thus applied to test all of the parameters:

```
parmchk -i lig.mol2 -f mol2 -o lig.frcmod
```

"tleap" is used to generate the parameters for the ligand:

```
echo "source leaprc.gaff
LIG = loadmol2 lig.mol2
loadamberparams lig.frcmod
loadamberprep lig.prep
saveamberparm LIG LIG.PRMTOP LIG.INPCRD
saveoff LIG lig.lib
quit" > tleap.in
tleap -f tleap.in
```

The ACPYPE package [31] would then be employed to convert LIG.PRMTOP AND LIG.INPCRD files to GROMACS format according to a simple way:

```
acpype.py -p LIG.PRMTOP -x LIG.INPCRD
```

The parameters of the ligand in GROMACS format are generated and stored in two files: MOL_GMX.top and MOL_GMX.gro. We need to delete the first and last six lines of the MOL_GMX.top file before it can be called in the complex topology file.

The N1 neuraminidase and neutralized ions were described via the Amber99SB-ILDN force field [33]. The TIP3P was consumed to feature the water model [34]. It should be noted that the combination of these force fields was selected since it is one of the most appropriate ones for free energy calculations [35, 36]. The N1-ligand system was then placed in a rectangular periodic boundary condition (PBC) box with the size of $7.16 \times 7.74 \times 9.90$ nm. Two Cl^- atoms were added to the solvated system to neutralize the one. The initial conformation with ~54,500 atoms totally of the solvated N1-inhibitor complex is described in Fig. 2.

2.2 Generating Initial Conformation for Umbrella Sampling Simulations Using Steered MD

The GROMACS version 5.1.5 [29] was employed to replicate the solvated N1-inhibitor system. In particular, the MD parameters were chosen following the methodology presented previously [21]. The integration timestep is set to 2 fs. The non-covalent pair affects within the radius of 0.9 nm. The van der Waals (vdW) interaction was cutoff with the same range, while the electrostatic interaction was imitated by the Particle Mesh Ewald method [37]. The solvated complex was primarily minimized with the steepest descent approach. The optimized conformation of the solvated N1-Zanamivir system was then relaxed over canonical (NVT) and isobaric-isothermal (NPT) imitations. The NVT and NPT simulations were of length 0.1 and 2.0 ns, respectively. It should be noted that the C_α atoms were constrained during the NVT and NPT imitations.

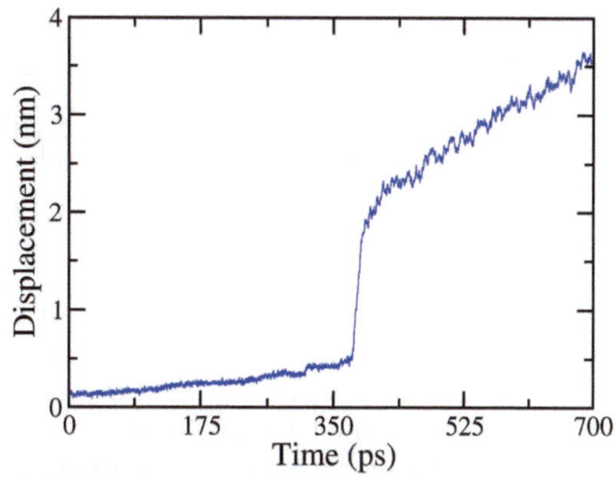

Fig. 4 The displacement of the ligand during FPL calculation

The relaxed shape of the solvated system was utilized as the starting structure of the SMD simulations. The ligand was then turned to be displaced utilizing the pulling code of GROMACS, in which a harmonic force with cantilever spring constant, $k = 600$ kJ mol^{-1} nm^{-2}, was applied on the ligand center of mass to carry out this task. The pulling velocity was chosen as $v = 0.005$ nm ps^{-1} as in the previous study [38]. The SMD simulation was carried out during 0.5 ns, and the systemic coordinates were recorded every 0.1 ps for generating the initial input of US calculations.

2.3 Umbrella Sampling Simulations

The conformations of the N1-ligand system were extracted along with FPL calculations upon the displacement of the ligand along the reaction coordinate, ξ (cf. Figure 4). The structure of N1 complex was recorded at every step of ~0.1 nm and then applied as commencing shapes for US evaluations (Fig. 4). Twenty-one N1 shapes were released from *bound* to *unbound* states along the unbinding pathway. A short NPT simulation with an interval of 0.1 ns was carried out to relax the system first, and the US simulation with an interval of 2 ns was then executed.

2.4 Computation of the Free Energy Profile Using the Potential Mean Force Method

The weighted histogram analysis method (WHAM) [22] was used to calculate a potential of mean force (PMF) value along the unbinding pathway via "gmx wham" scheme [39]. More precisely, the value along the unbinding pathway can be estimated as follows:

$$w(\xi) = -\beta \ln \left[\frac{P(\xi)}{P(\xi_0)} \right] \quad (1)$$

where $\beta = k_B T$, k_B is Boltzmann constant, T is absolute temperature, $P(\xi)$ is the probability density of ξ, and an arbitrary reference point ξ_0 corresponds to $w(\xi_0) = 0$. The value can be computed via the WHAM with the unbiased probability $P(\xi)$ with the implementation of auto-correlated time and robust error is as follows [39]:

$$P(\xi) = \frac{\sum_{i=1}^{N_w} g_i^{-1} h_i}{\sum_{j=1}^{N_w} n_j g_j^{-1} e^{-\beta\left(w_j(\xi) - f_j\right)}} \tag{2}$$

$$e^{-\beta f_j} = \int d\xi e^{-\beta w_j(\xi)} P(\xi) \tag{3}$$

where h_j is the histogram with total values n_j, $g_i = 1 + 2\tau_i$ is the statistical ineffectiveness with an auto-correlated time τ_i analogous to US windows i, and free energy constant f_i.

The PMF profile of N1 + zanamivir system was computed using "gmx wham"—a GROMACS tool with an example of the command:

```
gmx wham -it tpr.dat -if force.dat -unit kCal -ac -nBootstrap
50 -bins 50
```

where *tpr.dat* and *force.dat* are the lists of TPR and pulling force files. The list is described in Table 1.

The binding free energy of zanamivir to neuraminidase N1 can be estimated as the difference in PMF between *bound* and *unbound* regions. However, it is complicated to determine the PMF value in *bound* states (cf. Figure 5). Therefore, we assumed that the ΔG value approximately equals the free energy barrier of the binding, the binding free energy is roughly determined as the difference between the lowest and the largest PMF values as mentioned in Fig. 5 [21]. The obtained binding free energy thus is $\Delta G_{US} = -11.51 \pm 1.05$ kcal mol^{-1}, which is an appropriate result in comparison with the respective experiment $\Delta G_{EXP} = -12.38$ kcal mol^{-1} [40]. Moreover, because the ΔG value is estimated via the difference energy of two points, the calculated binding free energy probably does not depend on the kinetic binding rate.

3 Notes

1. The PRODRUG option *Chirality*, selected as **Yes**, means chiral centers of the input molecule would be maintained.
2. The PRODRUG option *Charges*, selected as **Full**, means atomistic full charges are considered for condensed-phase systems.
3. The PRODRUG option *EM*, selected as **No**, means the energy minimization would not be performed. The ligand coordinates would not be changed during the PRODRG calculation.

Table 1
An example of tpr.dat and force.dat content

No	tpr.dat	force.dat
1	*md_1.tpr*	*fx_1.xvg*
2	*md_2.tpr*	*fx_2.xvg*
3	*md_3.tpr*	*fx_3.xvg*
4	*md_4.tpr*	*fx_4.xvg*
5	*md_5.tpr*	*fx_5.xvg*
6	*md_6.tpr*	*fx_6.xvg*
7	*md_7.tpr*	*fx_7.xvg*
8	*md_8.tpr*	*fx_8.xvg*
9	*md_9.tpr*	*fx_9.xvg*
10	*md_10.tpr*	*fx_10.xvg*
11	*md_11.tpr*	*fx_11.xvg*
12	*md_12.tpr*	*fx_12.xvg*
13	*md_13.tpr*	*fx_13.xvg*
14	*md_14.tpr*	*fx_14.xvg*
15	*md_15.tpr*	*fx_15.xvg*
16	*md_16.tpr*	*fx_16.xvg*
17	*md_17.tpr*	*fx_17.xvg*
18	*md_18.tpr*	*fx_18.xvg*
19	*md_19.tpr*	*fx_19.xvg*
20	*md_20.tpr*	*fx_20.xvg*
21	*md_21.tpr*	*fx_21.xvg*

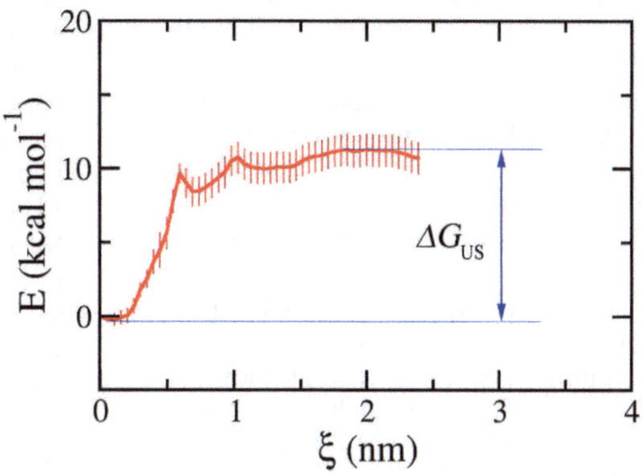

Fig. 5 The free energy profile of Zanamivir unbinding N1 via WHAM calculations

References

1. Woo H-J, Roux B (2005) Calculation of absolute protein–ligand binding free energy from computer simulations. Proc Natl Acad Sci U S A 102:6825–6830. https://doi.org/10.1073/pnas.0409005102
2. Siebenmorgen T, Zacharias M (2019) Evaluation of predicted protein–protein complexes by binding free energy simulations. J Chem Theory Comput 15:2071–2086. https://doi.org/10.1021/acs.jctc.8b01022
3. Yu W, MacKerell AD (2017) Computer-aided drug design methods. In: Sass P (ed) Antibiotics: methods and protocols. Springer New York, New York, NY, pp 85–106
4. Cavasotto CN, Aucar MG, Adler NS (2019) Computational chemistry in drug lead discovery and design. Int J Quantum Chem 119: e25678. https://doi.org/10.1002/qua.25678
5. Decherchi S, Cavalli A (2020) Thermodynamics and kinetics of drug-target binding by molecular simulation. Chem Rev 120 (23):12788–12833. https://doi.org/10.1021/acs.chemrev.0c00534
6. Mobley DL, Gilson MK (2017) Predicting binding free energies: frontiers and benchmarks. Annu Rev Biophys 46:531–558
7. Gapsys V, Perez-Benito L, Aldeghi M, Seeliger D, Van Vlijmen H, Tresadern G et al (2020) Large scale relative protein ligand binding affinities using non-equilibrium alchemy. Chem Sci 11:1140–1152. https://doi.org/10.1039/c9sc03754c
8. Trott O, Olson AJ (2010) Improving the speed and accuracy of docking with a new scoring function, efficient optimization, and multithreading. J Comput Chem 31:455–461
9. Aqvist J, Medina C, Samuelsson J-E (1994) A new method for predicting binding affinity in computer-Aidded drug design. Protein Eng 7:385–391. https://doi.org/10.1063/1.1740409
10. Srinivasan J, Cheatham TE, Cieplak P, Kollman PA, Case DA (1998) Continuum solvent studies of the stability of DNA, RNA, and Phosphoramidate−DNA helices. J Am Chem Soc 120:9401–9409. https://doi.org/10.1021/ja981844+
11. Ngo ST, Hung HM, Nguyen MT (2016) Fast and accurate determination of the relative binding affinities of small compounds to HIV-1 protease using non-equilibrium work. J Comput Chem 37:2734–2742. https://doi.org/10.1002/jcc.24502
12. Kirkwood JG (1935) Statistical mechanics of fluid mixtures. J Chem Phys 3:300–313. https://doi.org/10.1063/1.1749657
13. Zwanzig RW (1954) High-temperature equation of state by a perturbation method. I. nonpolar gases. J Chem Phys 22:1420–1426
14. Jarzynski C (1997) Nonequilibrium equality for free energy differences. Phys Rev Lett 78:2690–2693
15. Ryde U, Soderhjelm P (2016) Ligand-binding affinity estimates supported by quantum-mechanical methods. Chem Rev 116:5520–5566. https://doi.org/10.1021/acs.chemrev.5b00630
16. Marchi M, Ballone P (1999) Adiabatic bias molecular dynamics: a method to navigate the conformational space of complex molecular systems. J Chem Phys 110:3697–3702. https://doi.org/10.1063/1.478259
17. Mark AE, WFV G, HJC B (1991) Calculation of relative free energy via indirect pathways. J Chem Phys 94:3808–3816. https://doi.org/10.1063/1.459753
18. Laio A, Parrinello M (2002) Escaping free-energy minima. Proc Natl Acad Sci U S A 99:12562. https://doi.org/10.1073/pnas.202427399
19. Sugita Y, Okamoto Y (1999) Replica-exchange molecular dynamics method for protein folding. Chem Phys Lett 314:141–151
20. Ngo ST, Nguyen TH, Tung NT, Nam PC, Vu KB, Vu VV (2020) Oversampling free energy perturbation simulation in determination of the ligand-binding free energy. J Comput Chem 41:611–618. https://doi.org/10.1002/jcc.26130
21. Ngo ST, Vu KB, Bui LM, Vu VV (2019) Effective estimation of ligand-binding affinity using biased sampling method. ACS Omega 4:3887–3893. https://doi.org/10.1021/acsomega.8b03258
22. Ferrenberg AM, Swendsen RH (1989) Optimized Monte Carlo data analysis. Phys Rev Lett 63:1195–1198. https://doi.org/10.1103/PhysRevLett.63.1195
23. Ngo ST, Quynh Anh Pham N, Le LT, Pham D-H, Vu VV (2020) Computational determination of potential inhibitors of SARS-CoV-2 Main protease. J Chem Inf Model 60 (12):5771–5780. https://doi.org/10.1021/acs.jcim.0c00491
24. van der Vries E, Collins PJ, Vachieri SG, Xiong X, Liu J, Walker PA et al (2012) H1N1 2009 pandemic influenza virus: resistance of

the I223R neuraminidase mutant explained by kinetic and structural analysis. PLoS Pathog 8: e1002914. https://doi.org/10.1371/journal.ppat.1002914

25. Petrek M, Otyepka M, Banas P, Kosinova P, Koca J, Damborsky J (2006) CAVER: a new tool to explore routes from protein clefts, pockets and cavities. BMC Bioinform 7:316

26. Schüttelkopf AW, van Aalten DMF (2004) PRODRG: a tool for high-throughput crystallography of protein-ligand complexes. Acta Crystallogr D Biol Crystallogr 60:1355–1363. https://doi.org/10.1107/s0907444904011679

27. Bayly CI, Cieplak P, Cornell W, Kollman PA (1993) A well-behaved electrostatic potential based method using charge restraints for deriving atomic charges: the RESP model. J Phys Chem 97:10269–10280

28. Frisch MJ, Trucks GW, Schlegel HB, Scuseria GE, Robb MA, Cheeseman JR, et al. Gaussian~09 Revision E.01. 2016

29. Abraham MJ, Murtola T, Schulz R, Páll S, Smith JC, Hess B et al (2015) GROMACS: high performance molecular simulations through multi-level parallelism from laptops to supercomputers. SoftwareX 1–2:19–25. https://doi.org/10.1016/j.softx.2015.06.001

30. Case DA, Ben-Shalom IY, Brozell SR, Cerutti DS, Cheatham TEC III, Cruzeiro VWD, Darden TA et al (2018) AMBER 18. University of California, San Francisco

31. Sousa da Silva AW, Vranken WF (2012) ACPYPE—AnteChamber PYthon parser interfacE. BMC Res Notes 5:1–8. https://doi.org/10.1186/1756-0500-5-367

32. Wang J, Wolf RM, Caldwell JW, Kollman PA, Case DA (2004) Development and testing of a general amber force field. J Comput Chem 25:1157–1174. https://doi.org/10.1002/jcc.20035

33. Aliev AE, Kulke M, Khaneja HS, Chudasama V, Sheppard TD, Lanigan RM (2014) Motional timescale predictions by molecular dynamics simulations: Case study using proline and Hydroxyproline sidechain dynamics. Proteins 82:195–215. https://doi.org/10.1002/prot.24350

34. Jorgensen WL, Chandrasekhar J, Madura JD, Impey RW, Klein ML (1983) Comparison of simple potential functions for simulating liquid water. J Chem Phys 79:926–935. https://doi.org/10.1063/1.445869

35. Zhang H, Yin C, Jiang Y, van der Spoel D (2018) Force field benchmark of amino acids: I. Hydration and diffusion in different water models. J Chem Inf Model 58:1037–1052. https://doi.org/10.1021/acs.jcim.8b00026

36. Zhang H, Jiang Y, Cui Z, Yin C (2018) Force field benchmark of amino acids. 2. Partition coefficients between water and organic solvents. J Chem Inf Model 58:1669–1681. https://doi.org/10.1021/acs.jcim.8b00493

37. Darden T, York D, Pedersen L (1993) Particle mesh Ewald: an N·log(N) method for Ewald sums in large systems. J Chem Phys 98:10089–10092

38. Mai BK, Viet MH, Li MS (2010) Top-leads for swine influenza a/H1N1 virus revealed by steered molecular dynamics approach. J Chem Inf Model 50:2236–2247

39. Hub JS, de Groot BL, van der Spoel D (2010) g_wham—A Free weighted histogram analysis implementation including robust error and autocorrelation estimates. J Chem Theory Comput 6(12):3713–3720. https://doi.org/10.1021/ct100494z

40. Williams MA, Lew W, Mendel DB, Tai CY, Escarpe PA, Laver WG, et al. (1997) Structure-activity relationships of carbocyclic influenza neuraminidase inhibitors. Bioorganic Med Chem Lett 7(14):1837–1842. https://doi.org/10.1016/S0960-894X(97)00333-8

Chapter 15

Creating Maps of the Ligand Binding Landscape for Kinetics-Based Drug Discovery

Tom Dixon, Samuel D. Lotz, and Alex Dickson

Abstract

Simulations of ligand–protein interactions can be very useful for drug design and to gain biological insight. Full pathways of ligand–protein binding can be used to get information about ligand binding transition states, which form the rate-limiting step of the binding and release processes. However, these simulations are typically limited by the presence of large energy barriers that separate stable poses of interest. Here we describe a simulation protocol for exploring and analyzing landscapes of ligand–protein interactions that makes use of molecular docking, enhanced molecular simulation with the weighted ensemble algorithm, and network analysis. It can be accomplished using a modest cluster of graphics processing units and freely accessible software. This protocol focuses on the construction and analysis of a network model of ligand binding poses and provides links to resources that describe the other steps in more detail. The end result of this protocol is a map of the ligand–protein binding landscape that identifies transition states of the ligand binding pathway, as well as alternative bound poses that could be stabilized with modifications to the ligand.

Key words Molecular dynamics, Network analysis, Ligand binding, Kinetics, Transition paths, Weighted ensemble

1 Introduction

Generally, the binding affinity has been used to indicate a ligand's efficacy. However, this is not always the case. It has recently been shown that residence time (RT) is a better indicator of efficacy than the binding affinity for some systems [1, 2]. However, RT, unlike binding affinity, is a function of the transition path ensemble, making the unbinding pathway critical to designing more efficacious drugs. It is impossible to observe this pathway experimentally at the required level of detail; however, recent studies have shown that molecular dynamics (MD) can be used to simulate protein–ligand interactions and provide hypotheses on how ligands unbind from proteins [3–8].

Here we describe methodology to generate such pathways using a weighted ensemble MD algorithm [9] using a p-xylene-

Quentin Vanhaelen (ed.), *Computational Methods for Estimating the Kinetic Parameters of Biological Systems*, Methods in Molecular Biology, vol. 2385, https://doi.org/10.1007/978-1-0716-1767-0_15,

lysozyme system. Using our method, we were able to simulate unbinding trajectories of this protein–ligand complex. To characterize the free energy landscape generated from the ensemble, we cluster the resulting frames and generate a Markov State Model (MSM). To help understand the pathway, we can calculate observables such as ligand root mean square deviation (RMSD), and committor probabilities across the network. Cluster centers with committor probabilities at or near 0.5 can be identified as transition states of the ligand binding process, as they have an equal probability of reaching the bound or unbound basins. Finally, we can determine structural information for each network node and visualize it on the network using colors. These maps provide a holistic view of the ligand binding landscape allowing for deep insight into the binding and unbinding processes. Examples of ligand binding landscapes, including those that synthesize sampling from multiple starting poses, can be found in previous works [10–15].

This protocol includes an example script for the construction and analysis of a network model of the ligand binding landscape. Readers can follow along with these steps in the attached Python script (lysozyme_csn.py in tutorial_code/source) or download the 9.7 Gb dataset from the Zenodo repository (DOI: https://doi.org/10.5281/zenodo.3995599) and perform the analyses themselves on their local machine. Note that this protocol focuses on analysis of trajectory data, and provides only a brief description of the procedure for setting up and running the protein–ligand simulations. Typically, enhanced sampling algorithms for molecular dynamics are required to generate trajectories of ligand binding and/or unbinding, and we recommend using one of the weighted ensemble-based algorithms included in the wepy [16] package (e.g., WExplore [17] or REVO [18]). More detailed information on conducting simulations with wepy, including worked examples, can be found in Ref. [19].

2 Setup

Here we provide details for installing the software necessary for creating ligand binding landscapes following this protocol. Note that these steps work as described at the time of publication and might need to be modified in the future (*see* **Note 1**). We recommend using a Conda or Miniconda package management system [20], to setup a new conda environment with necessary tools, using python version 3.7:

```
> conda create --name LBmaps python=3.7
> conda activate LBmaps
```

Note that lines beginning with ">" indicate commands that should be typed into a UNIX-style command prompt. The necessary programs can then be installed inside the LBmaps environment:

```
> pip install wepy
> conda install -c omnia openmm
> pip install openmm_systems
> pip install git+https://github.com/ADicksonLab/CSNAnalysis@v0.5
> conda install scikit-learn
> conda install -c conda-forge backports.functools_lru_cache
```

3 Methods

1. Prepare one or more ligand-bound poses. This is typically done using either existing ligand poses from X-ray crystal structures or NMR, or by docking the ligand to a protein of interest (*see* **Note 2**). The researcher could start with a single pose, or alternatively could prepare a set of structures with large differences in ligand orientation that would not be expected to quickly interconvert along an MD trajectory. In the attached example, we use a single starting pose of p-xylene bound to T4 lysozyme.
2. Obtain force field parameters for the ligand that are compatible with the protein forcefield. In this example, we used AMBER ff96 to parameterize the protein and GAFF [21] to parameterize the ligand (*see* **Note 3**).
3. Prepare system for simulation. This includes minimizing the energy of the system as well as equilibrating the system to the desired starting conditions (temperature, pressure, or box volume, etc.) before running trajectory segments to analyze and calculate observables (*see* **Note 4**).
4. Run a series of unbiased trajectory segments that sample different regions of the ligand (un)binding pathway. (While this can be done using a variety of methods, in the example below the simulations were of weighted ensemble type and that the data is stored in a WepyHDF5 file format) (*see* **Note 5**). Here we briefly describe the protocol for running ligand unbinding simulations with a bin-less weighted ensemble algorithm called REVO ("Resampling Ensembles by Variation Optimization"). Please refer to the previous work [18, 19] for more information on running ligand unbinding simulations with REVO.
   (a) *Choosing a resampler and a distance metric:* The resampler is the algorithm that determines how walkers are merged and cloned. In our example, we use the REVO resampler

[12, 15, 18], a bin-less resampler that uses ligand RMSD between walkers as a distance metric (*see* **Note 6**). The resampler encourages cloning of walkers that are the farthest on average from all other walkers and merges the closest walkers together.

(b) *Boundary conditions:* We need to set up a boundary to determine when a walker has done the rare event of interest. Once a walker has crossed the boundary, we reset the walker back to the simulation's initial positions, which is called warping. The warped walker maintains its weight and continues performing resampling and dynamic steps to find new unbinding pathways. In our simulations, we use a 10 Å minimum distance between protein and ligand as the unbinding boundary condition which is consistent with the non-bonded force cutoff used in the simulation.

(c) *Number of walkers:* Having more walkers is beneficial because we have more opportunities to climb large energy barriers. However, the number of walkers is also proportional to the computational cost of the simulation. The simulations here use 48 walkers which are distributed across eight GPUs, which is in line with previous work [4, 22].

(d) *Minimum and maximum probabilities:* We use a maximum walker probability (p_{max}) to prevent the accumulation of weight in a single walker, and a minimum walker probability (p_{min}) to prevent computational time being wasted on trajectories that will not contribute significantly to observables we want to calculate. Here we use a p_{max} of 0.5 and a p_{min} of 10^{-12}.

(e) *Number of dynamic steps before resampling:* There should be enough dynamic steps between cycles so the trajectories can significantly diverge, but not so many that you could miss important fluctuations that could lead to barrier crossings. We find that 20 ps cycle time is a good compromise between these objectives.

5. Compute features that will be used for clustering (*see* **Note 7**). A commonly used set of features to describe ligand (un)-binding is a large list of distances between ligand and protein atoms (*see* **Note 8**). The attached example computes all possible distances between 18 ligand atoms and 470 protein binding site atoms—defined as those within an 8 Å proximity to any ligand atom in the starting pose—for a total of 8460 features.
6. Cluster the featurized data. Here we use a k-means clustering algorithm with 500 clusters (*see* **Note 9**).
7. Save the cluster assignments for each structure. These are attached to each structure in the WepyHDF5 database object using the `add_observable` function in wepy.

8. Calculate a counts matrix describing the number of transitions between the clusters. This is done using a lag time that defines the size of the time interval for each transition (*see* **Note 10**). For branched data structures, this must be done carefully so as to properly determine the beginning and ending points of each transition (*see* **Note 11**). The attached example uses wepy to define a `ContigTree` object that contains information needed to make the branched tree such as merging and cloning. This object is then used in tandem with the cluster identities to make the counts matrix using the `MacroStateNetwork` object.
9. Make a conformation space network using CSNAnalysis (*see* **Note 12**). This is initialized using the counts matrix generated by the `MacroStateNetwork`.
10. Calculate observables that are averaged over all frames in a given cluster (*see* **Note 13**). This is done by writing a function that calculates the observable of interest for all of the structures in all of the trajectories using the `network.wepy_h5.get_trace_fields` method. In our example we calculated: (1) free energy, (2) ligand RMSD to the starting structure, and (3) the minimum protein–ligand distance for each cluster.
11. Identify clusters that are members of the bound and unbound basins. In the example, we use ligand RMSD to the starting structure for all frames and use a cutoff of 5 Å to identify bound states. The unbound states are defined by the minimum protein–ligand distance, with a cutoff of 10 Å.
12. Determine committor probabilities between the bound and unbound basins for each cluster. This is done using the `calc_committors` function that is built into `csn` objects in CSNAnalysis and a list of lists containing the bound and unbound cluster labels.
13. Determine the steady state weights for each cluster. This is the sum of the weights of all trajectories that are assigned to a given cluster, normalized such that the weights all sum to 1. This is determined using the `MacroStateNetwork` `set_macrostate_weights` function.
14. Add observables to the MacroStateNetwork. This step uses the `network.set_nodes_attribute` function and helps visualize observables on the network. It requires a string to label the attribute and a dictionary containing the values to give each node. The attached example adds the ligand RMSD to the starting structure ("rmsd"), the minimum distance between the protein and ligand atoms ("min_dist"), the negative log probability of each state ("-log(p)"), and the committor probability to the unbound state.

15. Export the csn graph object to a file in GEXF format. This uses the `MacroStateNetwork` method `write_gexf`. It requires a string that gives the file path.
16. Open the GEXF file for visualization in the Gephi program.
17. Determine the layout of the network. Run ForceAtlas2 under the layout window in Gephi to visualize the graph topology. Set the node sizes to use the steady state weights. We used a maximum value of 50 and a minimum value of 2 for the node sizes.
18. Visualize observables on the graph as colors. Figure 1 shows four different network colorings using ligand RMSD to starting structure (top left), minimum protein–ligand distance (top right), −log(p) (bottom left), and the unbinding committor probability (bottom right).
19. Choose nodes of interest to visualize. Network representations are useful to identify states of interest. Figure 2 shows particular states along the unbinding pathway that were identified from the network.

4 Notes

1. This example points to specific versions of python packages that are currently supported as of the date of publication. This support could change in the future, and we encourage the readers to refer to the latest documentation for the wepy software as well as other related tools (OpenMM, CSNAnalysis).
2. If using existing structures, high-resolution (below 2.5 Å) structures give a better initial starting point to launch MD trajectories. Caution should be taken when interpreting the binding poses from lower low resolution structures.
3. Force field parameters can be quickly generated using a matching algorithm such as CGenFF [23] or GAFF [24]. These algorithms also predict the uncertainty in the assigned partial charges, bond parameters, and torsions. Caution, or additional molecular-specific QM calculations, should be undertaken for ligands with large uncertainties.
4. Periodic boundaries for the system should be large enough to allow the ligand to diffuse away from the protein and not simultaneously interact with multiple copies of the protein.
5. A large number of software packages are available to simulate molecular dynamics trajectories. In this example, we use OpenMM [25, 26] with the CUDA platform to run trajectories on graphics processing units. OpenMM is called as a Python library by the Wepy simulation toolkit.

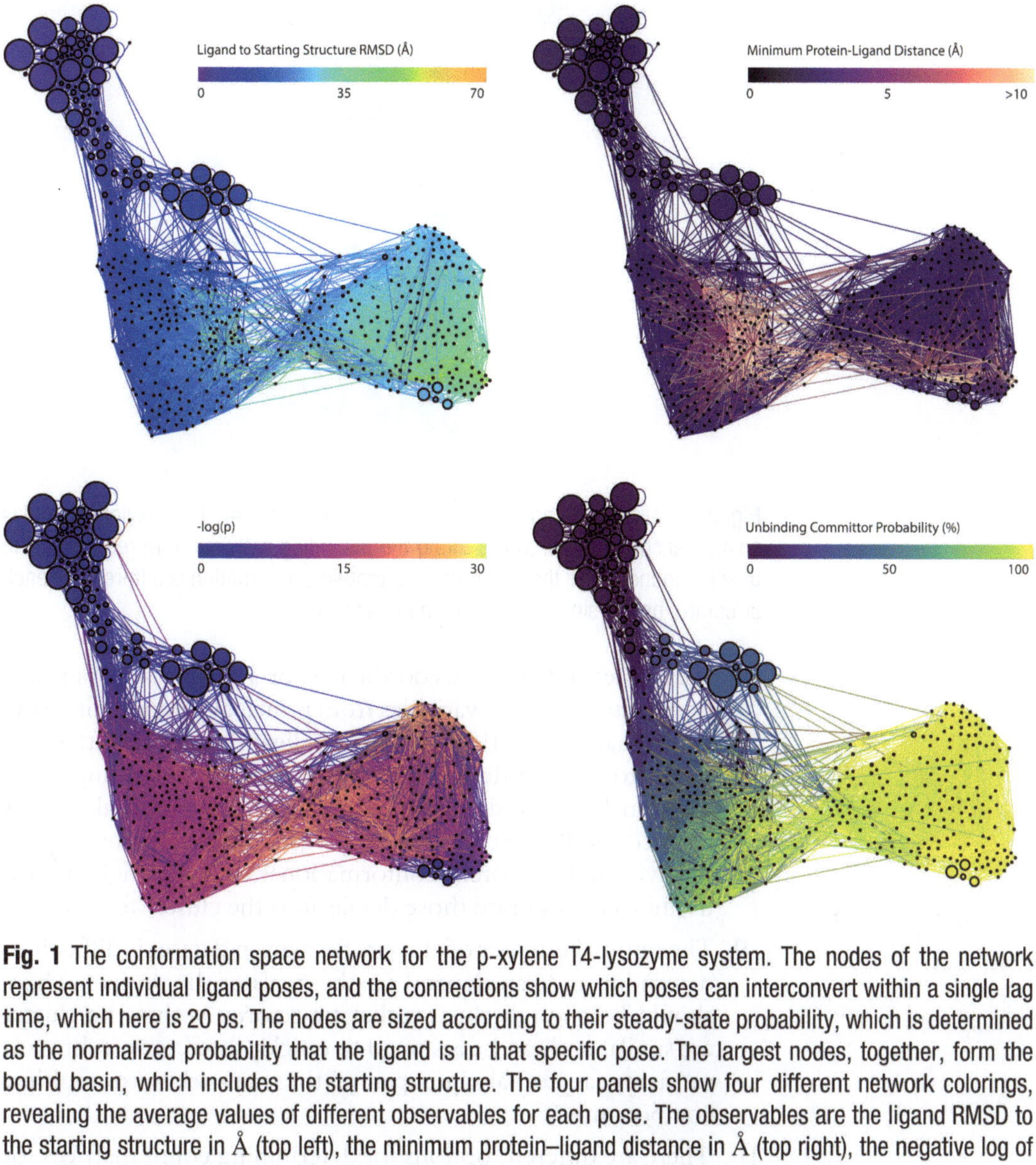

Fig. 1 The conformation space network for the p-xylene T4-lysozyme system. The nodes of the network represent individual ligand poses, and the connections show which poses can interconvert within a single lag time, which here is 20 ps. The nodes are sized according to their steady-state probability, which is determined as the normalized probability that the ligand is in that specific pose. The largest nodes, together, form the bound basin, which includes the starting structure. The four panels show four different network colorings, revealing the average values of different observables for each pose. The observables are the ligand RMSD to the starting structure in Å (top left), the minimum protein–ligand distance in Å (top right), the negative log of the node probability (bottom left), and the unbinding committor probability (bottom right)

6. The distance metric between walkers should compute meaningful differences between the ligand-protein conformations of walkers and be able to identify walkers for which the desired rare event is more likely to occur. For ligand unbinding, there are several suitable distance metrics such as ligand–protein atomic distances and RMSD between ligands after alignment to the binding site.
7. Care should be taken with regard to periodic boundary conditions when computing features.
8. Cluster features should be chosen to describe the phenomena of interest. For instance, a study of protein dynamics could use

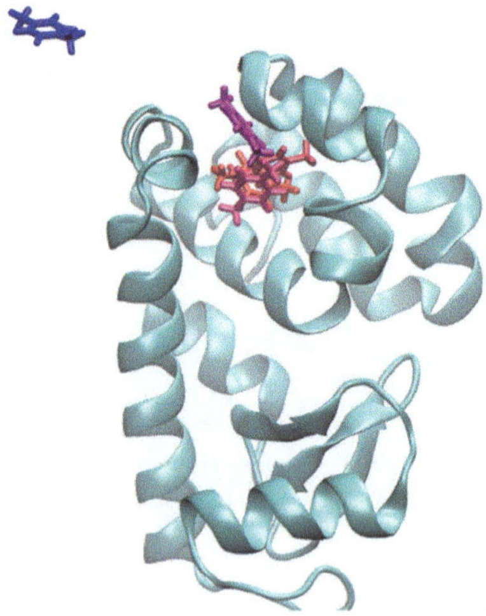

Fig. 2 Visualization of poses along the unbinding pathway. The p-xylene ligand is colored according to progress along the unbinding pathway from red (bound) to blue (unbound). Note that the lysozyme protein conformation is different for each structure, only a single conformation is shown

the full set of Cα x–y–z coordinates, or a smaller set of projections onto collective variables from principal component analysis. We have found that a large set of ligand–protein distances is useful to capture details relevant to ligand (un)binding pathways, such as ligand position, orientation, and internal degrees of freedom. Researchers studying ligand binding pathways that involve specific protein conformational changes might find it useful to incorporate those details into the cluster features.

9. The optimal number of clusters for a given dataset is difficult to determine and can depend on the goals of the project. There should be enough clusters that trajectories within the cluster are similar and can be assumed to quickly interconvert, but not so many as to falsely separate the network into multiple components.
10. There are different options for determining count matrices for weighted ensembles. One could simply count the number of transitions between the two states, but it is more accurate to make a weighted counts matrix using walker weights.
11. To combine multiple WepyHDF5 datasets, the researcher should cluster all data concurrently, make a separate counts matrix for each dataset, then add each counts matrix to combine for one overall counts matrix.
12. When building the csn object with CSNAnalysis, symmetrization of the transition counts (e.g., $c^s_{ij} = c^s_{ji} = 1/2\left(c_{ij} + c_{ji}\right)$) is not recommended as the system is typically out-of-equilibrium.
13. Computing observables for clusters can be done by calculating observables for each frame and averaging over all frames in a

cluster. The attached script provides examples of calculating average ligand RMSD and the minimum protein–ligand distances using network.node_fields_map. Once again, we caution to make sure that boundary conditions are properly considered during analysis.

References

1. Copeland RA (2016) The drug-target residence time model: a 10-year retrospective. Nat Rev Drug Discov 15:87–95. https://doi.org/10.1038/nrd.2015.18
2. Costa B, Da Pozzo E, Giacomelli C et al (2016) TSPO ligand residence time: a new parameter to predict compound neurosteroidogenic efficacy. Sci Rep 6:18164. https://doi.org/10.1038/srep18164
3. Uyar A, Karamyan VT, Dickson A (2018) Long-range changes in Neurolysin dynamics upon inhibitor binding. J Chem Theory Comput 14:444–452. https://doi.org/10.1021/acs.jctc.7b00944
4. Lotz SD, Dickson A (2018) Unbiased molecular dynamics of 11 min timescale drug unbinding reveals transition state stabilizing interactions. J Am Chem Soc 140:618–628. https://doi.org/10.1021/jacs.7b08572
5. Pan AC, Borhani DW, Dror RO, Shaw DE (2013) Molecular determinants of drug-receptor binding kinetics. Drug Discov Today 18:667–673. https://doi.org/10.1016/j.drudis.2013.02.007
6. Casasnovas R, Limongelli V, Tiwary P et al (2017) Unbinding kinetics of a p38 MAP kinase type II inhibitor from Metadynamics simulations. J Am Chem Soc 139:4780–4788. https://doi.org/10.1021/jacs.6b12950
7. Capelli R, Carloni P, Parrinello M (2019) Exhaustive search of ligand binding pathways via volume-based Metadynamics. J Phys Chem Lett 10:3495–3499. https://doi.org/10.1021/acs.jpclett.9b01183
8. Kokh DB, Amaral M, Bomke J et al (2018) Estimation of drug-target residence times by τ-random acceleration molecular dynamics simulations. J Chem Theory Comput 14:3859–3869. https://doi.org/10.1021/acs.jctc.8b00230
9. Huber GAGA, Kim S (1996) Weighted-ensemble Brownian dynamics simulations for protein association reactions. Biophys J 70:97–110. https://doi.org/10.1016/S0006-3495(96)79552-8
10. Huang D, Caflisch A (2011) The free energy landscape of small molecule unbinding. PLoS Comput Biol 7:e1002002. https://doi.org/10.1371/journal.pcbi.1002002
11. Dickson A (2018) Mapping the ligand binding landscape. Biophys J 115:1707–1719. https://doi.org/10.1016/j.bpj.2018.09.021
12. Dixon T, Lotz SD, Dickson A (2018) Predicting ligand binding affinity using on- and off-rates for the SAMPL6 SAMPLing challenge. J Comput Aided Mol Des 32:1001–1012. https://doi.org/10.1007/s10822-018-0149-3
13. Kirberger SE, Ycas PD, Johnson JA et al (2019) Selectivity, ligand deconstruction, and cellular activity analysis of a BPTF bromodomain inhibitor. Org Biomol Chem 17:2020–2027. https://doi.org/10.1039/c8ob02599a
14. Shukla D, Hernández CX, Weber JK, Pande VS (2015) Markov state models provide insights into dynamic modulation of protein function. Acc Chem Res 48:414–422. https://doi.org/10.1021/ar5002999
15. Dixon T, Uyar A, Ferguson-Miller S, Dickson A (2020) Membrane-Mediated Ligand Unbinding of the PK-11195 Ligand from TSPO. Biophys. J 120:158-167. https://doi.org/10.1016/j.bpj.2020.11.015
16. Dickson A, Lotz S, Uyar A, et al (2018) Wepy. Github Repository, Available at: https://github.com/ADicksonLab/wepy
17. Dickson A, Brooks CL (2014) WExplore: hierarchical exploration of high-dimensional spaces using the weighted ensemble algorithm. J Phys Chem B 118:3532–3542. https://doi.org/10.1021/jp411479c
18. Donyapour N, Roussey NM, Dickson A (2019) REVO: resampling of ensembles by variation optimization. J Chem Phys 150:244112. https://doi.org/10.1063/1.5100521
19. Lotz SD, Dickson A (2020) Wepy: a flexible software framework for simulating rare events

with weighted ensemble resampling. ACS Omega 5(49):31608–31623

20. Anon, 2017. Anaconda Software Distribution, Anaconda Inc. Available at: https://docs.anaconda.com/

21. Wang J, Wang W, Kollman PA, Case DA (2006) Automatic atom type and bond type perception in molecular mechanical calculations. J Mol Graph Model 25:247–260. https://doi.org/10.1016/j.jmgm.2005.12.005

22. Dickson A, Lotz SD (2017) Multiple unbinding pathways and ligand-induced destabilization revealed by WExplore. Biophys J 112:620–629

23. Vanommeslaeghe K, Hatcher E, Acharya C et al (2010) CHARMM general force field (CGenFF): a force field for drug-like molecules compatible with the CHARMM all-atom additive biological force fields. J Comput Chem 31:671–690. https://doi.org/10.1002/jcc.21367.CHARMM

24. Wang J, Wolf RM, Caldwell JW et al (2004) Development and testing of a general Amber force field. J Comput Chem 25:1157–1174. https://doi.org/10.1002/jcc.20035

25. Eastman P, Pande VS (2010) OpenMM: a hardware-independent framework for molecular simulations. Comput Sci Eng 12:34–39. https://doi.org/10.1109/MCSE.2010.27

26. Eastman P, Swails J, Chodera JD et al (2017) OpenMM 7: rapid development of high performance algorithms for molecular dynamics. PLoS Comput Biol 13:1–17. https://doi.org/10.1371/journal.pcbi.1005659

Chapter 16

Prediction of Protein–Protein Binding Affinities from Unbound Protein Structures

Alberto Meseguer, Patricia Bota, Narcis Fernández-Fuentes, and Baldo Oliva

Abstract

Proteins are the workhorses of cells to carry out sophisticated and complex cellular processes. Such processes require a coordinated and regulated interactions between proteins that are both time and location specific. The strength, or binding affinity, of protein–protein interactions ranges between the micro- and the nanomolar association constant, often dictating the molecular mechanisms underlying the interaction and the longevity of the complex, i.e., transient or permanent. In consequence, there is a need to quantify the strength of protein–protein interactions for biological, biomedical, and biotechnological applications. While experimental methods are labor intensive and costly, computational ones are useful tools to predict the affinity of protein–protein interactions. In this chapter, we review the methods developed by us to address this question. We briefly present two methods to comprehend the structure of the protein complex derived by either comparative modeling or docking. Then we introduce BADOCK, a method to predict the binding energy without requiring the structure of the protein complex, thus overcoming one of the major limitations of structure-based methods for the prediction of binding affinity. BADOCK utilizes the structure of unbound proteins and the protein docking sampling space to predict protein–protein binding affinities. We present step-by-step protocols to utilize these methods, describing the inputs and potential pitfalls as well as their respective strengths and limitations.

Key words Protein–protein interactions, Protein docking, Binding affinity, Protein structures, Protein interfaces

1 Introduction

1.1 Protein–Protein Interactions as Basis of Cellular Functions

Almost, if not all, cellular processes, and thus all the mechanisms that make life possible, are underpinned by highly coordinated and regulated communications between proteins. Protein–protein interactions (PPIs) are both space and time dependent. Two proteins will not be able to interact unless they find each other, and the time span of the interaction will be dictated, principally, by the strength of such association otherwise known as its binding affinity [1, 2]. The determination of protein binding affinities and analysis

Quentin Vanhaelen (ed.), *Computational Methods for Estimating the Kinetic Parameters of Biological Systems*, Methods in Molecular Biology, vol. 2385, https://doi.org/10.1007/978-1-0716-1767-0_16,

of the PPIs are therefore challenges that impact different fields of research within the life and biomedical science and biotechnology where PPI is the subject of study [3].

Determining the binding affinity of PPIs by experimental means can be achieved with biophysical methods such as isothermal titration calorimetry [4], surface plasmon resonance [5], fluorescence [6, 7], and stopped-flow analyses [8]. These approaches are usually time consuming and costly [3, 9]. Moreover, these different experimental methods can generate affinity measures which might vary as the estimation of binding affinity is dependent on the method and on the experimental context [10]. Therefore, computational approaches can help, either by providing a useful guidance on the context of the interactions or by giving an approximation of the affinity. The computational methods are classified into two large groups according to the type of information required as an input. On the one hand, there are sequence-based methods. These methods rely on mathematical and machine-learning models to infer the binding affinity from the protein primary structure [11, 12]. On the other hand, there are structure-based methods, which require the 3D coordinates of the protein complex to estimate the binding affinity [10, 13, 14]. Structure-based methods are more accurate, but these methods are limited by the fact that the structures of the complexes are known for a few numbers of interactions. To overcome this limitation, we developed a program, called BADOCK [15], that takes advantage of the accumulated knowledge on the role played by non-interface regions in the formation of the protein complexes.

1.2 Relevance on Nonnative Binding Interface Regions in the Formation of Protein Complexes

It has been estimated that the concentration of proteins in a cell goes from 2 to 4 million proteins per cubic micron [16]. In such a highly crowded environment, proteins are continually in contact with each other. This increases the likelihood of nonspecific interactions between proteins [17–20]. A paradox in protein–protein interactions is how the unbound proteins of a binary complex recognize each other in such a crowded environment and how they find their best interacting interface in a short period of time [21]. If finding the best interacting interface were the result of protein collisions with only one successful outcome (i.e., the native conformation of the binary complex), the formation of PPIs would require a large amount of time [21]. Nevertheless, interacting proteins find each other and form a complex in a relatively short time, and this is another paradox to solve after Levinthal's paradox in protein folding [22]. Recent works have proposed a plausible solution to this paradox by considering that all faces of the protein surfaces, and not only their binding sites, contain information to distinguish binding from nonbinding partners [15, 21, 23, 24].

First, it has been shown that distributions of docking poses can be used to distinguish interacting proteins from noninteracting proteins [15, 23]. Second, it was demonstrated that protein surfaces are under evolutionary pressure to recognize efficiently their binding partners (also known as positive design) and to avoid nonbinding partners (also known as negative design) [25–29]. This is in line with the work of Schweke et al. [24], proving that surface propensity to interact with ligands is shared by homologous proteins. Finally, it has been shown that protein-binding sites have similar features [30–33]: they tend to be flat, hydrophobic surfaces and provide a nonspecific binding patch that is suitable for many ligands. This suggests that binding sites may not be enough to define the specificity ruling protein–protein recognition.

All these findings agree with the funnel-like intermolecular energy landscape used to describe PPIs [34, 35]. According to this model, PPIs reach their native-state conformation after going through several intermediate conformations [2]. These conformations have lower values of free energy if they get close to the native state, resembling the energy funnel model for protein folding [36]. Therefore, albeit intermediate conformations involve nonnative assembly regions, they are still relevant for achieving the native conformation.

1.3 BADOCK: Estimating Binding Affinity from Unbound Proteins

The availability of structures of PPIs is scarce; consequently, in silico methods to predict binding affinities that require the structure of the complex cannot be widely used in many interactions. We developed BADOCK [15] to solve this problem, a computational method to predict the binding affinity of a PPI using the unbound structures of the partners. Specifically, we studied the formation and binding affinity of binary complexes of globular soluble proteins. BADOCK uses the results of docking to generate an ensemble of interacting poses between two proteins. The docking poses are generated with PatchDock (using the default parameters) [37], refined and rescored with FiberDock [38], and it includes additional scores from statistical potentials [39]. In most poses, the proteins interact through nonbinding sites rather than the correct binding interface. We used the poses resulting from the protein–protein docking search to scout the conformational space of potential encounter complexes. The analysis of this study proved the feasibility to predict binding affinity with the unbound protein structures. We classified the poses of the docking and affinity benchmark datasets [40] into four classes, depending on the orientation of the binding sites of the protein partners: two productive (near-native and face-face) and two nonproductive (face-back and back-back) (Fig. 1).

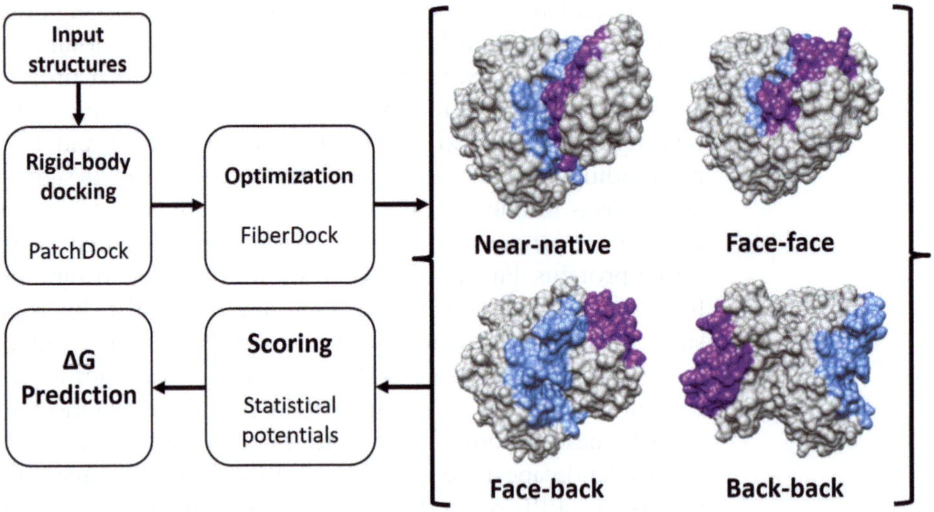

Fig. 1 BADOCK workflow. BADOCK takes as input two unbound structures. The method uses PatchDock to sample by rigid-body docking the interaction space. Docking poses are optimized using FiberDock. All docking poses are scored and used to predict ΔG. The orientations of the poses are classified as: (1) Near-Native, docking poses with a ligand-RMSD <10 Å [15]; (2) Face-Face, poses where binding sites of both protein-partners face each other (i.e., they are inaccessible to other proteins); Face-Back, poses where only one binding site interacts with the protein partner (i.e., the binding site of only one of the proteins is accessible; and Back-Back, when both binding sites are accessible to interact with other protein units (*see* details in [15])

Interestingly, the scores in the four classes of poses were correlated: low binding affinities in the native pose corresponds to the low energy-binding scores in the other classes, while high binding affinities in native conformations also correspond to large energy-binding scores in the other classes. Because of this feature, we could compare binding affinities of different PPIs using the distribution of scores from their docking space search, without relying on our knowledge of the native or near-native conformations. Using this principle, BADOCK applies a linear regression model to predict the difference of free energy, ΔG, between the binary complex and the two unbound proteins, using any of the docking scores. The difference of free energy is also known as binding free energy, and we will use it along this chapter as a measure of the binding affinity between two proteins. By default, the method uses the ES3DC statistical potential (the name of this potential was defined in Ref. 31) because this had the best correlation with the affinities in the benchmark [40]. The method was tested using a tenfold validation protocol in the affinity benchmark [40], achieving an average Pearson's correlation of 0.36 ($P < 0.05$) with an error around 2.84 kcal/mol [15].

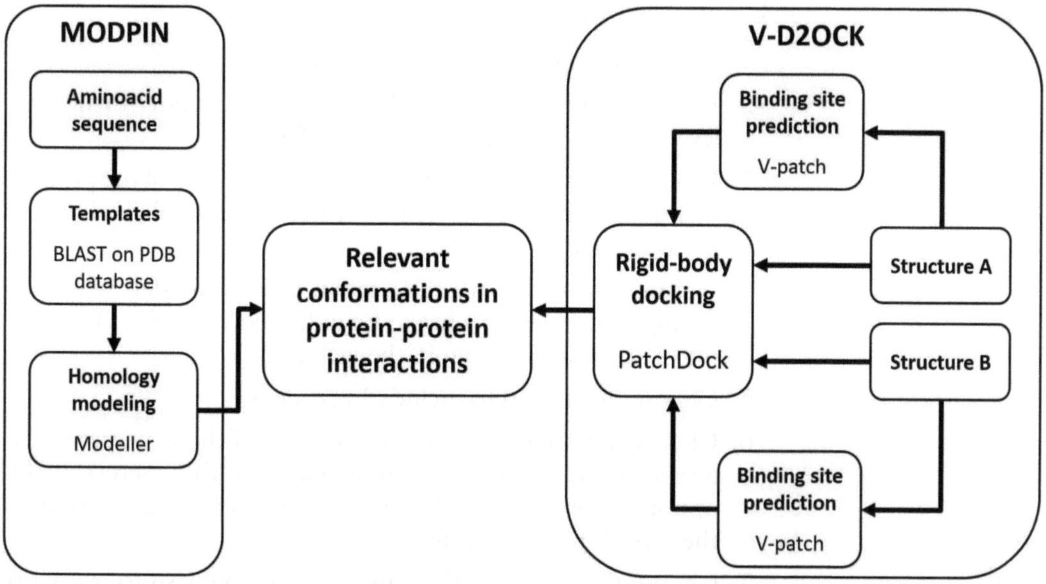

Fig. 2 MODPIN and V-D2OCK workflows. Both workflows produce conformations of PPIs. MODPIN uses the amino acid sequence of both partners as input, and it uses BLAST to search homologs of both sequences with a known structure of the interaction (PPI templates). One or more structural models are obtained with MODELLER [67] for each PPI template. V-D2OCK uses two unbound structures as input, predicts the interface of their interaction, and applies a rigid-body docking with PatchDock restricted to the predicted interface

2 Methods

2.1 Predicting the Structure of Protein Complexes by Comparative Modeling (MODPIN)

2.1.1 Background

The binding energy of PPIs can be predicted from the structures of their complexes, and we have discussed several methods that are available to carry out such calculations. If the structure of the protein complex is not available, several methods [41, 42] can be used to obtain a structural model by comparative modeling [43]. We have also developed a method, MODPIN, to model several conformations for each protein–protein interaction [44] (Fig. 2). This method can be used to explore the conformational space of an interaction and study the corresponding protein association [45, 46]. Since PPIs can have different conformations, a set of models in the form of an ensemble of conformations is more representative [47, 48]. MODPIN provides a range of metrics to account for the strength or quality of the interaction in the form of energy or docking scores [15, 40, 49, 50]. We have used MODPIN to assess the effect of mutations on the binding energy of complexes [51, 52]. Structural models of macro-complexes have been used by other authors to predict the binding affinity using a large variety of methods [53, 54]. However, in our study, we used MODPIN to understand and improve the prediction by our approach, BADOCK, without having to use the complex structure (*see* in Subheading 2.3).

2.1.2 Protocol

Using the server

1. Go to http://galaxy.interactomix.com.
2. In the section INTERACTOMIX TOOLS under Structural-level analyses, you can access MODPIN.
3. Upload a text file with the amino acid sequences of the proteins you want to model in FASTA format. An example is shown in the web page.
4. Upload a text file with the protein pairs to model. In case of testing mutations, this has to be indicated in the file. An example of an input is shown in the web page.
5. Choose the option to include hydrogens.
6. Choose the option to check the quality of the structure of the interaction (i.e., the structure is relaxed with Rosetta). If this option is not selected, the modeling is fast, but the quality of the model may be affected.
7. Choose the option to force modeling if you wish to overwrite the previous models.
8. Choose the option to include templates from the 3DiD database.
9. Choose the number of models to be obtained per template.
10. Choose the option to renumber the structure according to the amino acid sequence; it starts with 1 otherwise.
11. Submit your query.
12. The execution returns a set of structural models of PPIs.
13. To further analyze the models, the user is invited to the "Analysis" tool window. In this window, the user is requested to upload the input files as before. The program assigns the structural models and scores the interactions by docking potentials or physical-chemical energies (the program uses the docking scores of ZRANK by default [55]). Other available scoring methods are: InterfaceAnalyzer of Rosetta [56], FoldX [49], and Statistical Potentials such as ES3DC [39, 57].
14. Structural models with similar interface are clustered, and the server returns the files with the structures and scores of each cluster.

Local installation:

MODPIN can also be installed locally.

The scripts are available at https://github.com/structuralbioinformatics/MODPIN.

2.2 Predicting the Structure of Protein Complexes by Docking-Based Approaches (V-D2OCK)

2.2.1 Background

When neither the structure of the complex nor an acceptable template is available, the structure of the protein complex can be inferred using docking-based approaches [58]. Docking consists on modeling the structure of the complex starting from the structure of each separated molecule [37]. Docking algorithms produce many conformations (i.e., docking poses) that are subsequently scored to find the most preferable conformation(s) [59]. Wass et al. showed that the sets of docking poses could be used to distinguish interacting from noninteracting proteins using the distribution of docking scores [23]. Therefore, docking scores provide indications of the binding affinity between two proteins. In a previous work, we developed V-D2OCK (VORFFIP-driven DOCKing), a data-driven docking algorithm [60] that uses VORFFIP [61] to identify and delineate the likely interfaces of unbound proteins and direct the docking. V-D2OCK returns a list of possible conformations of protein complexes ranked using three different scoring functions (see below). Although these scores cannot be directly used as binding energies or affinities, they can be used nonetheless as qualitative measures to select the most likely conformation(s) and subsequently used to understand and complement the prediction of the binding energy with BADOCK. V-D2OCK's workflow is shown in Fig. 2.

2.2.2 Protocol

1. Go to http://galaxy.interactomix.com/.
2. In the section INTERACTOMIX TOOLS under Structural-level analyses, you can access V-D2OCK.
3. V-D2OCK requires two PDB coordinate files to perform the docking. Protein A will remain static during the docking. Submit both proteins.
4. Select the Scoring functions. Three choices are available to users: Patchdock score that takes into account geometrical features [37]; ZRANK scoring function that combines both physics-based and knowledge-based terms [55]; and (iii) ES3DC scoring function based on coarse-grained knowledge-based potentials [39].
5. Select the tolerance of the clustering algorithm.
6. Submit the query.
7. The server will return a compressed file (.tar) with the information on the rotation and translation matrices and a file with the scores for each of the docking poses.

2.3 Estimation of Binding Affinities from Unbound Proteins: BADOCK

2.3.1 Background

As discussed in the Introduction, BADOCK is a method that predicts the binding affinity of binary protein–protein interactions by means of the scores after sampling the docking space with the structures of the unbound proteins. Decoys for the docking simulation are binned in four classes depending on the orientation of their respective binding sites. The correlation between energetic terms and docking scores is then used to derive a linear regression model to predict the binding affinity. BADOCK's workflow is shown in Fig. 1.

2.3.2 Protocol

Using the server:

1. Go to http://sbi.upf.edu/BADOCK.
2. Upload two PDB files of the unbound protein partners (Fig. 3a). These structures may correspond to a multidomain structure or the user can split the structure by domains and test separated domains at a time.
3. Bookmark the page for later accession or wait until the program finishes (Fig. 3b).
4. Ultimately the server will return the predicted binding affinity (Fig. 3c).
5. By searching in the results page with the job ID, you can access to the BADOCK results corresponding to executions (Fig. 3d).
6. BADOCK is also accessible through the Galaxy InteractoMIX platform at http://galaxy.interactomix.com/tool_runner?tool_id=interactomix_BADOCK.

Local installation:

The scripts and parameters of BADOCK are available at: https://github.com/BADOCKsbi/BADOCK [15].

3 Notes

1. In the case the user attempts to model protein complexes with multiple domains but no suitable templates can be found for all the domains involved, it is suggested to divide the different monomers in domains (e.g., using PFAM [62]) and try the prediction with them (*see* further in **Notes 6** and 7).
2. V-D2OCK server can be accessed through the two following URLs: http://galaxy.interactomix.com/ or http://www.bioinsilico.org/VD2OCK. The latter provides richer 3D visualization capabilities and docking poses are visualized with a Jmol applet.
3. Unless there is an error with the input files of the coordinates of the proteins (i.e., usually this occurs if the input files are not in any of the standard PDB formats), V-D2OCK and BADOCK

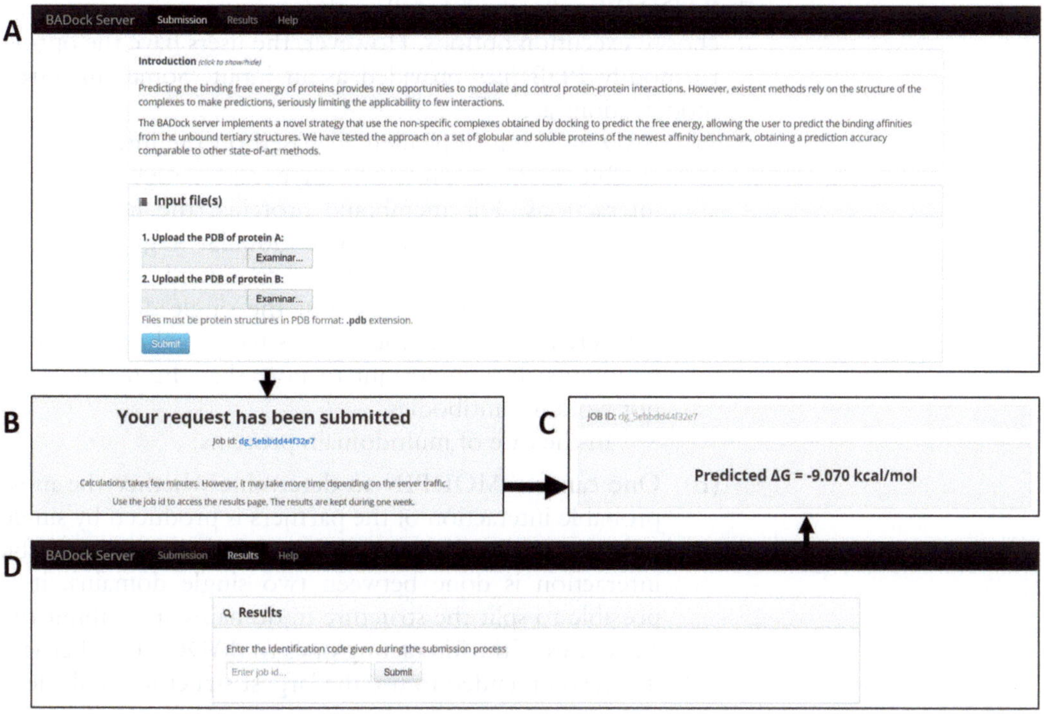

Fig. 3 BADOCK web service. (**a**) Submission page of the BADOCK web server. (**b**) Message sent after submission, it includes the job identification (JOB ID) and a link to access the results. (**c**) Results page of BADOCK web server shows the predicted ΔG for the interaction of the submitted proteins. (**d**) Menu of the results page: by introducing a JOB ID, results can be retrieved from a previous execution

will always produce docking poses of the complex and consequently a prediction of the binding energy. If they fail, the user is advised to verify that the coordinates of the proteins agree with the PDB standards.

4. Optimize the structure prior to submission; there are several methods available in Rosetta [56] (e.g., commands *fixbb* and *relax*) or in FoldX [49] (e.g., command *repair_pdb*).
5. Users can add hydrogens before optimization, using for example the program Reduce [63].
6. When docking multidomain proteins, the analysis of docking solutions will show the preferences for specific domains of one or both proteins. In this case, the protein could be decomposed into domains prior to the prediction of binding affinity with BADOCK. We must note that most protein–protein interactions are produced by the interaction between two single domains, as seen in Interactome3D [41]. Therefore, attempting to calculate the interaction with single domains is more suitable.

7. BADOCK server does not allow the user to tune parameters or choose execution options. However, the users have the option to modify PDB files provided as an input. Some suggested modifications are:
   (a) BADOCK was implemented based on the interaction of soluble globular proteins and except antigen–antibody interactions. For membrane proteins, the method can only be applied on the soluble globular fragment. Thus, for membrane proteins, we recommend splitting the structures by domains and test the single domains with BADOCK or restrict the analysis to the soluble globular fragments. It is important to note that the method will not work for antibodies.

       In the case of multidomain proteins:
   (b) One can use MODPIN to determine whether the most probable interaction of the partners is produced by single domains or imply multidomain conformations. If the interaction is done between two single domains, it is possible to split the structure in domains and submit the structures of the interacting ones to BADOCK, otherwise it is recommended to use the largest structure available.
   (c) If MODPIN cannot be applied, but the structures of the unbound partners are available, it is possible to use V-D2OCK to check the type of interaction (single or multidomain) and proceed accordingly as before.
8. iFrag can be used to predict the regions of the proteins that are more likely to interact and proceed following the instructions given in **Notes 6** and 7. iFrag is a web service that predicts regions of interaction in PPIs based on sequence information; it is available at http://sbi.imim.es/web/index.php/research/servers/iFrag [64].
9. BADOCK is also accessible through the Galaxy InteractoMIX platform: https://galaxy.interactomix.com in "Structure-based" menu.

4 Examples

4.1 Example to Complement BADOCK with MODPIN

MODPIN can be used to observe and analyze the conformational space of the interaction of two proteins through known interactions of homologs. It also complements BADOCK when one or both proteins of the interaction are composed of several domains. We use MODPIN to sample the conformational space of the interaction based on the homologs and to recognize the implication of domains and finally select those relevant on performing the interaction.

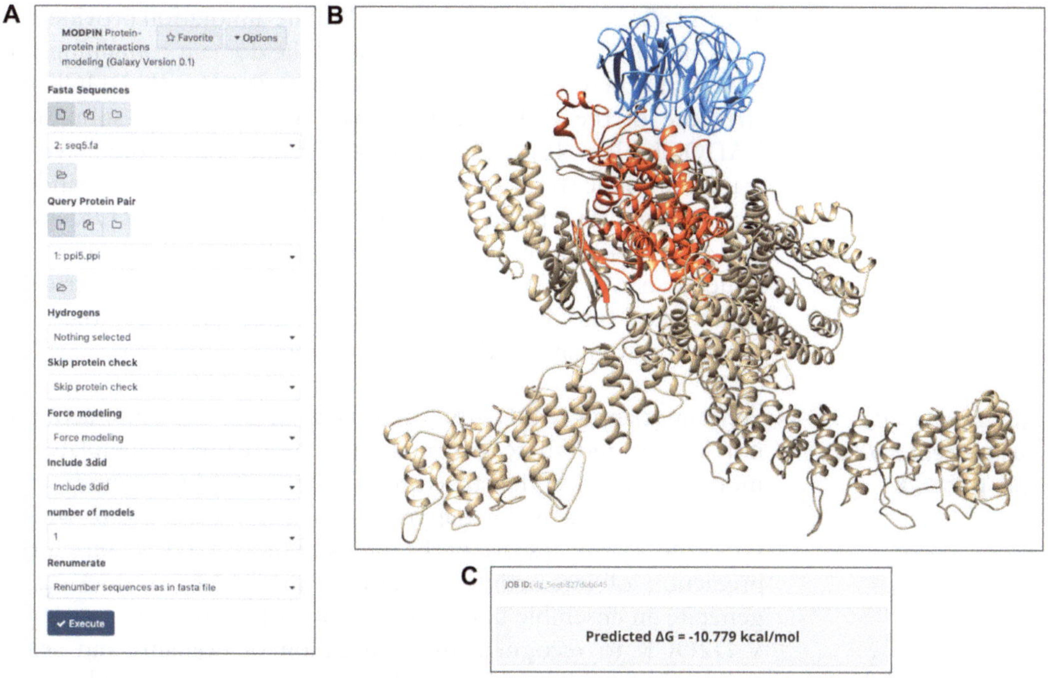

Fig. 4 Example of the use of BADOCK complemented with MODPIN. (**a**) Submission page of MODPIN in the galaxy InteractoMIX server. (**b**) Structure of a model of mTOR (orange and gray) interacting with the LST8 subunit of the target of rapamycin complex (blue). The region of mTOR involved in the interaction with the LST8 subunit in the majority of models obtained with MODPIN is highlighted in orange. (**c**) Result of BADOCK obtained with the structures of LST8 subunit and the highlighted region of mTOR

Here we show as an example the interaction between mTOR and the target of rapamycin complex LST8 subunit (also known as the mammalian lethal with SEC13 protein 8). LST8 is a subunit of both mTORC1 and mTORC2 complexes formed with Raptor (mTOR). While the LST8 subunit is a small protein consisting of a single domain, mTOR is a large multi-domain protein. We retrieve the amino acid sequences of human mTOR and LST8 subunit from the Uniprot database [65] (codes P42345 for mTOR and Q9BVC4 for LST8 subunit). We construct a single file with both sequences in FASTA format and input them to run MODPIN in the galaxy interactoMIX server (Fig. 4a).

MODPIN produces a collection of six models of the PPI. We assess the structural similarity of each partner of the interaction on the set of models using the superposition with MATCHMAKER from the Chimera package [66]. We observe that the structural models of the proteins involved in the interaction are highly similar between them (mean RMSD smaller than 1.5 Å and 1 Å for mTOR and LST8 subunit models, respectively). In addition, we observe that five out of six models share the same interface, where the region of mTOR in the interface involves amino acids 2262–2294 (Fig. 4b). This region belongs to the PI3K/PI4K domain (amino acids 2182–2516) according to the ProSite annotation provided by the Uniprot database [65]. We isolated the domain containing this region in a new PDB file. Then, we submitted to BADOCK the structure of the PI3K/PI4K domain and the LST8 subunit in BADOCK (Fig. 4c) and compared with the affinity predicted by using the whole structure of mTOR. The prediction of the binding free energy ΔG using the specific domain of mTOR gives $\Delta G = -11.429$ kcal/mol, while the prediction using the whole structure is $\Delta G = -10.779$ kcal/mol. Given an acceptable error of about 3 kcal/mol with BADOCK, both values are within the same range, and we may use the average of both.

4.2 Example to Complement BADOCK with V-D2OCK

We cannot apply MODPIN when the structure of an interaction is not known and there are no homologs to use as template for modeling. Then, if the structure of the unbound proteins is available, we can use a docking approach, such as V-D2OCK, to sample the conformational space of the interaction. V-D2OCK combines a prediction of interacting interfaces with rigid-body docking to generate an ensemble of conformers of a PPI. Therefore, we apply V-D2OCK to recognize the implication of domains and select those relevant on performing the interaction as above, albeit this approach is based on the sample of poses derived from docking, which may be less rigorous.

We use as an example the interaction between the cyclin-dependent kinase 9 (CDK9) and the La-related protein 7 (LARP7). The unbound structures are retrieved from the PDB for CDK9 (entry 3blh chain A) and two structures that correspond to two different domains for LARP7, one at the N-tail (Nt domain, formed by PFAM domains La and RRM-1, entry 4wkr chain A) and the other at the C-tail (Ct domain, formed by PFAM domain RRM-2, entry 6d12 chain A). We use V-D2OCK with the default parameters to sample the docking space and obtain several potential interacting interfaces between CDK9 and the Nt domain of LARP7 (Fig. 5a) and between CDK9 and the Ct domain of LARP7 (Fig. 5c). Then, we use BADOCK to predict the binding energy using the structures of both potential interactions (Fig. 5b, d). The predicted binding energies are respectively $\Delta G = -8.83$ kcal/mol and $\Delta G = -9.74$ kcal/mol. This is an interesting example where the full structure of LARP7 is unknown, but several domains have been independently crystallized. Therefore, we know the structure of individual domains but not their arrangement.

4.3 Example to Complement BADOCK with iFrag

To complement the previous use of V-D2OCK, we can also use a sequence-based approach to predict the regions involved in an interaction. Here, we propose the use of iFrag, a method that predicts the regions of the interface between two proteins based on the recursion of several interacting homologs (*see* Fig. 6). iFrag is available in the sequence based methods at https://galaxy.interactomix.com and at http://sbi.imim.es/iFrag [64].

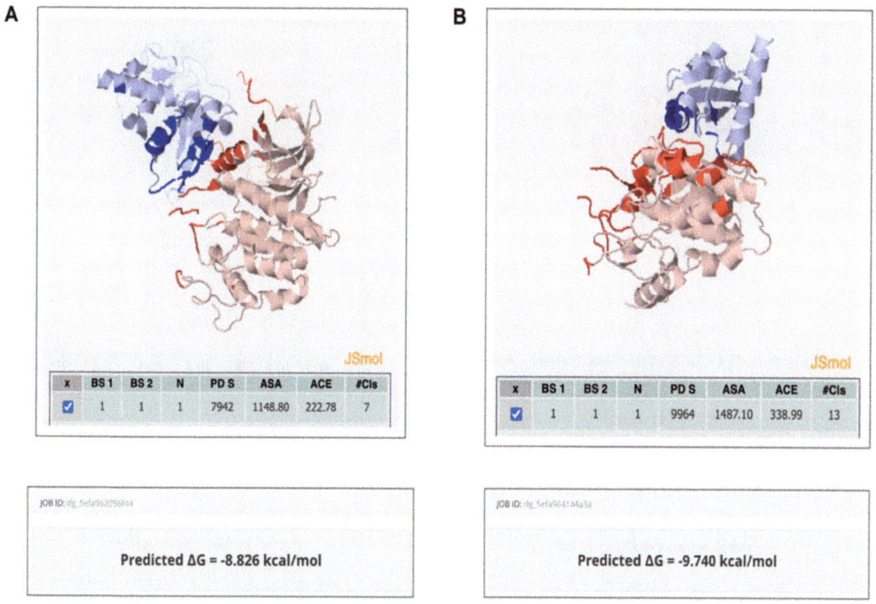

Fig. 5 Example of the use of BADOCK complemented with V-D2OCK. (**a**) Example of one of the predicted poses of the interaction between the Cyclin-dependent kinase 9 (CDK9, in red) and the Nt fragment of La-related protein 7 (LARP7, in blue). The prediction of affinity with BADOCK is shown at the bottom. (**b**) Example of one of the predicted poses of the interaction between the cyclin-dependent kinase 9 (CDK9, in red) and the Ct fragment, composed of the RRM-2 domain, of LARP7 (in blue). The prediction of affinity with BADOCK is shown at the bottom

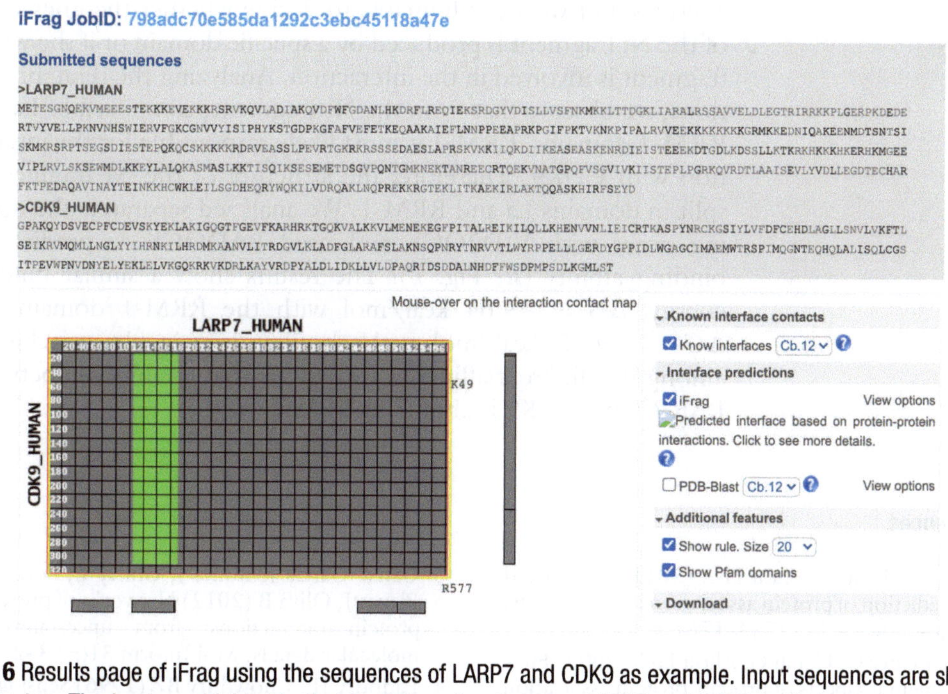

Fig. 6 Results page of iFrag using the sequences of LARP7 and CDK9 as example. Input sequences are shown at the top. The heatmap highlights that the interaction between LARP7 and CDK9 is likely produced by the region of domain RRM-1 of LARP7

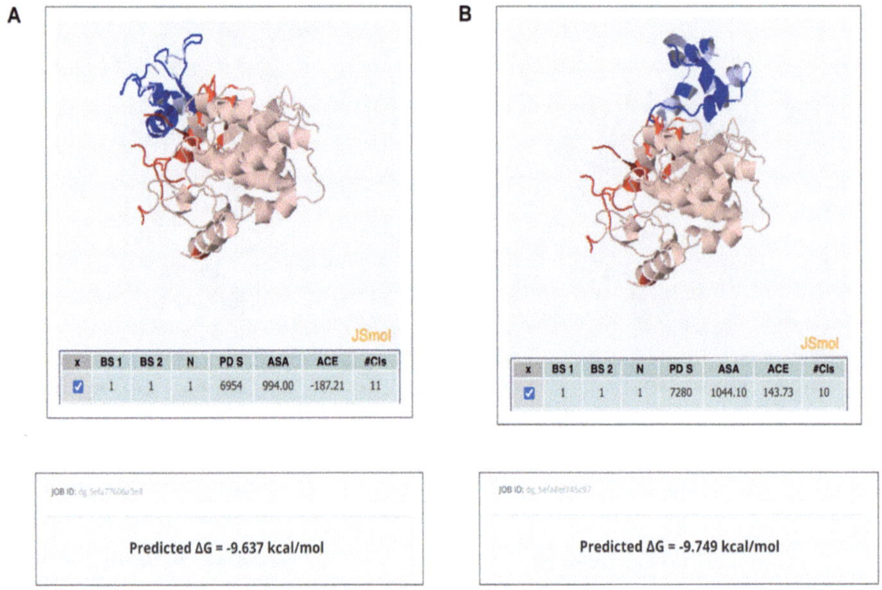

Fig. 7 Example of the use of BADOCK complemented with V-D2OCK after iFrag selection of domains. (**a**) Predicted pose involving the interface between the cyclin-dependent kinase 9 or CDK9 (CDK9, in red) and the RRM-1 motif of La-related protein 7 (in blue). The prediction of affinity with BADOCK is shown at the bottom. Predicted pose involving the interface between the cyclin-dependent kinase 9 or CDK9 (CDK9, in red) and the La domain of La-related protein 7 (in blue). The prediction of affinity with BADOCK is shown at the bottom

As shown with the use of V-D2OCK and BADOCK, the fragments Nt and Ct of the structure produce comparable binding energies, but we are still unable to decide whether the interaction of the Nt fragment is produced by a specific domain or if the whole fragment is involved in the interaction. Analyzing the iFrag predictions, we observe that from the Nt fragment, composed of La and RRM-1 domains, RRM-1 is most likely responsible for the interaction with CDK9. Therefore, the structure of the Nt fragment is split in domains La and RRM-1. We analyzed separately the docking poses with V-D2OCK and used BADOCK to predict the binding affinity (*see* Fig. 7). The results show a similar binding energy, $\Delta G = -9.64$ kcal/mol with the RRM-1 domain and $\Delta G = -9.75$ kcal/mol with the La domain. In conclusion, among the different affinities, we assume that the binding between LARP7 and CDK9 is about −9.64 kcal/mol.

References

1. Qin S, Pang X, Zhou H-X (2011) Automated prediction of protein association rate constants. Structure 19(12):1744–1751
2. Schreiber G, Haran G, Zhou H-X (2009) Fundamental aspects of protein-protein association kinetics. Chem Rev 109(3):839–860
3. Garcia-Garcia J, Bonet J, Guney E, Fornes O, Planas J, Oliva B (2012) Networks of protein–-protein interactions: from uncertainty to molecular details. Mol Inform 31(5):342–362
4. Ladbury JE, Chowdhry BZ (1996) Sensing the heat: the application of isothermal titration

calorimetry to thermodynamic studies of biomolecular interactions. Chem Biol 3 (10):791–801

5. Willander M, Al-Hilli S (2009) Analysis of biomolecules using surface plasmons. Methods Mol Biol 544:201–229
6. Lin T, Scott BL, Hoppe AD, Chakravarty S (2018) FRETting about the affinity of bimolecular protein-protein interactions. Protein Sci 27(10):1850–1856
7. Ciruela F (2008) Fluorescence-based methods in the study of protein-protein interactions in living cells. Curr Opin Biotechnol 19 (4):338–343
8. Dries DR, Newton AC (2008) Kinetic analysis of the interaction of the C1 domain of protein kinase C with lipid membranes by stopped-flow spectroscopy. J Biol Chem 283 (12):7885–7893
9. Li M, Simonetti FL, Goncearenco A, Panchenko AR (2016) MutaBind estimates and interprets the effects of sequence variants on protein-protein interactions. Nucleic Acids Res 44(W1):W494–W501
10. Vangone A, Bonvin AM (2015) Contacts-based prediction of binding affinity in protein-protein complexes. Elife 4:e07454
11. Abbasi WA, Asif A, Ben-Hur A, Minhas FUAA (2018) Learning protein binding affinity using privileged information. BMC Bioinformatics 19(1):425
12. Yugandhar K, Gromiha MM (2014) Feature selection and classification of protein-protein complexes based on their binding affinities using machine learning approaches. Proteins 82(9):2088–2096
13. Horton N, Lewis M (1992) Calculation of the free energy of association for protein complexes. Protein Sci 1(1):169–181
14. Moal IH, Agius R, Bates PA (2011) Protein-protein binding affinity prediction on a diverse set of structures. Bioinformatics 27 (21):3002–3009
15. Marín-López MA, Planas-Iglesias J, Aguirre-Plans J, Bonet J, Garcia-Garcia J, Fernandez-Fuentes N et al (2018) On the mechanisms of protein interactions: predicting their affinity from unbound tertiary structures. Bioinformatics 34(4):592–598
16. Milo R (2013) What is the total number of protein molecules per cell volume? A call to rethink some published values. BioEssays 35 (12):1050–1055
17. McGuffee SR, Elcock AH (2010) Diffusion, crowding & protein stability in a dynamic molecular model of the bacterial cytoplasm. PLoS Comput Biol 6(3):e1000694
18. Yu I, Mori T, Ando T, Harada R, Jung J, Sugita Y et al (2016) Biomolecular interactions modulate macromolecular structure and dynamics in atomistic model of a bacterial cytoplasm. Elife 01:5
19. Mika JT, Poolman B (2011) Macromolecule diffusion and confinement in prokaryotic cells. Curr Opin Biotechnol 22(1):117–126
20. Ellis RJ (2001) Macromolecular crowding: an important but neglected aspect of the intracellular environment. Curr Opin Struct Biol 11 (1):114–119
21. Planas-Iglesias J, Marin-Lopez MA, Bonet J, Garcia-Garcia J, Oliva B (2013) iLoops: a protein-protein interaction prediction server based on structural features. Bioinformatics 29(18):2360–2362
22. Levinthal C (1968) Are there pathways for protein folding? J Chem Phys 65:44–45
23. Wass MN, Fuentes G, Pons C, Pazos F, Valencia A (2011) Towards the prediction of protein interaction partners using physical docking. Mol Syst Biol 7:469
24. Schweke H, Mucchielli M-H, Sacquin-Mora S, Bei W, Lopes A (2020) Protein interaction energy landscapes are shaped by functional and also non-functional partners. J Mol Biol 432(4):1183–1198
25. Richardson JS, Richardson DC (2002) Natural beta-sheet proteins use negative design to avoid edge-to-edge aggregation. Proc Natl Acad Sci U S A 99(5):2754–2759
26. Pechmann S, Levy ED, Tartaglia GG, Vendruscolo M (2009) Physicochemical principles that regulate the competition between functional and dysfunctional association of proteins. Proc Natl Acad Sci U S A 106(25):10159–10164
27. Deeds EJ, Ashenberg O, Gerardin J, Shakhnovich EI (2007) Robust protein protein interactions in crowded cellular environments. Proc Natl Acad Sci U S A 104(38):14952–14957
28. Karanicolas J, Corn JE, Chen I, Joachimiak LA, Dym O, Peck SH et al (2011) A de novo protein binding pair by computational design and directed evolution. Mol Cell 42(2):250–260
29. Garcia-Seisdedos H, Empereur-Mot C, Elad N, Levy ED (2017) Proteins evolve on the edge of supramolecular self-assembly. Nature 548(7666):244–247
30. Lo Conte L, Chothia C, Janin J (1999) The atomic structure of protein-protein recognition sites. J Mol Biol 285(5):2177–2198
31. Chakrabarti P, Janin J (2002) Dissecting protein-protein recognition sites. Proteins 47 (3):334–343
32. Li X, Keskin O, Ma B, Nussinov R, Liang J (2004) Protein-protein interactions: hot spots

and structurally conserved residues often locate in complemented pockets that pre-organized in the unbound states: implications for docking. J Mol Biol 344(3):781–795

33. Keskin O, Ma B, Nussinov R (2005) Hot regions in protein--protein interactions: the organization and contribution of structurally conserved hot spot residues. J Mol Biol 345 (5):1281–1294
34. McCammon JA (1998) Theory of biomolecular recognition. Curr Opin Struct Biol 8 (2):245–249
35. Tsai H-HG, Reches M, Tsai C-J, Gunasekaran K, Gazit E, Nussinov R (2005) Energy landscape of amyloidogenic peptide oligomerization by parallel-tempering molecular dynamics simulation: significant role of Asn ladder. Proc Natl Acad Sci U S A 102 (23):8174–8179
36. Wolynes PG (2015) Evolution, energy landscapes and the paradoxes of protein folding. Biochimie 119:218–230
37. Schneidman-Duhovny D, Inbar Y, Nussinov R, Wolfson HJ (2005) PatchDock and SymmDock: servers for rigid and symmetric docking. Nucleic Acids Res 33:W363–W367
38. Mashiach E, Nussinov R, Wolfson HJ (2010) FiberDock: flexible induced-fit backbone refinement in molecular docking. Proteins 78 (6):1503–1519
39. Feliu E, Aloy P, Oliva B (2011) On the analysis of protein-protein interactions via knowledge-based potentials for the prediction of protein-protein docking. Protein Sci 20(3):529–541
40. Vreven T, Hwang H, Pierce BG, Weng Z (2012) Prediction of protein-protein binding free energies. Protein Sci 21(3):396–404
41. Mosca R, Céol A, Aloy P (2013) Interactome3D: adding structural details to protein networks. Nat Methods 10(1):47–53
42. Zhang QC, Petrey D, Garzón JI, Deng L, Honig B (2013) PrePPI: a structure-informed database of protein-protein interactions. Nucleic Acids Res 41(Database issue): D828–D833
43. Poglayen D, Marín-López MA, Bonet J, Fornes O, Garcia-Garcia J, Planas-Iglesias J et al (2016) InteractoMIX: a suite of computational tools to exploit interactomes in biological and clinical research. Biochem Soc Trans 44(3):917–924
44. Meseguer A, Dominguez L, Bota PM, Aguirre-Plans J, Bonet J, Fernandez-Fuentes N et al (2020) Using collections of structural models to predict changes of binding affinity caused by mutations in protein-protein interactions. Protein Sci 29(10):2112–2130
45. Levy Y, Cho SS, Onuchic JN, Wolynes PG (2005) A survey of flexible protein binding mechanisms and their transition states using native topology based energy landscapes. J Mol Biol 346(4):1121–1145
46. Andrusier N, Mashiach E, Nussinov R, Wolfson HJ (2008) Principles of flexible protein-protein docking. Proteins 73(2):271–289
47. Stein A, Rueda M, Panjkovich A, Orozco M, Aloy P (2011) A systematic study of the energetics involved in structural changes upon association and connectivity in protein interaction networks. Structure 19(6):881–889
48. Goh C-S, Milburn D, Gerstein M (2004) Conformational changes associated with protein-protein interactions. Curr Opin Struct Biol 14 (1):104–109
49. Schymkowitz J, Borg J, Stricher F, Nys R, Rousseau F, Serrano L (2005) The FoldX web server: an online force field. Nucleic Acids Res 33(Web Server issue):W382–W388
50. Stranges PB, Kuhlman B (2013) A comparison of successful and failed protein interface designs highlights the challenges of designing buried hydrogen bonds. Protein Sci 22 (1):74–82
51. Jankauskaite J, Jiménez-García B, Dapkunas J, Fernández-Recio J, Moal IH (2019) SKEMPI 2.0: an updated benchmark of changes in protein-protein binding energy, kinetics and thermodynamics upon mutation. Bioinformatics 35(3):462–469
52. Orchard S, Ammari M, Aranda B, Breuza L, Briganti L, Broackes-Carter F et al (2014) The MIntAct project--IntAct as a common curation platform for 11 molecular interaction databases. Nucleic Acids Res 42(Database issue): D358–D363
53. Siebenmorgen T, Zacharias M (2020) Computational prediction of protein–protein binding affinities. WIREs Comput Mol Sci 10(3): e1448
54. Siebenmorgen T, Zacharias M (2019) Evaluation of predicted protein-protein complexes by binding free energy simulations. J Chem Theory Comput 15(3):2071–2086
55. Pierce B, Weng Z (2007) ZRANK: reranking protein docking predictions with an optimized energy function. Proteins 67(4):1078–1086
56. Alford RF, Leaver-Fay A, Jeliazkov JR, O'Meara MJ, DiMaio FP, Park H et al (2017) The Rosetta all-atom energy function for macromolecular modeling and design. J Chem Theory Comput 13(6):3031–3048
57. Aguirre-Plans J, Meseguer A, Molina-Fernandez R, Marín-López MA, Jumde G, Casanova K, Bonet J, Fornes O, Fernandez-Fuentes N, Oliva B (2018) SPServer: Split-

Statistical Potentials for the analysis of protein structures and protein-protein interactions. BMC Bioinformatics 22:4

58. Takemura K, Matubayasi N, Kitao A (2018) Binding free energy analysis of protein-protein docking model structures by evERdock. J Chem Phys 148(10):105101
59. Barradas-Bautista D, Moal IH, Fernández-Recio J (2017) A systematic analysis of scoring functions in rigid-body protein docking: the delicate balance between the predictive rate improvement and the risk of overtraining. Proteins 85(7):1287–1297
60. Segura J, Marín-López MA, Jones PF, Oliva B, Fernandez-Fuentes N (2015) VORFFIP-driven dock: V-D2OCK, a fast and accurate protein docking strategy. PLoS One 10(3): e0118107
61. Segura J, Jones PF, Fernandez-Fuentes N (2011) Improving the prediction of protein binding sites by combining heterogeneous data and Voronoi diagrams. BMC Bioinformatics 12:352
62. El-Gebali S, Mistry J, Bateman A, Eddy SR, Luciani A, Potter SC et al (2019) The Pfam protein families database in 2019. Nucleic Acids Res 47(D1):D427–D432
63. Word JM, Lovell SC, Richardson JS, Richardson DC (1999) Asparagine and glutamine: using hydrogen atom contacts in the choice of side-chain amide orientation. J Mol Biol 285 (4):1735–1747
64. Garcia-Garcia J, Valls-Comamala V, Guney E, Andreu D, Muñoz FJ, Fernandez-Fuentes N et al (2017) iFrag: a protein–protein Interface prediction server based on sequence fragments. J Mol Biol 429(3):382–389
65. UniProt Consortium (2019) UniProt: a worldwide hub of protein knowledge. Nucleic Acids Res 47(D1):D506–D515
66. Pettersen EF, Goddard TD, Huang CC, Couch GS, Greenblatt DM, Meng EC et al (2004) UCSF chimera--a visualization system for exploratory research and analysis. J Comput Chem 25(13):1605–1612
67. Webb B, Sali A (2016) Comparative protein structure modeling using MODELLER. Curr Protoc Protein Sci 86:2.9.1–2.9.37

Chapter 17

Parameter Optimization for Ion Channel Models: Integrating New Data with Known Channel Properties

Marco A. Navarro, Marzie Amirshenava, Autoosa Salari, Mirela Milescu, and Lorin S. Milescu

Abstract

Ion channels play a central role in membrane physiology, but to fully understand how they operate, one must have accurate kinetic mechanisms. Estimating kinetics is not trivial when the mechanism is complex, and a large number of parameters must be extracted from data. Furthermore, the information contained in the data is often limited, and the model may not be fully determined. The solution is to reduce the number of parameters and to estimate them in such a way that they not only describe well the new data but also agree with the existing knowledge. In a previous study, we presented a comprehensive formalism for estimating kinetic parameters subject to a variety of explicit and implicit constraints that define quantitative relationships between parameters and describe specific mechanism properties. Here, we introduce the reader to the QuB software, which implements this constraining formalism. QuB features a powerful visual interface and a high-level scripting language that can be used to formulate kinetic models and constraints of arbitrary complexity, and to efficiently estimate the parameters from a variety of experimental data.

Key words Ion channel, Kinetic mechanism, Kinetic parameters, Markov model, Model constraints, Prior knowledge

1 Introduction

Ion channels are membrane proteins that provide a controlled pathway for ions to cross the lipid bilayer of the cellular membrane [1]. Structurally, the ion channel molecule contains a pore-forming domain with specific ionic selectivity, and some control elements that open and close the pore with defined sensitivity and timing, in response to specific ligand molecules or physical variables, such as voltage, mechanical tension, or temperature. Functionally, an ion channel is fully characterized by its ionic selectivity and kinetic properties. The kinetic mechanism can be conveniently represented as a Markov model with discrete states that correspond to experimentally and/or statistically detectable molecular conformations

Quentin Vanhaelen (ed.), *Computational Methods for Estimating the Kinetic Parameters of Biological Systems*, Methods in Molecular Biology, vol. 2385, https://doi.org/10.1007/978-1-0716-1767-0_17,

[2]. The states can represent, for example, the channel closed, open, or inactivated. The transitions between states are quantified by rate constants that follow the Eigen rate theory and can be modulated by the concentration of a ligand or by the magnitude of a physical variable. For example, voltage-gated ion channels have rate constants that can be parameterized as follows:

$$k_{ij} = k_{ij}^{0} \times \exp\left(k_{ij}^{1} \times V\right), \tag{1}$$

where V is voltage, k_{ij} is the rate constant for the transition between states i and j, k_{ij}^{0} is the rate at zero voltage, and k_{ij}^{1} is a factor that denotes voltage sensitivity. An example of a kinetic mechanism is shown in Fig. 1, as proposed to describe the kinetic properties of voltage-gated sodium (Nav) channels [3].

To understand how an ion channel functions, dynamically, at the molecular and cellular levels, one must fully characterize its kinetic mechanism. For this, one has to know the number of states and their connectivity (the "structure" or "topology") and the parameters that quantify the rate constants and their sensitivity to external factors, such as expressed by Eq. 1. Several decades of ion channel research have seen the development of powerful parameter estimation algorithms that extract the rate constants for a given topology from a variety of experimental data types, such as single channel or whole-cell voltage-clamped currents [4–19] or current clamp recordings [20].

Generally, parameter estimation algorithms use an optimization engine that explores a multidimensional parameter space to find a set of values that best explain the data [21]. Finding these optimal parameters is not easy for ion channels, because they often have kinetic mechanisms with many states and transitions, which results in a large number of parameters that must be estimated. For example, the Nav model shown in Fig. 1 is rather minimalistic, yet it has no less than 64 kinetic parameters! Combined with the stochastic nature of the data and with the inevitable presence of recording artifacts, these many parameters would be very difficult to estimate. If we cut corners and try to use smaller models, with fewer states

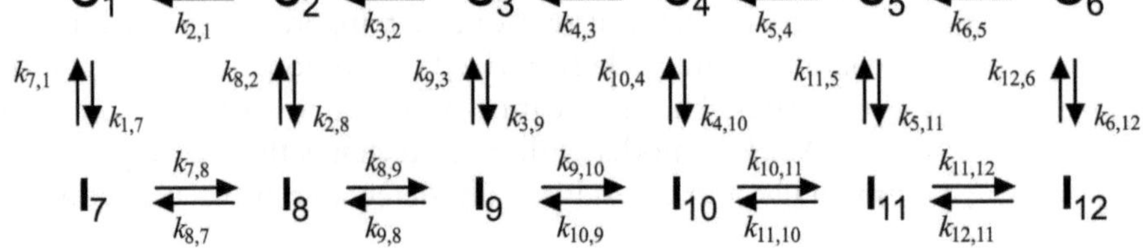

Fig. 1 State models for ion channel kinetic mechanisms. This example model was previously proposed to describe the kinetic properties of neuronal voltage-gated sodium (Nav) channels [3]. During an action potential, Nav channels undergo transitions between closed (C), open (O), and inactivated (I) states

and parameters, we may end up with a model that simply has not enough flexibility to describe the operation of the channel in a live cell [22].

However, if we are fortunate enough, we may know something about the channel that can help us not only to reduce the number of parameters to be estimated but also to guide the optimizer in its search for a solution. For example, we know that Nav channels have four voltage sensors that feature similar voltage sensitivities. Then, we can assume that both the forward rates along the $C_1 > C_2 > C_3 > C_4 > C_5$ pathway and the backward rates along the $C_5 > C_4 > C_3 > C_2 > C_1$ pathway are in a 4:3:2:1 ratio. The same assumptions can be made about the rates connecting the inactivated states $I_7 \ldots I_{11}$. We can further assume that the channel obeys microscopic reversibility, which means that for each loop in the kinetic scheme (e.g., $C_1 > C_2 > I_8 > I_7 > C_1$), the product of the rates in the clockwise direction is equal to the product of the rates in the counterclockwise direction. Altogether, these are reasonable assumptions that reduce the number of model parameters from 64 to around 20. Finally, we also know that Nav channels have certain kinetic properties, and these must be reflected in the mechanism. For example, the channel must fully recover from inactivation in 5–20 ms, so as to enforce a brief but necessary refractory phase for nerve conduction, yet be ready quickly for the next action potential [23].

How can we integrate new data with everything we already know about the kinetic mechanism? We addressed this question in two recent papers [24, 25], where we proposed a methodology for extracting kinetic parameters that not only explain new experimental data but also satisfy existing knowledge. We showed how prior knowledge can be enforced via two types of constraints: *linear parameter constraints* and *behavioral constraints.* Linear parameter constraints represent explicit mathematical relationships between model parameters, which include the rate constant parameters k_{ij}^0 and k_{ij}^1 in Eq. 1, and all other parameters or external variables that are needed to describe the experimental data and the recording conditions, such as the number of active channels in the recording. We implemented a unified mechanism that handles both equality and inequality linear parameter constraints, using linear algebra methods that transform the inter-dependent parameters of the model into a reduced set of independent parameters that are then optimized [24]. Linear relationships can be used to implement a variety of constraints, such as scaling one rate to another, parameterizing allosteric relationships, enforcing microscopic reversibility.

In contrast to linear relationships, behavioral constraints represent either specific model behaviors that result from implicit parameter relationships, or explicit parameter relationships that cannot be implemented via the linear algebra method [25]. For example, one may wish to enforce an estimated parameter to take values in a specific range. Although this constraint is both explicit and linear, it cannot be implemented via the linear constraint-type mechanism. As another example, one may want to enforce a set of parameters that produce a model with a specific maximum open probability. This constraint is implicit, in the sense that a mathematical relationship cannot be formulated between model parameters, at least not easily and not in the general case.

These constraints that cannot be handled by the linear algebra mechanism can be enforced by calculating a penalty that represents the degree by which the model deviates from that constraint, and then adding that penalty to the cost function that measures the error between the data and the prediction of the model (e.g., the sum of square errors or the likelihood). As the optimizer minimizes a cost function that now includes the penalty for behavioral constrains, it will generate a model that will not only fit the data but will also satisfy all the prior knowledge, formulated as a combination of linear parameter constraints and behavioral constraints. An interesting distinction is that linear parameter constraints are "hard" constraints that are implemented exactly, whereas behavioral constraints can be implemented with a certain degree of flexibility. Another difference is that linear parameter constraints reduce the number of free parameters (one for each equality relationship), whereas behavioral constraints effectively reduce the search space.

This chapter is meant as a companion to our previous studies [24, 25], where we presented extensively the mathematical and computational formalism for estimating ion channel kinetics with prior knowledge. The algorithms described there have been implemented in our freely available software, QuB—the MLab edition [26]. Here, we focus on demonstrating the practical aspects of using the QuB software to define model constraints and estimate kinetic parameters. To make it simpler for the reader, we use the same examples that were used in [25], where we explained the required calculations step-by-step, using detailed numerical examples.

2 Materials

2.1 Computer

You need a PC running a Windows version compatible with the QuB software (XP, 7, 8, or 10). QuB has no special requirements in terms of computer hardware, but computation speed is commensurate with the characteristics of the processor. As QuB can parallelize the computation, a multi-core machine is recommended.

3 Methods

3.1 Install the QuB Software

Go to the QuB website (https://milesculabs.org/qub.html) and follow the instructions for how to install QuB. Download the example files that will be used in this tutorial (data file *CCOC_-Sim.qdf*, model files *CCOC_Sim.qmf*, *CCOC_Fit.qmf*, and *CCOC_FitStart.qmf*, and library file *Constraints.slf*). You will also find materials that describe the scripting language that will be used to implement the behavioral constraints. If you encounter difficulties during the installation process, given the existing variety of Windows versions and user configurations, contact the authors for assistance.

3.2 Set up the Modeling Interface in QuB

For convenience, QuB has a number of predefined graphical user interfaces, each displaying a layout of windows appropriate for a given task—in this case, modeling the kinetic mechanism of ion channels. These predefined interfaces can be further customized, and new ones can be created. To turn on or off a window, go to *Main Menu* > *View* and check or uncheck that window. By default, QuB starts in a layout mode where windows are mutually resizable (to switch, select *Main Menu* > *Interface* > *Floating windows*). In this mode, to change the size of a window, use the spacer bar between that window and the adjacent one. To reposition a window, left-click on the caption bar of that window, hold the mouse button down, and move the mouse around. A rectangle will become visible to indicate where the window will be positioned when you release the mouse button.

For this demonstration, select the *Modeling* interface and customize it as shown in Fig. 2, with the *Data*, *Model*, *Scriptlets*, *Graph*, and *Report* windows visible (your windows will be empty until you load the content, as explained in the next sections). The *Data* window is where we load the data that we want to fit. The *Model* window is where we construct the ion channel kinetic mechanism and define the linear parameter constraints. The *Scriptlets* window contains user-defined snippets of code ("scriptlets"), which will be used here to implement the computation of behavioral constraints. The *Graph* window provides a simple way of displaying curves and other graphics, and it will be used here by the optimization engine to show the convergence of the fit. Finally, the *Report* window is a text interface used by the optimizer (and other functions) to communicate status and performance and to display results.

3.3 Prepare the Data for Fitting

QuB can handle a number of data file formats, including its own native files with extension ".qdf" (and the older format ".ldt"), Molecular Devices (Axon) pClamp binary files (".abf"), HEKA Pulse binary files (".pul"), and simple text files. QuB native data

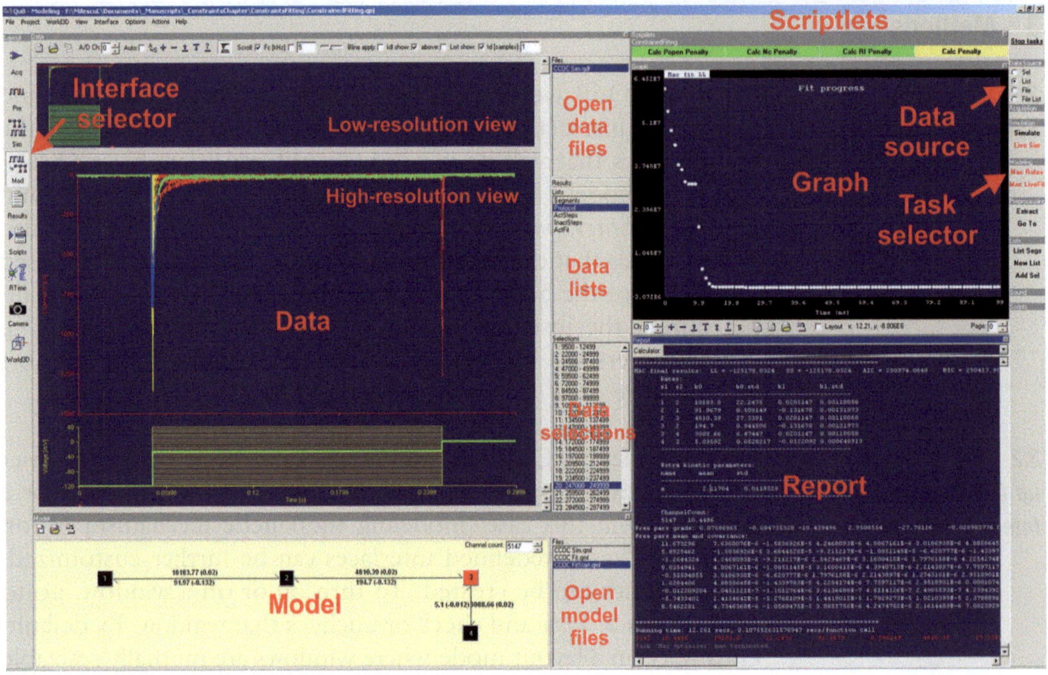

Fig. 2 Customizing the modeling interface in the QuB software. Different windows can be displayed and resized to create a convenient layout for visualization and analysis

files can be created by data acquisition, simulation, or by processing other files (e.g., by extracting a subset of data from a file). For this demonstration, we have already prepared a data file (*CCOC_Sim.qdf*) that can be downloaded from the QuB website. Load this file in the data window (*Main Menu > File > Open Data*).

This example file contains a stochastic simulation generated with the Nav channel model that will be discussed in the next section, in response to a voltage-clamp protocol typically used to record Nav macroscopic currents. The entire protocol contains a series of 33 sweeps, where each sweep first equilibrates the channels at −120 mV for 1 s, then "conditions" them for 200 ms at −120... +40 mV, and finally "tests" them for 50 ms at 0 mV. This protocol is typically used to extract information about the time course of supra-threshold activation and inactivation, and about the steady-state activation and inactivation.

A simple but powerful feature in QuB is that it allows the user to manually select different sections in a data file and assign them to a "data list," which can contain multiple "selections." For the so-called gap-free data files, QuB has the option of displaying the entire data as if it were a text that can be scrolled through line by line, or page by page. For data files that are structured in sweeps, QuB automatically creates one list for each stimulation protocol, with the selections in the list corresponding to the sweeps in the protocol. Whether created by hand or automatically, lists can be

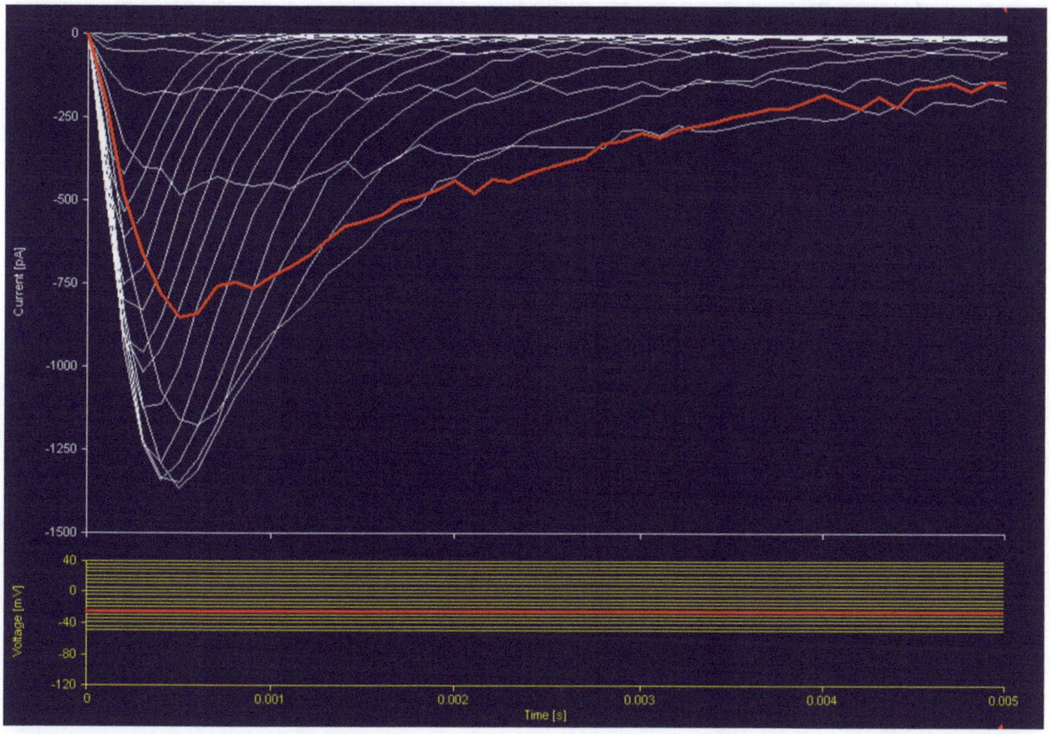

Fig. 3 Preparing the data for analysis. For this demonstration, a test data file (*CCOC_Sim.qdf*, downloadable from the QuB website) was previously generated and is loaded here in the *Data* window. The file contains stochastically simulated macroscopic sodium currents, generated with the Nav model shown in Fig. 4, in response to a typical voltage-clamp step protocol. To have these traces displayed in QuB as shown in this figure, select the *ActFit* selection list in the *Data* window

displayed individually, with the selections overlapped, as shown in Fig. 2. We will use this feature here.

For your convenience, we created five lists in the *CCOC_Sim.qdf* file, named *Segments*, *Protocol*, *ActSteps*, *InActSteps*, and *ActFit*. The *Segments* list points to all the 33 sweeps in the file, each in its entirety, whereas the other lists point to different subsections in the data. For example, the *Protocol* list, which is the one displayed in Fig. 2, points to the last 300 ms in each sweep. In general, you have the choice of analyzing the entire data file (or multiple data files), a single selection list with multiple selections, or a single stretch of data that you have just selected on the screen. For this demonstration, we will analyze the data referenced by the *ActFit* list, as shown in Fig. 3. The *ActFit* list points to the first 5 ms of data in the conditioning step and contains only the steps where the voltage is equal to, or greater than −50 mV. The other traces were not included in the list, because the channel is not activated at those negative voltages and we would gain nothing by fitting them.

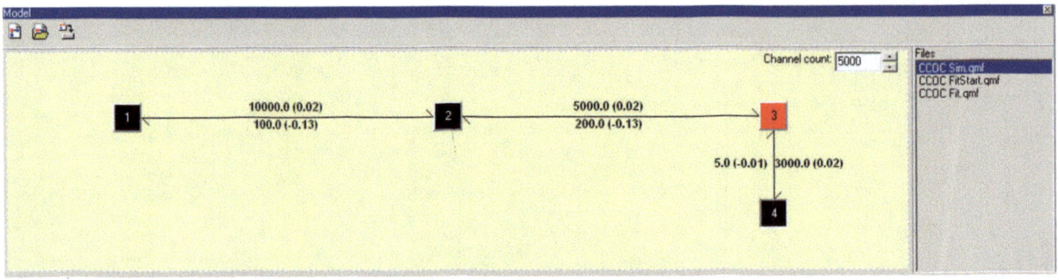

Fig. 4 A simple Nav model. For this demonstration, a test model file (*CCOC_Sim.qmf*, downloadable from the QuB website) was previously created and is loaded here in the *Model* window. This model was previously used to simulate the data shown in Fig. 3. Each number represents the value of that rate constant at 0 mV, equal to k_{ij}^{0} in Eq. 1. The voltage sensitivity, equal to k_{ij}^{1} in Eq. 1, is shown in parentheses

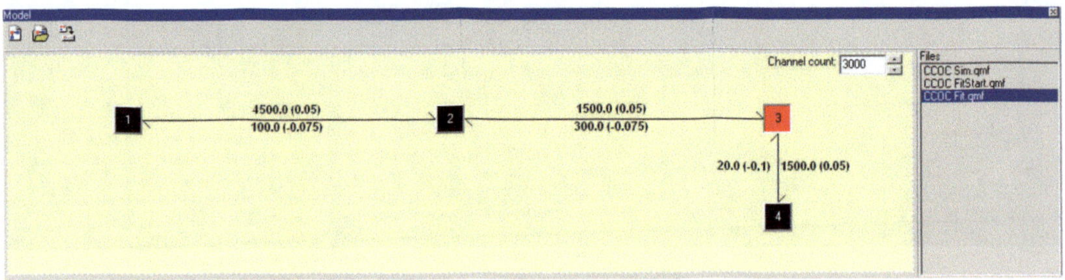

Fig. 5 The model used as the starting point of optimization. This model (*CCOC_Fit.qmf*, downloadable from the QuB website) is the same as the model shown in Fig. 4, but the parameters were changed to generate different predictions (Fig. 6)

3.4 Prepare a Kinetic Model

QuB can be used to create and analyze ion channel kinetic mechanisms of arbitrary complexity. For this demonstration, we have already prepared several model files (*CCOC_Sim.qmf*, *CCOC_FitStart.qmf*, and *CCOC_Fit.qmf*) that can be downloaded from the QuB website. Load these files in the *Model* window.

The first of these models, *CCOC_Sim.qmf* (Fig. 4), was used to simulate the data contained in the *CCOC_Sim.qdf* file, and it is the same as the model used in [25]. We must emphasize that this simple model is not a realistic representation of the Nav mechanism. Nevertheless, it will serve well for this demonstration, as it is smaller and easier to use than the model shown in Fig. 1 but still captures the essential properties of Nav channels, as illustrated in Fig. 3. The other two models, *CCOC_FitStart.qmf* and *CCOC_Fit.qmf* (Fig. 5), are identical with each other and have the same topology as the *CCOC_Sim.qmf* model, but their parameters have been modified from the "true" values that were used to simulate the data and result in a very different set of curves (Fig. 6). The

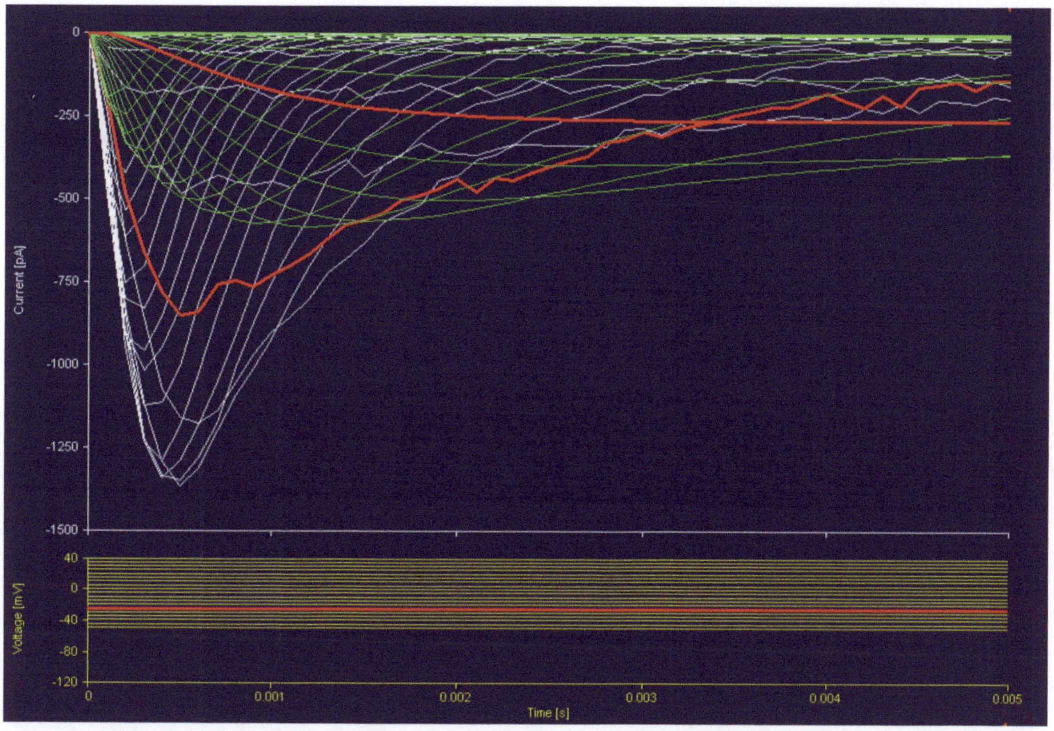

Fig. 6 The starting model makes predictions that significantly deviate from the data. The yellow curves were simulated with the *CCOC_Fit.qmf* model. Will the optimizer find a good solution, if the search is started with these parameters?

CCOC_FitStart.qmf model will be used here simply to mark the starting point of the optimization, whereas the *CCOC_Fit.qmf* will be used to actually fit the data, and it will contain the estimated parameters produced by the optimizer. When you want to repeat an optimization using the same initial values, you can right-click on the model name in the list of opened model files, and select *Revert to saved*. If you accidentally saved the *CCOC_Fit.qmf* model with the wrong parameters and cannot revert, you can copy and paste the original model from the *CCOC_FitStart.qmf* model, or download the file again from the QuB website.

3.5 Define Linear Parameter Constraints

The model provided for this demonstration, *CCOC_FitStart.qmf* (Fig. 5), has already been edited to feature the same linear parameter constraints as in [25]. These constraints define the following mathematical relationships:

$$k_{1,2} = a_1 \times k_{2,3}, \tag{2}$$

$$k_{3,2} = a_1 \times k_{2,1}, \tag{3}$$

$$k^1_{3,4} = k^1_{2,3}, \tag{4}$$

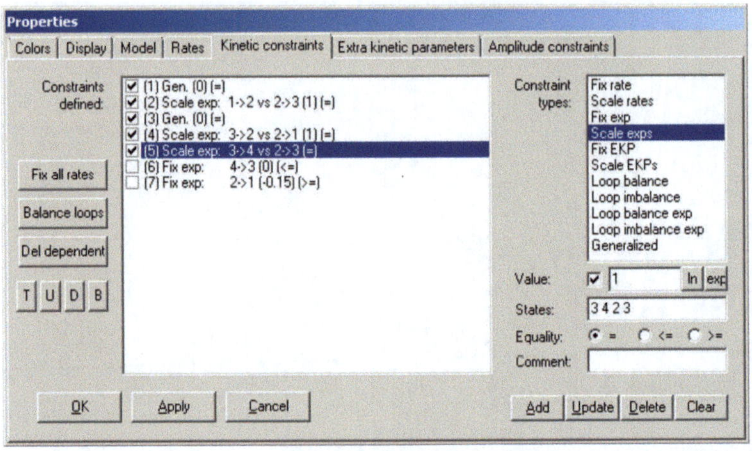

Fig. 7 Defining linear parameter constraints. To bring this dialog in view, right-click on the model and select *Properties*. The constraints prefixed by a check-mark are active and will be enforced during optimization

$$k^1_{4,3} \leq 0, \tag{5}$$

$$k^1_{2,1} \geq -0.15, \tag{6}$$

where a_1 is a factor that defines relationships between rates, and which itself must be estimated.

To see how these linear parameter constraints are implemented in QuB, right-click on the model and select *Properties*, which brings in view the model editor. Select the *Kinetic constraints* panel, as shown in Fig. 7. This interface provides a list of *Constraint types*, from which you can select and create your list of *Defined constraints*. Note that you can create any number of *Defined constraints* but have only a subset of these active and enforced by the optimizer, which is convenient for running tests with different combinations of constraints.

As discussed in [24], each linear parameter constraint requires one or two mathematical relationships, with each of this implemented as a *Defined constraint*. For example, the constraint defined by Eq. 2 is implemented by *Defined constraints* #1 and #2, which take the following mathematical forms:

$$\log(k^0_{1,2}) - \log(k^0_{2,3}) - \log(a_1) = 0 \tag{7}$$

$$k^1_{1,2} - k^1_{2,3} = 0. \tag{8}$$

The constraint defined by Eq. 3 is similarly implemented by *Defined constraints* #3 and #4, Eq. 4 by #5, Eq. 5 by #6, and Eq. 6 by #7. For convenience, the common types of constraints are

Fig. 8 Defining behavioral constraints. The *Scriptlets* window contains user-defined snippets of code that can be used to implement behavioral constraints

predefined (e.g., *Fix rate* or *Scale rate*), but arbitrary relationships can also be created using the *Generalized* type. For the predefined constraints, you only need to specify the transitions involved, the value of the ratio or the fixed value, and whether you have equality or inequality. For example, to specify the constraint that enforces Eq. 4, you select *Scale exps* as the *Constraint type*, enter "3 4 2 3" for *States* to indicate that you want to scale $k_{3,4}^1$ to $k_{2,3}^1$, and enter 1 for *Value* and "=" for *Equality*, to indicate that the two parameters are equal to each other. For generalized constraints, you have a table where you provide the coefficients of the parameters present in that relationship. For example, to enforce Eq. 7, you enter coefficients values of 1 for $\log\left(k_{1,2}^0\right)$, -1 for $\log\left(k_{2,3}^0\right)$, and -1 for $\log(a_1)$, a free term value of 0, and "=" to indicate equality.

3.6 Formulate Behavioral Constraints

Unlike linear parameter constraints, which are predefined and can be created easily via the graphical constraint editor shown in Fig. 7, behavioral constraints are by necessity arbitrary and must be defined as custom code. This offers full flexibility but requires you to write code. Fortunately, QuB provides a high-level scripting language that implements many useful functions for manipulating ion channel models. This scripting language is accessible in two ways: as a program-like script that can be edited and run in the *Script* window, or as (smaller) snippets of code that can be defined in the *Scriptlets* window. These scriptlets can be run interactively, by clicking on their corresponding buttons, or can be called programmatically as functions, from the main script or from other scriptlets. For this demonstration, we have already prepared a scriptlet library file, *Constraints.slf*, which can be downloaded from the QuB website. Load this file in the *Scriptlets* window, as shown in Fig. 8.

The library contains three scriptlets that define the set of behavioral constraints that were used in [25] to enforce the number of channels (*Calc Nc Penalty*), the maximum open probability (*Calc Popen Penalty*), and the recovery from inactivation (*Calc RI Penalty*). The code associated with each scriptlet can be accessed and modified by right-clicking with the mouse on the button. Once you made a change in the code, click the *Compile* button. If you made a coding error, the program will highlight its place.

The *Calc Nc Penalty* scriptlet contains the following code:

```
/*
Enforce a range constraint on Nc:
  Nc_Low <= Nc <= Nc_High
*/

//Nc range limits
var dbl
  Nc_Low  = 6000,
  Nc_High = 8000;

//Declare some variables
var dbl
  Nc,
  Nc_PenaltyLow  = 0,
  Nc_PenaltyHigh = 0;

//Get Nc from the optimized model
Nc = macionchannel.channelcount;

//Calculate penalty
if Nc < Nc_Low then {
  Nc_PenaltyLow = ((Nc - Nc_Low)/Nc_Low)^2;
};
if Nc > Nc_High then {
  Nc_PenaltyHigh = ((Nc_High - Nc)/Nc_High)^2;
};

//Return result
result = Nc_PenaltyLow + Nc_PenaltyHigh;
```

As coded, this constraint will force the estimate of N_C to be between 6000 and 8000. The code is very simple, as it only needs to retrieve N_C from the model being optimized by the *Mac Rates* algorithm (the *macionchannel* object in the code) and then calculate the penalty.

The *Calc Popen Penalty* scriptlet contains the following code:

```
/*
Enforce an equality constraint on Popen:
  Popen = Popen_Value
*/

//Popen value
var dbl
  Popen_Value  = 0.5;

//Local copy of the optimized model
ionchannel icMyModel;

//Declare some variables
var dbl
  V,
  V0 = -120.0, //mV
  V1 = 0.0,
  tp = 5.0, //ms
  dt = 0.01,
  Popen_Max;

//Get the kinetics from the optimized model
icMyModel.copyfrom(macionchannel);
icMyModel.setvoltage(V);

//Calculate the equilibrium state occupancies under V = V0
icMyModel.calcpeq(voltage = V0, update = true);

//Simulate response to voltage step (V = V1, t = tp), and retrieve maximum Popen
V = V1;
icMyModel.calcq;
icMyModel.calca(dt = dt);
Popen_Max = icMyModel.advance(time = tp, return = RETURN_MAX_POPEN);

//Calculate penalty and return result
result = (Popen_Max - Popen_Value)^2;
```

The effect of this constraint is to force the optimizer to find a set of parameters that satisfy the condition that the maximum open probability reached by the channel during a depolarizing voltage step has a value of 0.5. This code is still simple, as it takes advantage of some high-level functions that first calculate the equilibrium probabilities (*calcpeq*), the rate constant matrix (*calcq*), the transition probability matrix (*calca*), and then simulate the model in response to a voltage step (*advance*). If the stimulation protocol contained more sections, you would only need to change the value assigned to *V*, and repeat the sequence of *calcq*, *calca*, and *advance* calls, as shown next.

The *Calc RI Penalty* scriptlet contains the following code:

```
/*
Enforce an equality constraint on RI fraction:
  RI = RI_Value
*/

//RI value
var dbl
  RI_Value  = 0.8;
//Local copy of the optimized model
ionchannel icMyModel;
//Declare some variables
var dbl
  V,
  V0 = -80.0, //mV
  V1 = 0.0,
  tp = 5.0, //ms
  tr = 50.0,
  dt = 0.01,
  Popen_Max1, Popen_Max2, RI;
//Get the kinetics from the optimized model
icMyModel.copyfrom(macionchannel);
icMyModel.setvoltage(V);
//Calculate the equilibrium state occupancies under V = V0
icMyModel.calcpeq(voltage = V0, update = true);

//Simulate response to first voltage step (V = V1, t = tp), and retrieve maximum
Popen
V = V1;
icMyModel.calcq;
icMyModel.calca(dt = dt);
Popen_Max1 = icMyModel.advance(time = tp, return = RETURN_MAX_POPEN);
//Simulate response to recovery interval (V = V0, t = tr)
V = V0;
icMyModel.calcq;
icMyModel.calca(dt = dt);
icMyModel.advance(time = tr);
//Simulate response to second voltage step (V = V1, t = tp), and retrieve maximum
Popen
V = V1;
icMyModel.calcq;
icMyModel.calca(dt = dt);
Popen_Max2 = icMyModel.advance(time = tp, return = RETURN_MAX_POPEN);
//Calculate RI fraction
RI = Popen_Max2 / Popen_Max1;

//Calculate penalty and return result
result = (RI - RI_Value)^2;
```

This constraint forces the optimizer to find a set of parameters that satisfy the condition that the fraction of channels that recover from inactivation after 50 ms is equal to 0.8. As explained in [25], these three constraints enforce values that are intentionally different from the "true" values used to simulate the data contained in the *CCOC_-Sim.qdf* file, to demonstrate that arbitrary values can be enforced. The true values were as follows: the number of channels $N_C = 5000$, maximum $P_{Open} \approx 0.4175$, and the recovered fraction ≈ 0.4292.

The three constraints above are just examples that illustrate how the scripting language in QuB can be used to create code that calculates the response of the model to a certain protocol. To implement any other constraints, you can take these examples as templates and further customize them. You should also be aware that there are more sophisticated ways in QuB for defining and optimizing models with constraints, but we leave that as an exercise, using the examples provided on the QuB website.

3.7 Calculating the Total Penalty

The three user-defined behavioral constraints implemented in this example are managed by the *Calc Penalty* scriptlet, which simply calls the individual constraint functions, sums up the calculated penalties, and returns the result to the optimization engine, as follows:

```
//Calculate total Penalty


var dbl

  p1 = 0, p2 = 0, p3 = 0;


//Call individual constraint functions

p1 = runscript("Calc Nc Penalty");

p2 = runscript("Calc Popen Penalty");

p3 = runscript("Calc RI Penalty");


//Sum up the values and return the result

result = p1 + p2 + p3;
```

Within this code, individual constraints can be simply turned off by commenting out their calls. For example, to apply only the channel count constraint, the code can be modified as follows:

```
...

p1 = runscript("Calc Nc Penalty");

//p2 = runscript("Calc Popen Penalty");

//p3 = runscript("Calc RI Penalty");

...
```

Of course, it would be possible to have the code that calculates the three constraints embedded in the Calc Penalty scriptlet. However, having separate code for each constraint makes it easier to manage the analysis.

3.8 Setting Up the Optimization

We have so far prepared the data, the model, the linear parameter constraints, and the code that calculates the behavioral constraints. The last step before fitting the data is to set up the optimization engine. QuB provides algorithms for analyzing both whole-cell [19] and single channel [17, 18] data, but the constraint mechanism is universal. As our example uses macroscopic currents, we will use the *Mac Rates* algorithm. The user interface for this type of optimization is shown in Fig. 9. There is an extensive (and perhaps a little intimidating) array of options available, but the defaults work well, and there are only a few settings that must be tweaked for enforcing behavioral constraints. As highlighted in the figure, these are *Constraints*, *Cycles*, *Weight*, and *Factor*.

For this demonstration, you must simply instruct the program that there are behavioral constraints to be enforced, which you indicate by checking the *Constraints* option and by providing the name of the scriptlet that calculates them, which is *Calc Penalty*. The optimizer will fit the model iteratively, in several *Cycles*. During the search for optimum parameters, the optimizer will not only

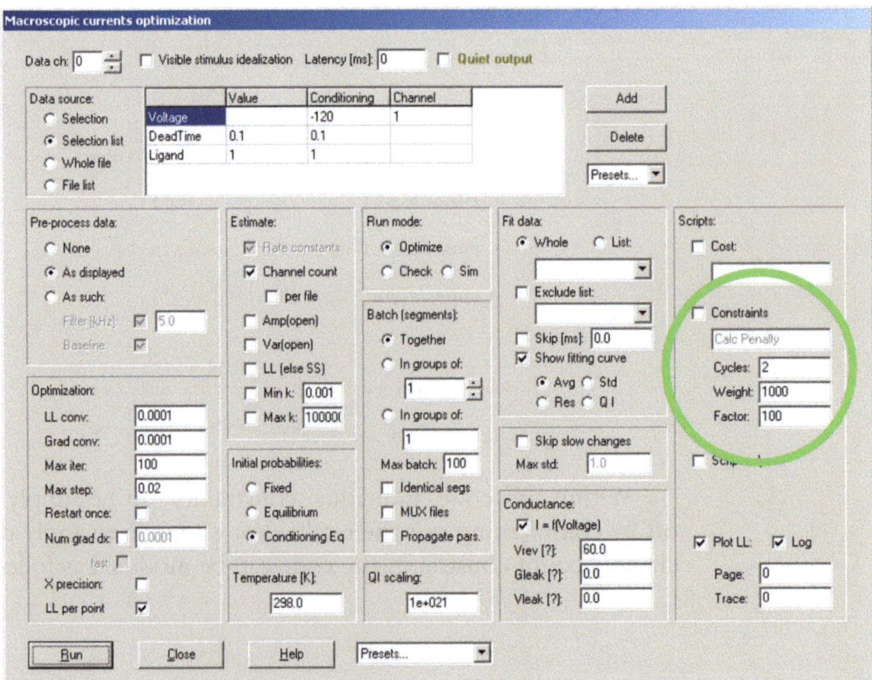

Fig. 9 Setting up the optimization. For macroscopic currents, we use the *Mac Rates* algorithm. To raise this interface, right-click on the *Mac Rates* button in the *Task selector* area

calculate the error between data and model prediction, but will also run the *Calc Penalty* scriptlet to calculate the constraints. The penalty value returned by this scriptlet will be multiplied by a *Weight* factor, and then the error and the weighted penalty will be added together. It is this sum that the optimizer will minimize during each cycle, in multiple iterations per cycle. The parameter values reached at the end of one cycle will become the starting parameters for the next cycle. However, the weight of the penalty will be increased by a *Factor*.

Through these options, the constraints can be enforced more loosely at the beginning of the optimization, but more and more tightly with each optimization cycle. This iterative approach has a reason: if the behavioral constraints are too strong in the beginning, the optimizer may get stuck with a suboptimal solution. In practice, we find that a few cycles—or sometimes, depending on the problem, just one—are enough for a good fit with tightly enforced behavioral constraints. To better understand these options and the general workflow of optimization with constraints, consult Fig. 1 in [25].

3.9 Run the Optimization

3.9.1 Overview of the Method

In [25], we tested the constrained optimization algorithm in several runs, with different combinations of constraints applied in each run. For the sake of simplicity, here we run only two tests: in the first run, we fit the data without enforcing any constraint; in the second run, we fit the data with all the linear parameter constraints and the behavioral constraints enforced. Nevertheless, when you are done with this demonstration, you can experiment with other combinations. Furthermore, we only fit the data shown in Fig. 3, without including the steady-state activation and inactivation curves, as in the previous study. Considering these procedural differences, we expect to obtain different parameter estimates than in [25]. The prediction of the model corresponding to the starting parameter values is shown in Fig. 6.

3.9.2 Run I: No Constraints Applied

To turn off the linear parameter constraints, select the *CCOC_Fit-Start.qmf* model file, right-click on the model and select *Properties*, then go to the *Kinetic constraints* panel and uncheck all the seven *Defined constraints*. Next, select the *CCOC_Sim.qdf* data file, and make sure the *FitAct* list is selected (click on it). Also make sure that *Data source* (Fig. 2) is set to *Selection list*. Then, right-click on the *Mac Rates* button to raise the optimization interface, and check that you have the same configuration as in Fig. 9, with the exception of the *Constraints* option, which should be unchecked. Hit *Run* and watch magic happen.

The optimizer will run for 100 iterations, reporting its progress via text messages in the *Report* window, and graphically, in the *Graph* window. During optimization, you will notice that the rate constants of the model change continuously, as the optimizer

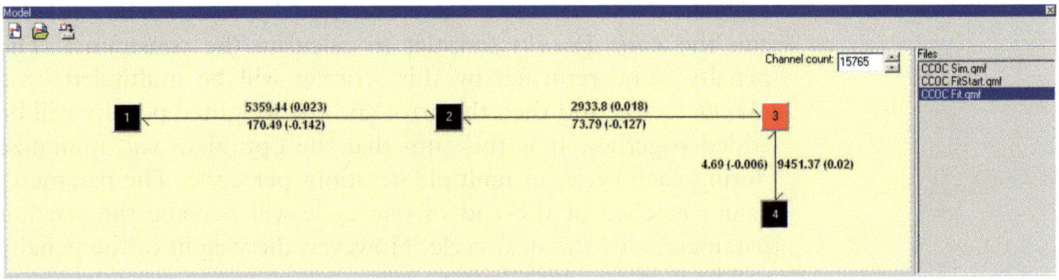

Fig. 10 Model parameters—no constraints applied. The rate constants shown here represent the solution found by the optimizer

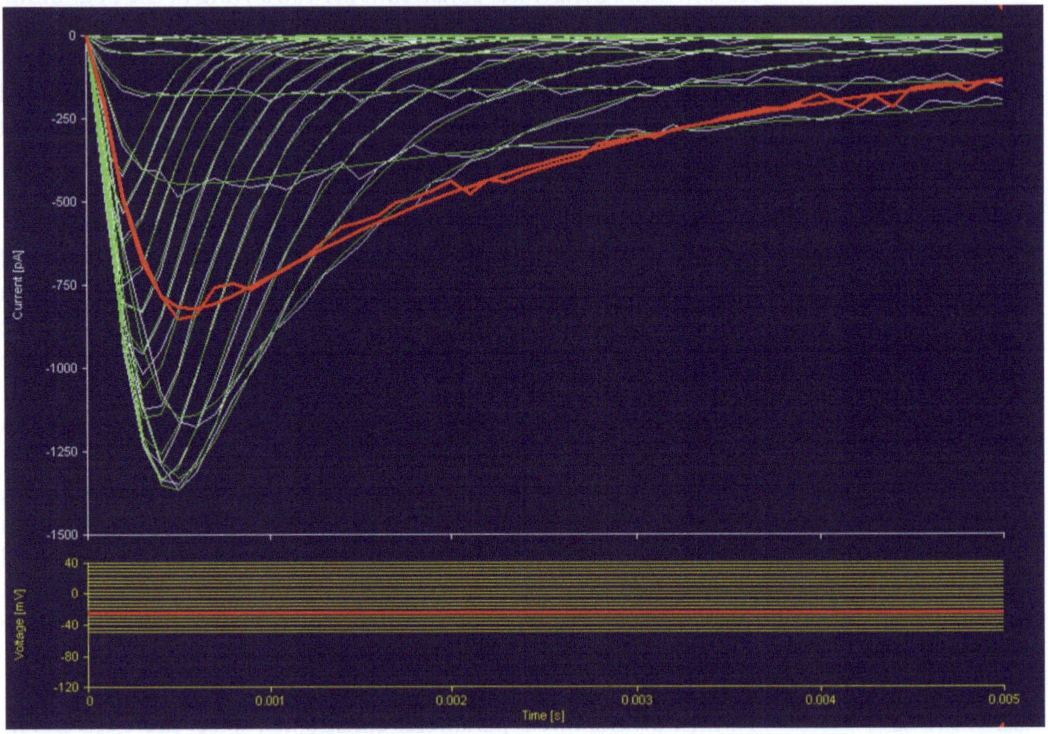

Fig. 11 Model predictions—no constraints applied. These fit curves were generated with the parameters shown in Fig. 10

searches the multidimensional parameter space. You will also notice that the four scriptlets are called each time the fit curves are recalculated.

The estimated model is shown in Fig. 10, the fit curves are shown in Fig. 11, and the progress of the optimization is shown in Fig. 12. The results demonstrate the convergence of the optimizer when started with the "wrong" parameters. Nevertheless, be aware that the initial model must be good enough, qualitatively. Otherwise, the optimizer, which only searches locally in the parameter

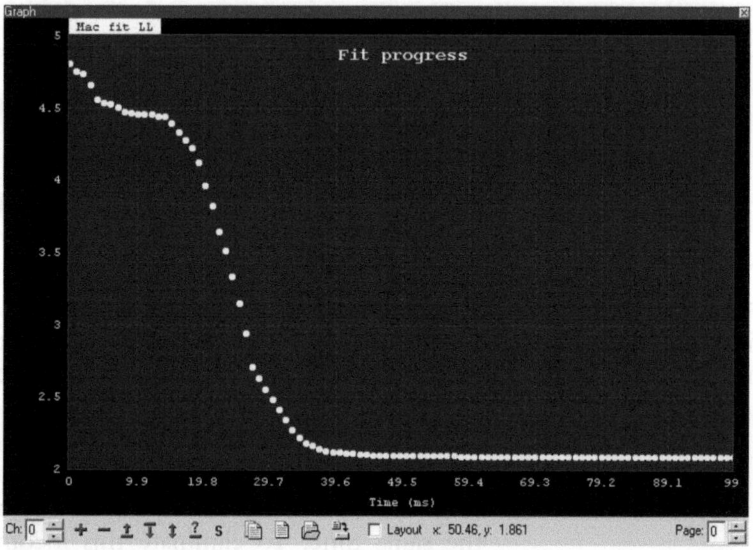

Fig. 12 Optimization progress—no constraints applied. The points represent the logarithm of the sum of square errors, as reduced by the optimizer at each iteration, during its search for a solution. We recommend you try different values for the *Max step* parameter (Fig. 9), which can make the search slower but more stable (e.g., with *Max step* = 0.01), or faster but possibly unstable (e.g., with Max step = 0.1)

space, may not find a good solution. Clearly, the estimated parameters are not a good match for the "true" parameters, yet the fit is "perfect" (sum of square errors ≈ 127.73). The explanation is that the model simply has too many parameters and multiple solutions exist. If this were a real modeling project, and not just a simple demonstration, the parameters of an undetermined model would be of little use.

3.9.3 Run II: All Constraints Applied

To optimize with all the constraints enforced, you simply repeat the same steps as described above, but turn on all the linear parameter constraints in the *Kinetic Constraints* editor, as well as the *Constraints* option in the *Mac Rates* interface. Also set the *Max step* to 0.1, *Cycles* to 2, *Weight* to 1000, and *Factor* to 100. You are now ready to run the optimization. However, before you hit *Run*, you must take a moment and ask yourself: are the three behavioral constraints compatible with each other and with the data?

In fact, they are not. Specifically, the two constraints that enforce $6000 \leq N_C \leq 8000$ and maximum $P_{Open} = 0.5$ are mutually incompatible, given the actual data and the conductance properties of the model. Thus, the maximum peak current of −1290 pA reached during the step to 0 mV corresponds to approximately 2150 channels open, given that the conductance of the open state is 10 pS and the reversal voltage is +60 mV. Therefore, we can have either $N_C = 6000$, which gives a maximum $P_{Open} \approx 0.358$, or a

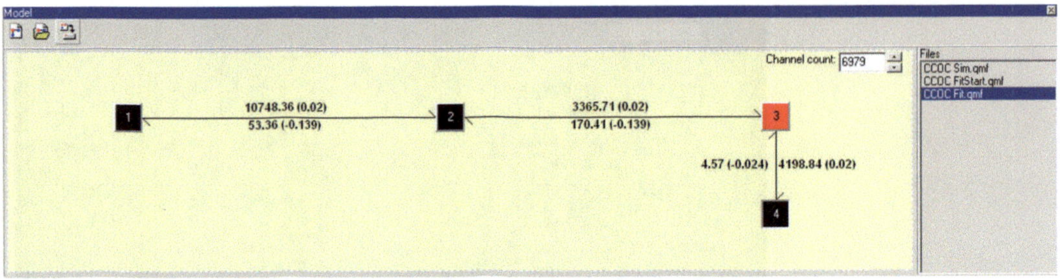

Fig. 13 Model parameters—all constraints applied. As in Fig. 10, but with all the linear parameter and model behavioral constraints applied. The values of the enforced model properties (number of channels, maximum open probability, and fraction of channels recovered from inactivation) are given in the main text

maximum P_{Open} of 0.5, which gives $N_{\text{C}} = 4300$. However, we cannot have both at the same time. In [25], this issue has not been raised, because these two constraints were not enforced at the same time. To eliminate this incompatibility here, you can either disable the N_{C} constraint, or change the maximum P_{Open} constraint to enforce a compatible value, such as 0.3. In this case, the *Calc Popen Penalty* scriptlet will be modified as follows:

```
...

//Popen value

var dbl

  Popen_Value   = 0.3;

...
```

Now run the optimization. As expected, at the end of the two cycles, the estimated parameters are quite different from the previous run, due to all the enforced constraints (Fig. 13). The good news is that the fit curves match the data just as well (Fig. 14), which means that the constraints are all mutually compatible, and also compatible with the data. The behavioral constraints were already enforced quite tightly at the end of the first cycle, which ran for only 70 iterations before convergence criteria were satisfied, with the maximum $P_{\text{Open}} \approx 0.282$, instead of 0.3, and the fraction recovered ≈ 0.799 instead of 0.8. The sum of square errors was ≈ 129.47, just a little higher than in the unconstrained run, on account of the many constraints that reduce the degrees of freedom (Fig. 15, green points). At the end of the second cycle, both maximum P_{Open} and the recovered fraction were within 0.0001 of their constraining values, and the sum of square errors was ≈ 129.79 (Fig. 15, magenta points). In conclusion, you have successfully integrated new data with known kinetic properties!

As an exercise, you can to try to run the optimization with mutually incompatible behavioral constraints. What happens then? Another exercise would be to enforce the "correct" model

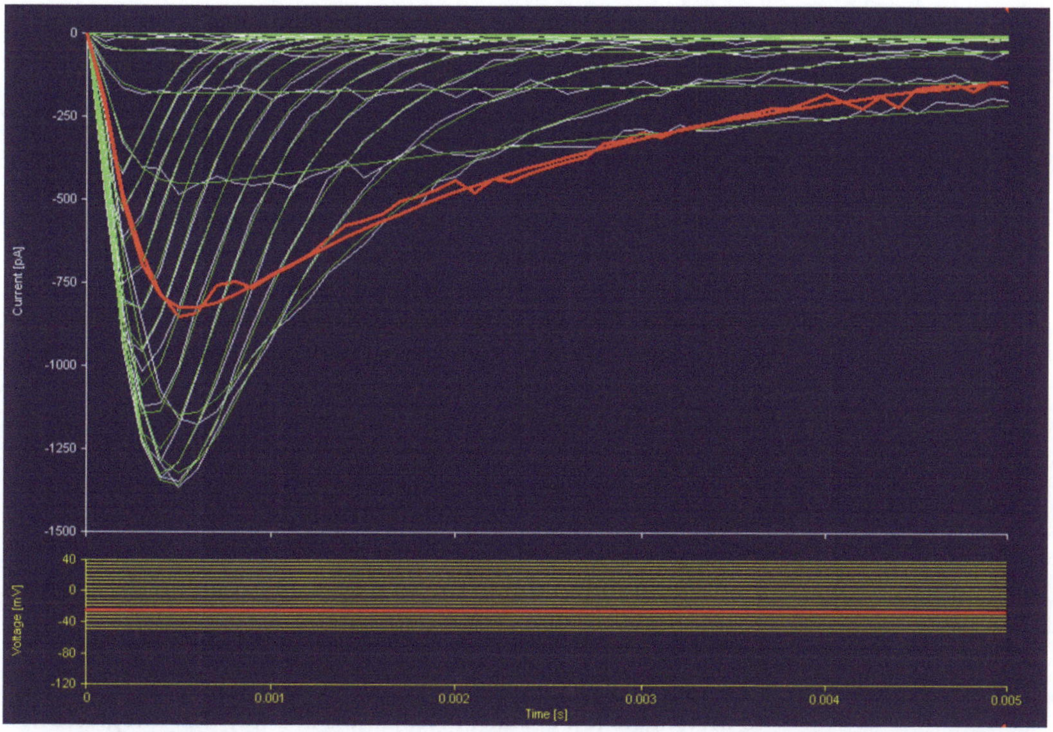

Fig. 14 Model predictions—all constraints applied. These fit curves were generated with the parameters shown in Fig. 13 and are just as good as the fit curves shown in Fig. 11. However, all the constraints were enforced now, which has resulted in a set of parameters (Fig. 13) that differ from the parameters used to simulate the data. This demonstrates that the optimization algorithm has successfully integrated new data with known kinetic properties

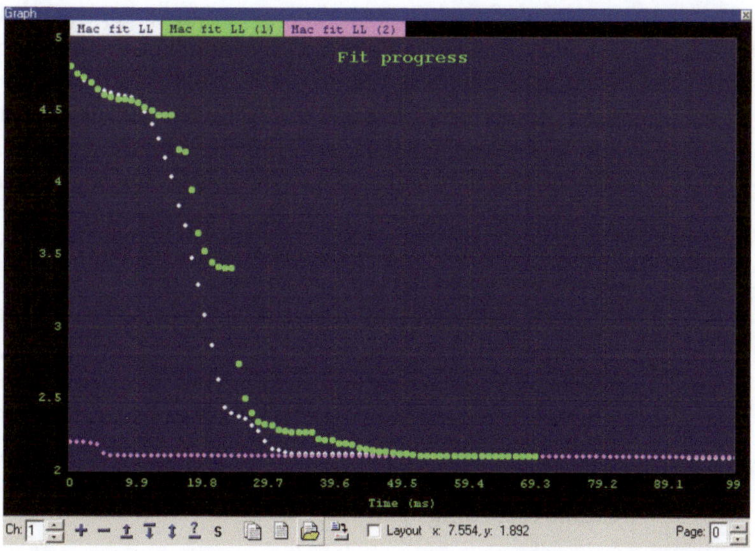

Fig. 15 Optimization progress—all constraints applied. The white points are the same as in Fig. 12, shown here for comparison. The green and the magenta points represent the sum of square errors during the first and second optimization *Cycles*, respectively, when all constraints were enforced. The first cycle has terminated in 70 iterations, instead of 100, because the convergence criteria were satisfied sooner

behavior, i.e., a channel count of 5000, a maximum open probability of 0.4175, and a recovered fraction of 0.4292. Will the estimates in this case be closer to the "true" values?

An important test you should always run is to check whether the constrained model is fully determined or not, in which case you have too many degrees of freedom relative to the information contained in the data. You can run this test easily, by defining an additional linear parameter constraint that fixes any of the model parameters to a slightly modified value, relative to the value obtained in the above fit. If the optimizer can still find a solution that fits the data just as well, even with that additional constraint that keeps it away from the previous solution, it means that the model is not fully determined. In other words, the solution you found in Run II is not unique. Alternatively, if the fit is slightly worse, you know that the solution is unique, at least in the local space surrounding it. Be aware that the optimizer engine in QuB relies on a local search algorithm that is not meant to look for a global solution, although it can be restarted by the user from different initial points in the parameter space.

3.10 Conclusions

Our objective in this chapter was to explain how QuB can be used to investigate ion channel mechanisms in a way that integrates new experimental data with known channel properties. We demonstrated this idea with stochastically simulated macroscopic currents, which were fitted with a Nav model that featured a variety of linear parameter constraints and model behavioral constraints. When you want to analyze real data, you simply repeat the same steps described here: prepare the data you want to fit, create a state model, define the constraints you want to apply, and then optimize the model. If you have single channel (or single molecule) data, then you can use the MIL or MPL single-channel algorithms [17, 18]. For macroscopic data, use the Mac algorithm [19].

If you only need to apply linear parameter constraints, then you do not have to write code: you simply use the *Kinetic constraint* editor (Fig. 7). Of course, all the difficulties inherent to mechanistic kinetic analysis still apply. However, if there are specific model behaviors that you want to enforce, you will have to write some code. If you have even minimal experience using any scientific computation software, such as Matlab, the scripting language in QuB should present no difficulty at all, and you can easily reuse the example code available on the QuB website. Using scripts, you can also implement more complex analysis, where you can, for example, combine multiple types of data and different experimental paradigms recordings [20].

References

1. Hille B (2001) Ion channels of excitable membranes, vol 507, 3rd edn. Sinauer Associates, Inc., Sunderland, MA
2. Colquhoun D, Hawkes AG (1995) A Q-matrix cookbook: how to write only one program to calculate the single-channel and macroscopic predictions for any kinetic mechanism. In: Sakmann B, Neher E (eds) Single-channel recording. Plenum Press, New York, pp 589–636
3. Kuo CC, Bean BP (1994) Na+ channels must deactivate to recover from inactivation. Neuron 12(4):819–829
4. Colquhoun D, Sigworth F (1995) Fitting and statistical analysis of single-channel records. In: Sakmann B, Neher E (eds) Single-channel recording. Plenum Press, New York, pp 483–587
5. Venkataramanan L, Sigworth F (2002) Applying hidden Markov models to the analysis of single ion channel activity. Biophys J 82 (4):1930–1942
6. Csanady L (2006) Statistical evaluation of ion-channel gating models based on distributions of log-likelihood ratios. Biophys J 90 (10):3523–3545
7. Stepanyuk AR, Borisyuk AL, Belan PV (2011) Efficient maximum likelihood estimation of kinetic rate constants from macroscopic currents. PLoS One 6(12):e29731
8. Stepanyuk A, Borisyuk A, Belan P (2014) Maximum likelihood estimation of biophysical parameters of synaptic receptors from macroscopic currents. Front Cell Neurosci 8:303
9. Moffatt L (2007) Estimation of ion channel kinetics from fluctuations of macroscopic currents. Biophys J 93(1):74–91
10. Ball FG, Sansom MS (1989) Ion-channel gating mechanisms: model identification and parameter estimation from single channel recordings. Proc R Soc Lond B Biol Sci 236 (1285):385–416
11. Celentano JJ, Hawkes AG (2004) Use of the covariance matrix in directly fitting kinetic parameters: application to GABAA receptors. Biophys J 87(1):276–294
12. Hawkes AG, Jalali A, Colquhoun D (1992) Asymptotic distributions of apparent open times and shut times in a single channel record allowing for the omission of brief events. Philos Trans R Soc Lond Ser B Biol Sci 337 (1282):383–404
13. Epstein M et al (2016) Bayesian statistical inference in Ion-Channel models with exact missed event correction. Biophys J 111 (2):333–348
14. Gurkiewicz M, Korngreen A (2007) A numerical approach to ion channel modelling using whole-cell voltage-clamp recordings and a genetic algorithm. PLoS Comput Biol 3(8): e169
15. Horn R, Lange K (1983) Estimating kinetic constants from single channel data. Biophys J 43(2):207–223
16. Menon V, Spruston N, Kath WL (2009) A state-mutating genetic algorithm to design ion-channel models. Proc Natl Acad Sci U S A 106(39):16829–16834
17. Qin F, Auerbach A, Sachs F (1996) Estimating single-channel kinetic parameters from idealized patch-clamp data containing missed events. Biophys J 70(1):264–280
18. Qin F, Auerbach A, Sachs F (2000) A direct optimization approach to hidden Markov modeling for single channel kinetics. Biophys J 79(4):1915–1927
19. Milescu LS, Akk G, Sachs F (2005) Maximum likelihood estimation of ion channel kinetics from macroscopic currents. Biophys J 88 (4):2494–2515
20. Milescu LS et al (2008) Real-time kinetic modeling of voltage-gated ion channels using dynamic clamp. Biophys J 95(1):66–87
21. Salari A, Milescu LS (2016) Modeling the kinetic mechanisms of voltage-gated ion channels. In: Korngreen A (ed) Advanced patch-clamp analysis for neuroscientists, vol 113. Humana Press, New York, pp 267–304
22. Milescu LS et al (2010) Kinetic properties and functional dynamics of sodium channels during repetitive spiking in a slow pacemaker neuron. J Neurosci 30(36):12113–12127
23. Bean BP (2007) The action potential in mammalian central neurons. Nat Rev Neurosci 8 (6):451–465
24. Salari A et al (2018) Estimating kinetic mechanisms with prior knowledge I: linear parameter constraints. J Gen Physiol 150 (2):323–338
25. Navarro MA et al (2018) Estimating kinetic mechanisms with prior knowledge II: behavioral constraints and numerical tests. J Gen Physiol 150(2):339–354
26. Navarro MA et al (2015) 3D data mapping and real-time experiment control and visualization in brain slices. Biophys J 109(8):1521–1527

Index

Quentin Vanhaelen (ed.), *Computational Methods for Estimating the Kinetic Parameters of Biological Systems*, Methods in Molecular Biology, vol. 2385, https://doi.org/10.1007/978-1-0716-1767-0,

MIX
Papier aus verantwortungsvollen Quellen
Paper from responsible sources
FSC® C105338

If you have any concerns about our products,
you can contact us on
ProductSafety@springernature.com

In case Publisher is established outside the EU,
the EU authorized representative is:
Springer Nature Customer Service Center GmbH
Europaplatz 3, 69115 Heidelberg, Germany

Printed by Libri Plureos GmbH
in Hamburg, Germany